Where Is Science Leading Us?

Lars Jaeger · Michel Dacorogna

Where Is Science Leading Us?

And What Can We Do to Steer It?

Lars Jaeger
Baar, Switzerland

Michel Dacorogna
Zürich, Switzerland

ISBN 978-3-031-47137-7   ISBN 978-3-031-47138-4 (eBook)
https://doi.org/10.1007/978-3-031-47138-4

© The Editor(s) (if applicable) and The Author(s), under exclusive license to Springer Nature Switzerland AG 2023

This work is subject to copyright. All rights are solely and exclusively licensed by the Publisher, whether the whole or part of the material is concerned, specifically the rights of reprinting, reuse of illustrations, recitation, broadcasting, reproduction on microfilms or in any other physical way, and transmission or information storage and retrieval, electronic adaptation, computer software, or by similar or dissimilar methodology now known or hereafter developed.
The use of general descriptive names, registered names, trademarks, service marks, etc. in this publication does not imply, even in the absence of a specific statement, that such names are exempt from the relevant protective laws and regulations and therefore free for general use.
The publisher, the authors, and the editors are safe to assume that the advice and information in this book are believed to be true and accurate at the date of publication. Neither the publisher nor the authors or the editors give a warranty, expressed or implied, with respect to the material contained herein or for any errors or omissions that may have been made. The publisher remains neutral with regard to jurisdictional claims in published maps and institutional affiliations.

Cover illustration: The cover images illustrate the fact that sometime modern science is playing the alchemist

This Springer imprint is published by the registered company Springer Nature Switzerland AG
The registered company address is: Gewerbestrasse 11, 6330 Cham, Switzerland

Paper in this product is recyclable.

Contents

1	**Introduction**	1
2	**The Takeover of Scientific Leadership**	9
	Geniuses Create a New World	10
	The Shift of the Scientific Gravity Centre from Europe to the USA	12
	Philosophical Implications of Quantum Theory – The Concept of Reality Called into Question	14
	Highly Controversial Philosophical Discussions Among Physicists in the 1930s	16
	How Europe and Philosophy Both Lost Their Dominance in Science	19
	Pragmatism Now Governing Science – With Consequences	22
	Philosophy for Quantum Physics - Still Essential Today	25
	Scientific Revolutions Beyond Physics	27
3	**Publicly Backed Science in Competition with Private Companies**	31
	The Transition from Fundamental Scientific Research to Technological Applications	31
	Relationship of Sciences and Technological Applications Today	33
	Who Finances Modern Science?	35
	How Do We Judge the Quality of Scientific Papers Today?	38
	Where Should We Go in Science?	41

	Three Past Examples from Fundamental Research to Revolutionary Technology - Penicillin, PCR and the Atomic Bomb	42
	Penicillin	42
	Atomic Bomb	44
	Polymerase Chain Reaction PCR	45
	The Most Important Technological Efforts Today - And Those in the Future	47
4	**Philosophy in Science Is Over**	51
	The Tradition of Science Interacting with Philosophy	51
	The Breakdown of Absolute Knowledge in Science Is a Profound Philosophical Challenge	56
	Does Philosophy Still Have an Importance for Science?	59
	Open Questions in Science Today Are as Well Open Philosophical Questions	64
	Why the Influence of Philosophy into Science, Besides All Its Needs, Is Still so Low Today	68
	A Key Mathematician Steps Out of Science Protesting Against Its Nature of Leaving Out Important Topics	70
	A Good Example for an Interaction of Philosophy and Science: Research on the Nature of the Human Ego-Consciousness	73
	More and More Important: The Relationship Between Science and Ethics	76
5	**Promising and Scary Developments in Future Technologies**	79
	The Future Technological Application Changing the World and Human Beings	80
	Artificial Intelligence - Improving or Controlling Our Lives?	80
	Quantum Computers – Millions of Times Faster Computation or Just a Dream by Physicists?	81
	CO_2-Neutrality – Can We Create Enough Alternative Energies in the Next Few Years to Prevent a Climate Catastrophe	83
	Nuclear Fusion – The Solution of Our Energy Problems or Just a Topic of a Century of Dreaming?	85
	Genetics - The Victory Over Cancer or Manipulation of Mankind?	85
	Internet of Things – New Industrial Technologies and Smart Fabrications or a Full Invasion of Privacy?	87

Neuro-Enhancements – Improving Our Thinking
and Acting or Move Away From Today's Reality? 88
Understanding Our Minds Through VR-Technologies –
Finding Our Ego Or Is It Unfindable for Scientists? 89
Digital Algorithms and Big Data – New Profiles for Our
Lives or Controlling Humans' Thinking and Acting? 90
Blockchain Technology - Is It a Groundbreaking
Innovation or Just a Passing Trend? 91
Cybersecurity – Is It a Consistently Significant Concern
or Merely an Occasional Problem? 91
Nanotechnology – Creation of Things from "Nothing"
or Just a Dream? 91
Stem Cells – Using Cells That Can Do Everything. Also
for Our Entire Body and Mind? 92
Biotechnology – From Frogs for Pharmacists
to Nano-Robots in Our Bodies, a Medical Dream
or Future Reality? 93
New Food Technologies - How We Will Provide Food
to 10 Billion People or Just a Science Dream Story? 94
Synthetic Life– When Humans Play God: Part I 95
Life Prolongation – When Humans Play God: Part II 96
Are These All the Technologies that Will Shape Our Future? 97

6 Physics from 1960 to Today 99
The New Quantum World – How to Deal With
An Uncountable Number of New Particles 99
A First Theory Integrating Various Particles and the Strong
Force 102
The Standard Model of Elementary Particles 107
"Chaos" Theory and Emergence Patterns in Today's Physics 109
Today's Situation in the Macrocosm – Will We Soon Get
Answers to the Fundamental Questions About the Universe? 114
How Realistic Is a Unifying Theory for Physics? 116
Research in Physics Today 118
Philosophy of Physics Today 119

7 Computers, Nanotechnology, Internet and Many Other
Technologies 125
What New Technologies Physics Has Brought Us:
A Tremendous Amount of Life Improvements and a Few
Important Open Questions 125

	Nuclear Technologies	126
	Electronics, Digital Technologies and the Miniaturisation of Processors	127
	Digital Revolution (also Known as the "Third Industrial Revolution" or "Microelectronic Revolution")	128
	Lasers	129
	Mobile Phones	130
	Internet	131
	Superconductivity and Superfluidity	132
	Satellites in and Beyond the Atmosphere	133
	New Materials That Do Not Exist in Nature	134
	Solid-State Physics	135
	Quantum Computer	136
	Nano Particles and Nanotechnologies	139
	Where Are We Going?	143
8	**Biology from 1953 to 2023: Major Breakthroughs and Their Ethical Issues**	**147**
	The Second Foundation of Biology: Genetics	147
	First Steps in "Genetic Engineering"	151
	The Development of Life on Earth	153
	The Origin of Life	155
	Genetics Since the 1970s	161
	Revolution of Genetics in 2012 – As Amazing as Scary New Technologies	163
	Synthetic Life	165
	Life Prolongation – When Humans Want to Play God II	167
	Ethics for Today's Biology	170
	Humans as a Bull in a China Shop	172
9	**Brain Research Since the 1990s**	**175**
	History of Brain Research Until 1990s – A Rather Short Story Compared with What Happened Thereafter	175
	Early Brain Research as of 1990 – First Insights and Many Problems Left	178
	Research About Our Consciousness - How the Brain Generates Our Mind	180
	About Our ("Ego-")consciousness – Fundamental Open Problems	183
	More Methods and Results of Research on Our Consciousness	185
	And Yes, It Does Change - The Plasticity of Our Brain	188

Key Technologies – "Improving" Our Minds
with Neuro-Enhancements 190
More Philosophical Questions 192
Our Inner Model as Virtual Reality 195
Our Mind and Self-Consciousness – More Empirical
Studies, Dramatic Applied Technologies and - yet Again -
Ethical Issues 197
Summary: Scientific Knowledge, Philosophical and Ethical
Questions, and Remaining Openness 201
New Questions on Social Relationships 203
What is a Human Being and What Should a Human Being Be? 204

10 Artificial Intelligence from Its Origins Via Today to the Future 207
History of Artificial Intelligence 208
History of Computers and Computer Science 210
Where AI Stands Today 214
The Current Interaction of AI and Our Brain – Does
that Eventually Lead to Superhuman Intelligence? 218
How AI Shapes Our Society 222
Who Should Deal with the Decline of Our Privacy? 225
The Development of Big Data 228
Artificial Intelligence's Possible Consciousness of (Strong) AI 232

11 The Path Towards Modern Mathematics 237
Mathematics Before 1920 237
The Crisis in Mathematics 240
The Revolution 243
The Path Towards Modern Mathematics – More and More
Abstraction 245
Dealing with Concrete Problems Through Numerical Methods 252
Mathematics Today and in the Future 255

12 Astronomical Research 259
A (Very) Brief History of Astronomy Prior to 1960 259
A Rather Recent Revolution in Observing the Universe 262
New Discoveries in the Last 25 years 265
Cosmology – The Origin of the Universe 269
Cosmology – How the Universe is Developing 270
The Current Unified Theory of the Universe – Many Open
Questions 273

13	**The Future of Sciences/Technologies?**	279
	More of the Promising and Challenging Areas in Science and Technology	280
	CO_2-Neutrality	280
	Nuclear Fusion	282
	Food Technology	286
	Synthetic Life and Life Prolongation	287
	Historical Issues	290
	"We Go Under" Versus "Yes, We Can" - Dystopian Pessimism Versus Utopian Optimism	294
	Social Drivers	298
	Us	299
14	**The Myth of the Optimally Functional Invisible Hand**	301
	The Legend of an Invisible Hand	301
	Who is Likely to Best Govern the Scientific Future? – I. The Side Actors: Cultural Figures, Journalists or the Church?	306
	Who is Likely to Best Govern the Scientific Future? – II. The - Democratically Elected - Government?	308
	Who is Likely to Best Govern the Scientific Future? – III. The Scientists Themselves?	311
	Who is Likely to Best Govern the Scientific Future? – IV. All of Us!	312
15	**Science, Technology and Spirituality**	319
	How New Technologies Shape Up the Economy – In the Right Direction?	319
	More Openness, Less Dogmatism	322
	Rationally Irrational	322
	How Can Broad Knowledge About Science and Technologies and Its Rational and Democratic Assessments Make the World a Safer and Better Place	325
	A New Way of Approaching "Spirituality"	328
	Summary: Ideas Instead of Ideologies	332
Name Index		335

1

Introduction

In around eighty years, from the end of the nineteenth to the middle of the twentieth century, science made advances that led to the greatest revolution in human thoughts of all time. These upheavals were more drastic than the revolutions of the Renaissance and the Enlightenment and happened roughly during the lifetime of an individual person. Moreover, these advances occurred after a period during which all sciences, including mathematics, had faced the greatest crises in their respective histories, crises that frustrated scientists like never before (and even led to a suicide of one scientific leader). The work of a whole series of scientific geniuses of breath-taking creativity then led all sciences out of their crises in ways that were as exciting as they were bizarre. In this process of completely restructuring of all its disciplines, science itself not infrequently reached the limits of reason and coherence, as these developments also called into question principles established over 2000 years ago in Western philosophy.[1]

Furthermore, during this process science experienced a significant change of its very foundation which is of equally important philosophical significance. The scientists were forced to drastically change a paradigm which their believes had been fundamentally based on until then: They had to abandon the idea that their job consists of seeking *eternal truths*, a scientific belief

---

[1] For more details about the scientific revolutions between 1870 and 1950, see: Lars Jaeger, *The Stumbling Progress of 20th Century Science - How Crises and Great Minds Have Shaped Our Modern World*, Springer (2023).

that had been strongly rooted in Western philosophy (and often even in religions). Science today has to be content with *relative truths*, i.e. truths that can change, once there exists enough empirical evidences for a better truth, without losing its ability to change our reality, quite the contrary.

As of 1960 all these developments strongly shaped human societies. As exciting and influential for the character of our societies the revolutions from 1870 to around 1950 were, the developments of sciences and – especially – of technological applications based on them profoundly changed the human society itself in the years after 1950 to today. Responsible for this were the following three developments:

- The *number of scientists* worldwide increased exponentially, and has grown by more than twenty times since then. Today, it stands at around nine million people.
- The *number of scientific articles* grew also extremely fast. In the same period, it has doubled every 9 years, i.e. a growth factor of about 250 times from 1950 to today.[2] There is a joke among physicists that says one thing can eventually grow faster than the speed of light and thus break Einstein's law: the list of research articles published by all these scientists, as the number published has become the sole measure of scientific success (the ironic answer of other physicists to this is that this event will not break Einstein's laws as no information is passed along any longer).
- Also, the pure spectrum of scientific areas (i.e. universities, public and private research institutes, etc.) grew equally exponentially to host all those scientists.

A consequence of this "explosion" of scientific research, the pure number of individuals working in this field since 1950, and their respective scientific articles published is that, compared to the 80-year phase before, not any individual scientist today is as well-known as Albert Einstein, Max Planck, Charles Darwin, Louis Pasteur or Robert Koch in the revolutionary period between 1870 and 1950. As of the 1960s even the greatest scientists became more and more experts in specific fields due to the equally immense growth of special scientific areas. The time of universal scientific geniuses is now over. This explains why today the names of Nobel Prize winners rarely ever remain known in the broader public's memory a few days past the award announcement. This makes the writing of an entertaining book about this period to

---

[2] For more details, see: Lutz Bornmann, Ruediger Mutz, *Growth rates of modern science: A bibliometric analysis based on the number of publications and cited references*, Journal of the Association for Information Science and Technology, 65, 11 (2015).

non-scientists more difficult than covering the period before. Furthermore, the different sciences today are much more connected than they were before, so they can barely be discussed separately. Accordingly, we refer with respective remarks to other chapters and sometimes even discuss the same topic from two different perspectives.

Nevertheless, it was the processes in (natural) sciences and the technologies they made possible that created our modern human life today beyond what society experienced with the first industrial revolution. Besides scientists getting deeper and deeper into areas such as particle (quantum) physics, functions of individual genes, or understanding more deeply the human brain, the *real shaping* entities that have formed our modern society and are continuing to do so - ever more dramatically and rapidly than ever before - are the technological application thereof. In this process, as we will see in this book, the usual distinction between science and technology blurs and it becomes difficult to separate them.

This book aims to provide an overview of the scientific progress since 1960. However, it will also discuss how technologies can in the next 30 years shape the *future life of humans and human societies* more than it has ever done in the past. We thereby assess some of the main problems today as well as in the near future (like the climate change, probable loss of personal data protection, potential genetic change of us, virtual reality) and opportunities (like addressing the climate change[3]). For this we also introduce and assess a philosophical context which is almost deserted by modern science. This combination of the historical context after the scientific revolution of the first half of the 20th until today, a discussion of the possible implications for our societies in the next 30 years, and providing a philosophical perspective on this is, as far as we know, the first assessment of this kind. The main question is hereby: What are the dramatic changes today and how should (and can) we address and deal with them from a social and society perspective?

In the last 250 years, women and men have already profoundly changed their environment and living conditions through science and new technologies from it (we refer to the development in the nineteenth century already as the "Industrial Revolution"). Yet the biological and psycho-spiritual foundation of mankind has so far remained largely untouched. But now, for the first time in history, the human being itself is becoming an object of technological developments. In the dynamics of scientific as well as technological progress, we are at a point where biotechnology, genetic engineering, quantum technology, neurotechnology and their interactions can transform the human

---

[3] For discussing climate change issues and possible paths towards addressing it: Lars Jaeger, *Ways Out of the Climate Catastrophe - Ingredients for a Sustainable Energy and Climate Policy*, Springer (2022).

being and human civilisation themselves in hitherto unimaginable ways, and this in only a fraction of a human life. What awaits us is not just another industrial revolution, of which there have already been a few, but we must prepare ourselves for the first *revolutionem humanam,* a revolution of being human in itself, an "inversion" of what makes us human in our innermost being and what we define ourselves as.

The technological developments could potentially be anything from just continuing in making us richer and richer and living better and better (unlikely) to changing or even reprograming humans themselves (much more likely). Developments will with highest probability be much more society shaping than what we have seen in the past. For this reason, it is not only the scientists and technologists who should shape the content of their doing. They rather should be joined - next to by economists - by politicians, social workers, and many other people co-deciding on sciences' technological effort - not on the fundamental research itself, but on the technologies grown out of it. For this, also the amount of public and private money that flows into scientific research requires that a - better than today - control of scientists' doing with respect to technological developments must be installed.

In fact, even if they have been quite distant from science over the past 85 years, in the future, philosophers in particular should be asked to critically assess the development of science and technologies given where science and especially technologies have developed towards. However, a philosophical discussion of science is not new: In the history of modern science since its early beginnings with Galilei, Kepler and Newton philosophy has always played a key role. Yet, it was mostly concentrated on what philosophers call *epistemology,* i.e. in what sense scientific results are influenced by our own given capability to observe and judge on natural processes. *Today,* a more significant philosophical field which needs to be applied to technologies is *ethics,* i.e. the questions what we can do without breaking given values. But what are the appropriate ethical values? How to perform scientific research ethically is exactly a sort of question that is philosophical and not scientific. Possibly the use of nuclear bombs by the US against Japan in late WW II can be seen as the first outcome of science having become ethically critical (with an outcome not so highly ethical).

In all this discussion, we need to make a highly important distinction between two types of science: the leading core scientific models on the one side and current scientific research on the other. The first is a body of knowledge upon which (almost) all scientists agree and there is currently no serious doubt about it. Classic examples of this are the facts that the Earth is round, that all life has gone through an evolution or that physics in the magnitudes of atoms is structurally different than in the range of our experiences. The second aims at answering a body of currently open questions raised within science or from new societal problems like climate change or pandemics. Regarding those, there is a lot of doubts, critics and controversies among scientists. Taking the second for the first is unfortunately a widespread confusion in non-scientific circles. Thus, people often challenge accepted knowledge like that humans are genetically quite close to monkeys or that quantum physics matches experiments very, very well, as they do not match with our daily perceptions. They therefore argue that science is relative. To state it very clearly: While our daily research is relative and questionable among the various scientists, the body of collected scientific knowledge is not. These two areas must be clearly separated and treated differently! (as we will discuss in more detail starting in Chap. 14).

An important fact of today's technological progress is: While in the past, there were at given times *single* technologies that evolved over periods of several generations, today, in contrast, there exists *an entire list of key technologies that develop at the same time* that will each have a similarly massive impact on our lives as the technological revolutions in the past (most likely even higher ones). Their list is as long and fascinating as it is frightening: Conversions occur from more than a dozen different directions with each one providing a fundamental change for humanity itself. And any of them develops at a speed that will create massive impacts on human lives within only a fraction of a generation. All of them together they create a depth of changing the very human life at today still unimaginable, uncontrollable, and by most humans likely unwanted steps. It is thus necessary that a *broad public* understands these dramatic changes. This is what we are aiming to achieve in this book.

We want to mention a last aspect of modern science before getting started. It concerns the appearance of new types of risks created by it that, other than single populations such as the South Americans after the European invaded their countries, can potentially lead to mankind dying out *as a whole*. An example of this is the power unleashed by nuclear bombs, but today also bioengineered pandemics (the frankenvirus[4]), genetical modifications, superintelligences and others. Thus, *risk management* is a new tool that must also be applied to science, including the assessment of such extreme risks. The key question is here: How do we evaluate scientific projects, with their highly complex questions and insights, in what their new knowledges and life benefits are, but also the risky and dangerous consequences of their technologies?

---

[4] Frankenviruses are so called from "Frankenstein" the famous hero of Mary Shelley and are viruses developed in laboratories by piecing together various parts of different viruses.

It is why, besides describing the history of science and technologies of the last 60 years, we cannot get by the important question to where we want to go with it, a question that is of philosophical nature and more important than ever in history.

On the structure of the book: The second and third chapters deal with the general development of the natural sciences from about 1960 onwards. After their centre had shifted from Europe to the USA after the Second World War (second chapter), the differentiation of the natural sciences increased strongly and - parallel to this - the financial dependence on third-party funding (third chapter). The fourth chapter then illustrates the detachment of the American-influenced sciences from philosophy, and how it needs to get back to science, this time also and especially for the ethics. The fifth chapter then presents the most significant future technologies, each of which will probably bring about a similarly profound social upheaval as the individual technological revolutions in the past (from mechanics and thermodynamics in the first, electrodynamics for the second, and quantum physics and information technology for the third industrial revolution). From the sixth chapter onwards, the focus is then on the individual scientific fields - physics (including the most recent astrophysics), chemistry (implied by physics on the one hand and providing the basis for biology on the other), biology (mainly the new genetics) and mathematics, and two core technologies that have emerged from them - in more detail:

- Chapters 6 and 7: Physics;
- 8th chapter: Genetics and Biology;
- 9th chapter: Brain Research;
- 10th chapter: Artificial Intelligence;
- 11th chapter: Mathematics;
- 12th chapter: Astrophysics.

In Chap. 13, we will outline scenarios of what potentially unimaginable social upheavals the sciences and technologies could bring in the (near) future, before the final two chapters, the 14th and 15th, that will deal with the social shaping of the future sciences and, in particular, the applications that emerge from them in order to manage the technological changes for the benefit of all humans, the development of their freedom rather than serving to their enslavement in whatever form.

So let us now dive into the for the future most fascinating and influential area that determines our future lives more than anything else. It is thus most mesmerizing and scary at the same time: We are on the one side in times of great opportunities and on the other side of great risks. It is only through reason and knowledge that we can shape an enriching future for mankind.

Lars Jaeger and Michel Dacorogna, Zug/Zürich in June 2023.

2

The Takeover of Scientific Leadership

How the Centre of Science Shifted from Europe to the USA in Just a Few Years - And How This Changed the Nature of Science Itself

As we saw in the introduction: The eighty years from 1870 to 1950, a blink of an eye in the history of mankind, was the period of the greatest revolution in thinking of all times. The drivers of these revolutions were processes in science, and the most dramatic of the scientific upheavals was what occurred in physics. However, fundamental revolutions also occurred in all other sciences, among them biology, organic chemistry, genetics, brain research, artificial intelligence, even in mathematics.

The period around 1960 to today then meant even more and more scientific revolutions, this time including and often lead by technologies that shaped up mankind itself like in an unprecedented manner and which constitute not only our modern society but will likely shape its future even more dramatically. Soon we might face a time when humanity in its biological foundation itself is changed.

In this chapter we will describe some of the key feature of science and technological development in the period from the 1960s onwards. However, in order to grasp this more deeply, it is necessary to first take a brief look back at the development of the period before, in particular the years from roughly 1900 to 1950.[1] There will be a somewhat natural focus on physics. However, we will also assess dramatic changes in other areas, like in Chaps. 8–11.

---

[1] For more details of this revolutionary developments, see: Lars Jaeger, *The Stumbling Progress of 20th Century Science - How Crises and Great Minds Have Shaped Our Modern World*, Springer (2022).

Geniuses Create a New World

Geniuses are generally characterised by more than just a particularly high talent or intelligence. As creators of something completely new out of very brilliant ideas they are particularly creative in the very meaning of the word. On top of this, they come with highest levels of perceptions and endurances in dealing with particularly difficult problems over a longer period of time, with the power to turn a domain fundamentally upside down. They thus enter realms that are not only beyond what has already been explored and accomplished, but that their contemporaries did not even imagine they existed. In this way, geniuses expand the boundaries of our knowledge and enlarge the playing field available to humanity.

In the history of science, progress has mostly been driven by solitary geniuses; Kepler, Galileo, Newton and Leibniz, or Gauß were such figures in their time (regarding Newton, see next chapter), as were James Clerk Maxwell, Ludwig Boltzmann, Georg Cantor, or Sofia Kowalewskaja in the second half of the nineteenth century. But from the first half of the twentieth century onwards, a surprisingly large number of exceptional minds worked together in all scientific disciplines and engaged in lively exchange with each other eventually (and after several crises) creating the most significant revolutions in science's history. In physics and mathematics, chemistry and biology, but also in philosophy or anthropology, and psychology or history, the genius scientists opened up possibilities for their disciplines to deal with fundamental contradictions that had come up. In some cases, they even succeeded in eliminating inconsistencies that had been known already by ancient thinkers. Their courage and achievements can thus not be overestimated.

Physicists such as Max Planck, Albert Einstein or Niels Bohr made a start when they left classical physics of the seventeenth to nineteenth century behind with relativity (combining time and location into a four-dimensional space) and first steps into quantum theory (behaviour of smallest particles), and even moved outside Western philosophy that had not been questioned for 2500 years. In fact, all sciences were until then actually founded on thoughts that had been shaped by ancient philosophers, especially by Plato and Aristotle, as well as the Catholic church, i.e. a strong belief in God, or in transcendence as far as Greek philosophers are concerned.[2]

Also in mathematics, geniuses such as Cantor, Hilbert, Poincaré, Lebesgue, Borel, Emmy Noether, von Neumann, Kolmogorov and Gödel left the safe

---

[2] See Lars Jaeger, *Sternstunden der Wissenschaften* (only in German), Südverlag (2020).

and familiar shores of the classical - rather concrete - version of their discipline and ventured out into new, much deeper waters. The first acts of genius were to mathematically tackle the (for many prior mathematicians) scary features of uncountable infinities as well as to search for ways to deal with probabilistic chances. They had to leave behind the clarity and classical logic that had given mathematicians a foothold and framework until the late nineteenth century and thus developed, respectively conquered areas of mathematics whose seeming absurdity and high degree of abstraction defy any everyday experience and conceptuality - but that did often (sometimes many years later) end up being very important for science, in particular physics. They even succeeded in "controlling chance" by transforming many uncertainties into measurable risks thus laying the ground for quantitative risk management.

In addition, from the early twentieth century onwards, the science of life also received decisive stimuli. It was not the least due to numerous physicists, such as Max Delbrück and even Erwin Schrödinger in his late life, whose focus had shifted to biology. With their mathematical and physical skills they solved riddles left behind by nineteenth century researchers Darwin and Mendel in biology and advanced the decoding of heredity by discovering the genetic code. The decisive breakthrough of the molecular gen structures came in 1953 with the deciphering of the DNA code by the physicist Francis Crick and the biologist James Watson - as well as by the chemist and X-ray crystallographer Rosalind Franklin who also played crucial roles in deciphering the helical structure of the DNA molecule but was not given the deserved credit for that. The resulting possibilities of applications in genetic engineering are affecting us today more than ever.

Another example of an ongoing revolution in science today, with potentially equally influential consequences, is the understanding of the human brain structure and our thinking, which had its origin in the discovery of neurons and synapses in our brain more than 100 years ago. Recent research in this field might even lead us to grasping the fundamentals of our mind (see Chap. 9). In addition to the scientific motivation to understand what is arguably one of the most complex structures known in the universe today, the expansion of neuroscience has been driven by two main factors. One is a growing awareness of the social and economic burden of brain disease, but also the old disease of depression (major depressive disorder), which in most industrialized countries grows with the increase of populations age. The second one is a rising confidence of researchers that brain diseases are becoming a tractable problem. Conditions such as depression, schizophrenia, bipolar disorder, stroke and age-related cognitive decline, once considered inevitable features of the human condition, are now seen as specific diseases whose causes can be identified, prevented and ultimately cured. E.g. the specific progress for brain tumours (glioblastoma), until today not treatable, are currently more and more achieved.

With the spreading of science into many different fields the list of ongoing scientific revolutions has gotten correspondingly longer. And the speed at which they have been evolving has been rising constantly - and are still rising such that we experience an ongoing acceleration of the subsequent changes in our societies.

The Shift of the Scientific Gravity Centre from Europe to the USA

In the late 1930s and early 1940s, especially driven by the Nazis kicking out key scientists from Germany and the outbreak of the Second World War in Europe, there occurred a significant geographical shift in the scientific gravity centre from Europe, particularly Germany, to the USA (which is presented in detail below). Here yet another group of geniuses brought itself onto the scene, such as Richard Feynman, who developed the first coherent quantum field theory, Robert Oppenheimer, (scientific) head of the Manhattan Project building the first atomic bomb, John Bardeen, who initiated the miniaturisation of computers with his semiconductor theory, for which he received the Physics Nobel Prize of 1956, and who later also explained superconductivity, for which he received a second Nobel Prize in Physics (1972), or, last but not least, Linus Pauling who also received two Nobel prices, one for Peace, the other in 1954 for chemistry for his work on chemical bonds.

With the US taking over the leading role in science something fundamental happened which we will elaborate on in this chapter: a major shift took place in the way science was understood and done. This change began first in physics but from there it spread into all other natural sciences, and it has transformed the face of scientific work to this day: Scientists left the field of philosophical reflection, which had played such a prominent role in European science from its very origin until then. The American scientific educational system gave much less importance to the general culture (in the spirit of Alexander von Humboldt). It placed much more emphasis on the hard sciences and their ability to solve practical problems than on philosophical questions. Thus, the researchers now devoted themselves much more to concrete application opportunities of the new scientific findings, while in Europe technological application had been advanced by engineers rather than scientist. Hence the triumphant advance of the technologies that followed scientific discoveries, which define our world to this day, had its core in the US (and has it there still today). The following list provides some important examples:

- Electronics together with the miniaturisation of processors (made possible by quantum physics) has led to the development of today's fast computers,
- Digital technologies,
- Lasers,
- Mobile phones,
- Satellites beyond the atmosphere,
- Television,
- Nuclear technology,
- Medical diagnostics and disease healing technologies,
- New materials with new functions that do not exist in nature (best example: plastic),
- Genetic engineering processes,
- Neuro-technologies,
- and last but not least, a whole new mathematics, the numerical methods of which have become so powerful (through ever and ever more powerful computers) that even world-wide hyper complex climate models have become realistic today.

The first and powerful manifestation of this development of US science and their applications was: the nuclear bomb (while the German scientists also working on it had still been in their thinking phase about it). It transformed the world in fundamental ways - changing forever the potential of global wars by making extinction of the entire mankind itself possible. Stalin, realizing after the war the importance of technological advances for the military, strongly increased the status of the academies of science in his country. Only four years after the US, his USSR had constructed the nuclear bomb themselves, with the help of Soviet spies in the US and the UK, in particular Klaus Fuchs (who emigrated from Germany to the UK and then, as of 1944, participated in the construction of the nuclear bomb in the US), sympathizing with Marxist ideology.

In the US as well as in the USSR, the new power of sciences was first directed towards technological developments mostly for military purposes. Other than in the US the scientific focus in the USSR remained largely in the military area, while in the West the triumphant advances of the technologies shaped the entire society. It improved everybody's daily life, and still does so today. To the broadening of wealth, another consequence of scientists leaving the field of philosophy and classical education was opening up the interest of and the access to sciences to all people, including those coming from less educated origin.[3] In a way, it brought a democratisation of the access to science, by introducing a way for gifted young people to climb up the social ladder.

The separation of science and philosophy had yet another consequence: Core philosophical questions remain unanswered by science. While still being influenced by its progress, questions such as the limits of reason, the implication of randomness in nature, the existence of reality outside the observer, and more fundamentally the very nature of life and consciousness were simply ignored. The human Ego remains largely a "terra incognita" today. This leads to strange contradictions, such as although we do not concretely understand what exactly and why something happens in the subatomic world, we can calculate it more exactly than anything else before, and accordingly also control it technologically. Let us thus make a quick excursion around the philosophical discussion in quantum physics.

Philosophical Implications of Quantum Theory – The Concept of Reality Called into Question

Philosophy is still fundamentally important for physics (and vice versa). For more than three hundred years, until the early part of the twentieth century, classical physics and classical philosophy, the later going back to the Hellenistic tradition - supplemented by modern philosophies, for example Immanuel Kant's - were inextricably interwoven in the Western world. During this time, it was natural for physicists and other scientists to participate in philosophical discussions, particularly as they saw their laws as absolutely given and thus being a solid fundament of philosophical discourse as well. Conversely, many philosophers were able to discuss deep connections in the natural sciences, as well; Kant, for example, also taught Newton's physics at his university. It was not until the middle of the nineteenth century that mathematical calculations in physics became so complex that only a few philosophers were still able to assess the advances in physics. The physicists, on the other hand, continued to know and apply philosophy very well.

---

[3] For the separation of science and philosophy, see: *The Origin of the Separation between Science and Philosophy*, Proceedings of the American Academy of Arts and Sciences, Vol. 80, No. 2 (May, 1952), pp. 115–139.

Heisenberg, for example, dealt intensively with Plato, Einstein with Spinoza, Leibniz and Berkeley[4] and Mach (he also had a long philosophical controversy with Bergson in 1922 about the concept of time), Langevin with Marx. One particular aspect of Western philosophical thought shaped by Plato and Aristotle was to cause physicists a lot of trouble in the first decades of the twentieth century: the dualism that separates the world in two parts, objective nature and human subjective experience and thinking. This dualism even became the basis of classical physics with two scientific principles:

- To find the material, unchanging building blocks of the world,
- To extract the substance of things from the confusion of subjective impressions in the form of eternally valid laws with the help of objective, repeatable experiments that provide us these laws of nature.

In the 1920s, however, it became known that strange conditions prevailed in the quantum world, in particular the simultaneous feature of electrons and other smallest particles as well as for electromagnetic waves of being *particles as well as waves* (however never at the same time) which in classical physics - as well as in our daily experiences - is impossible. It depends on the *subjective choice* to perform a particular experiment which one is obtained. What a strange thing this was, given the 2500 years tradition of decisively separating objects and subjects. However, the properties of quantum objects, electrons, photons, etc. are not only undetermined before the measurement, but they *do not exist* at all! To put it very clearly: a quantum object has no independent properties before the measurement! Only the measurement gives it such. They have no reality, but "only" a *potentiality*, as the physicists say. Consequently, the core of the description of a quantum object is the *wave function* (more about it below) that does not describe a "real" physical entity, but rather a mathematical, probabilistic one.

With quantum theory, central concepts of classical thinking disappeared from the map of physics (as well our everyday mental map)[5]:

1. Reality
2. Causality
3. Identity.

The last point, the indistinguishability of quantum particles, contradicts the classical philosophical principle that Gottfried Wilhelm Leibniz had

---

[4] In his book *De Motu* Berkeley argued against Newton's doctrine of absolute space, time and motion.
[5] For more details see: Lars Jaeger, *The second Quantum Revolution*, Springer (2018).

formulated 250 years earlier, but which Aristotle already knew: the *principium identitatis indiscernibilium (pii)*. This "theorem of the identity of the indistinguishable" says that two separate things that are completely the same in *every* respect cannot exist. For Immanuel Kant it is above all the factor of "place" that is significant for each particle: Even if the properties of two parts are of the same material and are ruled by the same mechanism and thus appear to be very much the same, they must at least be in different places. The fact that pii does not apply in the micro-world is not a theoretical gimmick, but the very reason why in quantum theory fundamentally different laws apply than in classical physics.

Highly Controversial Philosophical Discussions Among Physicists in the 1930s

But if very different laws apply to the micro- and the macroworld there needs to be an area in between them in which the laws transfer from one area to the other. This caused a subtle philosophical discussion between the physicists in the mid-1930s. Let us have a rather quick review of it.[6] Bohr had pragmatically stated that the laws of classical physics apply to "sufficiently large" systems and the laws of the quantum world apply to "sufficiently small systems". Where exactly, however, on the size scale the transition is located, he had to leave entirely open. However, he and his followers had to assume that the transition does not happen gradually, but that there must be a sharp split. The alternative would have been that there is a region in which two different types of laws apply simultaneously.

Schrödinger and Einstein recognised that Bohr's interpretation only concealed a central logical gap in his interpretation: Somewhere on the size scale the quantum world and the macroscopic world, in each of which completely different laws apply, must touch. Where exactly on the size scale is that transfer located?

In order to illustrate this issue Schrödinger described in his 1935 essay "The Present Situation in Quantum Mechanics"[7] his famous thought experiment with the cat. With it, he wanted to reveal the absurdity of the cut between quantum and macros laws. In a locked steel chamber, there is a cat

---

[6] For more details see: Lars Jaeger, *The Stumbling Progress of 20th Century Science - How Crises and Great Minds Have Shaped Our Modern World*, Springer (2022).

[7] Erwin Schrödinger, *Die gegenwärtige Situation in der Quantenmechanik* (The Present Situation in Quantum Mechanics), Naturwissenschaften. Volume 23 (1935), Pages 844–849 (English version originally translated in: Proceedings of the American Philosophical Society, 124 (1980), p. 332–338; available under http://hermes.ffn.ub.es/luisnavarro/nuevo_maletin/Schrodinger_1935_cat.pdf).

and a quantity of a radioactive substance calculated in such a way that statistically one of its atoms decays per hour. Since only an average is possible to steer, it may be that one, two, three or even no atom decay within an hour. Each decay of an atom is detected by a fine measuring device; if there is a decay, it shatters a container of deadly hydrogen cyanide. After an hour, an observer will not know whether one of the atoms has decayed and therefore whether the cat is alive or not. The genius of Schrödinger's thought experiment is that it causally couples the quantum and macro worlds directly with each other and thus makes the strict separation postulated by Bohr and Heisenberg permeable: Only one atomic nucleus, i.e. quantum particle has to decay for a macroscopic object, the cat, to be directly affected (here: killed).

- As long as no measurement takes place on the atomic nucleus, it is in a state of superposition of "decayed" and "not decayed". This is quantum physically possible but does not exist in classical physics.
- In Schrödinger's experimental set-up, also the state of the cat in the steel chamber must consist of a superposition of "dead" and "alive": As long as the door to the box is not opened, the cat is objectively in both states at the same time! This is macrocosmically not possible.

This cat paradox was precisely Schrödinger's intention. Nobody doubted that the quantum laws were valid in the microworld. The question was: how far does their power extend into the macro world? Where in the chain *atomic nucleus - measuring device - prussic acid ampoule - cat - observer* must one place the transition from the quantum world to the macro world, in which we may trust our usual views and objects have clearly defined and independent properties?

Physicists had quite different answers to where to make the cut between macro- und microworld. And Schrödinger himself? Where did he make the cut? He found his own way of approaching this question: He looked at the measurement process, because it is precisely there that a device from the macro world comes into contact with a quantum object. While other physicists strictly separated the macroscopic world from the quantum world, Schrödinger consistently applied the principle of the common wave function to macroscopic measurements as well, i.e. to the moment when the macroscopic measurement system comes into contact with the quantum object to be measured. Somewhat casually, he remarked:

> It [*the Ψ-function of the measured object*] has, according to the inevitable law of the total Ψ-function, become entangled with that of the measuring instrument […]

Schrödinger thus gave his interpretation its own name, which to this day can hardly be surpassed in importance: "entanglement":

> This property [*entanglement*] is not one, but *the* property of quantum mechanics, the one in which the entire deviation from the classical way of thinking manifests itself.[8]

The most bitter opponent of the quantum uncertainty interpretation was, however, Albert Einstein. In 1935, he and his American assistants Boris Podolsky and Nathan Rosen published the article: "Can the quantum mechanical description of physical reality be considered complete?"[9] The thought experiment therein, later called the Einstein-Podolsky-Rosen paradox (EPR), was as ingenious as it was simple: Two quantum particles in superposition with each other are parts of a common wave function. Their momentum cannot be determined exactly, but it is known that the momentum of one particle is directly related to the momentum of the other particle. Now the two particles are separated from each other and brought to two different places. Then one of them is measured for his momentum so that the common wave function collapses so that the momentum of the partner particle is now also (instantly) defined. That cannot happen instantly though, as it breaks Einstein's law of the speed of light being fastest possible progress of information.

This was a rather clever move, which Bohr had little to oppose against. Nevertheless, the shot backfired, albeit only after Einstein's and Bohr's death: Precisely the falsified and seemingly impossible instantaneous action at a distance is the basis of many technological applications today, not least a possible future quantum computer. French physicist Alain Aspect's experiment from almost 50 years later, in 1982, validated the quantum entanglement and locality principles.[10] It also offered an experimental answer to Albert Einstein, Boris Podolsky and Nathan Rosen's paradox. Hereby, the laws of Einstein are not broken, as no information can be transferred in the described process.[11] For his experiment, Alain Aspect received the Nobel Prize in Physics another 40 years later, in 2022.

---

[8] E. Schrödinger, Discussion of Probability Relations between separate systems. *Proceedings of the Cambridge Physical Society*, 31, (1935) S. 555.

[9] Albert Einstein, Boris Podolsky, Nathan Rosen, Can Quantum-Mechanical Description of Physical Reality be Considered Complete?, Physical Review. 47, 10 (1935) Pages 777–780.

[10] Alain Aspect, Philippe Grangier, Gérard Roger, *Experimental Realization of Einstein-Podolsky-Rosen-Bohm Gedankenexperiment: A New Violation of Bell's Inequalities*, Physical Review Letters, 49 (1982).

[11] In order to transfer information, one needs to know the initial state of either side. This can only obtained by a measurement. And these results have to be transferred by some kind of traditional transmission technology, not faster than the speed of light.

We see: Only when they had clarified these questions for themselves did the European developers of modern physics use mathematics to put their theory into suitable forms. Einstein and his colleagues understood their activity as part of a broader philosophical tradition in which they also saw their intellectual home.

How Europe and Philosophy Both Lost Their Dominance in Science

In 1935, the dispute about the philosophical background of quantum physics climaxed. Einstein and other researchers such as John von Neumann, however, had already emigrated to the USA, and thus, the discussions took place over great distances. With that year, the philosophical discussion about quantum physics abruptly broke off entirely. Hardly anyone was further interested in the inner contradictions that had disturbed Einstein and Bohr so deeply. This had to do with a generational as well as the above already mentioned geographic change of the scientific centre after which the vast majority of physicists were no longer bothered by the fact that they could not explain *why* quantum mechanics worked, but simply applied it mathematically and in experiments. Only four of the greatest physicists of the twentieth century - Einstein and Schrödinger as critics of the given quantum mechanics, Bohr and Heisenberg as their advocates - continued the search for an explanation until their deaths. Despite all their efforts, they had to leave open the questions about spooky remote effects, entangled particles and half-dead cats until the end of their lives. For a long time, they were the last leading physicists for whom the philosophical background of their subject had any meaning. For the others, the successes of mathematics - and gradually technological applications - had become more important than unanswered philosophical questions. The same is true for British science and mathematics, despite Paul Dirac.

What caused the dramatic shift in particular? Until the mid-1930s, the development of physics (quantum mechanics and relativity theory) as well in all other sciences and mathematics was almost entirely driven by German-speaking scientists. Göttingen, Zurich and Berlin were the world's centres of theoretical physics and mathematics. The Dane Bohr himself spoke and loved this language like his mother tongue. Even the Englishman Paul Dirac spoke fluent German.

The elaboration of the mathematical foundations of the new physics was also shaped by German-speaking mathematicians: David Hilbert, Emmy

Noether and Hermann Weyl worked in Göttingen (Weyl was in Zurich for a long period but then took over Hilbert's chair in 1930). The best students and young scientists in physics and mathematics flocked to Germany from all over the world to learn and do research there. Plus, major progress in chemistry, biology and medicine originated in Germany. Paris, a leading place for mathematics and science for centuries, still had Borel, the leading mathematician of probability and statistics, Hadamard, who made major contributions in number theory, complex analysis, differential geometry and partial differential equations, and Louis de Broglie, the discoverer of wave properties of quantum particles, but overall had a much lower position in mathematics and science than one generation earlier at the turn of the century.

However, in 1933 the Nazis took over Germany and immediately applied a ban on Jewish scientists and mathematicians. Many scientists (even non-Jewish ones, when they had their own mindsets and were thus threatened by the Nazis) had to leave Germany. For most of them the US was the best place to move to. Among them were important names, but also highly talented young researchers.

- Albert Einstein was German (and of Jewish ancestry) but despised the conservative atmosphere in the country of his birth. As a young man he was stateless for a few years and in 1901 he took the Swiss citizenship, which he kept until the end of his life. Because he accepted a professorship in Berlin in 1914, he had to apply for a German passport again. In 1933, he gave up this citizenship for good and left Germany with the firm intention of never returning (which he fulfilled for the rest of his life).
- German Jewish statistician Emil Gumbel was one of the first to leave the University of Heidelberg in 1933 where he was a constant source of controversy with his pacifism and his leftist positions. After a short stay in France, he settled in New York where he taught at Columbia University. He ended up being one of the pioneers in statistics of Extreme Value Theory of which he wrote the first book dedicated entirely to this subject.
- Wolf Price recipient (in its inaugural year in 1978, together with Israel Gelfand), Carl Ludwig Siegel, considered then as the greatest living mathematician was an antimilitarist at the end of the First World War. Although he was not Jewish, he left Göttingen in 1940 and went to the US via Norway, where he joined the famous Institute of Advance Study in Princeton with Einstein and Gödel. He only came back to Germany in 1951.
- The Hungarian John von Neumann (born János Jajos Neumann) had already attended the German-language grammar school in Budapest and

lived in Berlin, Zurich and Göttingen as one of the most brilliant mathematicians in the 1920s and early 1930s. Like Einstein, he emigrated to the USA in 1933 (and - also like Einstein - despised Germany until his death), where he played a crucial role in the development of the nuclear bomb.
- The equally Hungarian Edward Teller also emigrated to America and later played a significant role in the development of the American hydrogen bomb.
- Hans Bethe worked on the properties of the electron in the 1920s and early 1930s, emigrated first to England in 1933 and later moved to the US.
- Emmy Noether, arguably the greatest female mathematician in history, was forced to leave Germany because of her Jewish background and went to the USA.[12]
- Max Delbrück, a theoretical physicist who shifted to biology, in 1935 (before he emigrated to the US), together with two others, published a work on gene mutations, in which they were the first to propose that genes can be understood as complex atomic assemblies. This was the beginning of modern genetics.
- An example from the Nazi's fascist partner: Enrico Fermi, the creator of the first nuclear reactor, the Chicago Pile-1, left fascist Italy in 1938 to escape the Italian racial laws that affected his Jewish wife, Laura Capon.

Many more names of great physicists and mathematician, but also biologists and medical researchers could be cited here. The brain drain was exorbitant and left behind a devastated scientific landscape in Germany. When the (already retired) mathematician David Hilbert attended a banquet in 1934, where he sat next to the Nazi Minister of Culture Bernhard Rust, the latter asked whether "the Mathematical Institute had really suffered so much because of the departure of the Jews". Hilbert's answer was: "Suffered? Mathematics has not suffered, Herr Minister. There is no such thing anymore!"

The centre of scientific activity shifted within not more than five years from the German-speaking countries to the English-speaking USA. And this was as we saw more than a mere change of location or language: Because in the USA a completely different intellectual climate prevailed, a new kind of theoretical and experimental physics (as well as of other sciences) developed. Here, people were simply not interested in the question of "what holds the world together at its core", but rather "what can we do to solve specific problems". Mathematics was to take over the explanation of the world, not the

---

[12] For more details on Emmy Noether, see Lars Jaeger, *Emmy Noether – Her rocky path to the world's top of mathematics* (only in German) Springer, 2022.

human imagination. With its help, all problems of physics were supposed to be solved. The focus of science now became the technological application of scientific knowledge, and this at first primarily for military purposes during and after WW II.

Pragmatism Now Governing Science – With Consequences

The skill in dealing with complex variables and operators and thus with a more and more abstract mathematics was the primary focus of theoretical physicists in the US. Based on this the Standard Model of particle physics (read more in Chap. 6) was established largely in the US in the 1960s and 1970s - and is still valid today, although lacking any elegance (perhaps a consequence of the "fumbling" in complex spaces). Its development is a triumphal example of the corresponding particular way of doing theoretical physics: Pragmatic and sober, relying more on mathematical virtuosity than on the ability to reflect deeply through difficult conceptual and last but not least philosophical problems. To put it simple: While in the European tradition, physics had begun with an empirical observation and thereupon based theoretical and philosophical concepts, and only in a second step had been translated into mathematics, Americans physicists started with mathematics trying to find proper theories, and only then searched for experiments to verify the theories. Philosophy thereby plays no role. The physicist David Merlin summed up this new methodological style in theoretical physics: "Shut up and calculate".[13]

A particular abstractly thinking American physicists was Richard Feynman. He succeeded in developing a new interpretation of the Dirac equations so that it was possible to precisely calculate the electromagnetic interactions between quantum entities. In order not to lose track of the many individual calculation steps for a system - the number of calculations is in fact theoretically even infinite in quantum field theory - Feynman developed very abstract mathematical tools with which the complexity of the trajectories could be

---

[13] N. David Mermin, *What's Wrong with this Pillow?*, Physics Today (April 1989), page 9. This phrase is sometimes also assigned to Paul Dirac or to Edward Teller talking to Richard Feynman. Interestingly, already in the seventeenth century Leibniz was seeking a solution to some of the denominational quarrels that were plaguing his generation by envisioning a calculus ratiocinator that would make it possible for the quarrelling parties to "sit down and calculate"; see Wolfgang Lenzen, *Leibniz and the Calculus Ratiocinator*, Springer (2018), German original: *Calculus Universalis — Studien zur Logik von G. W. Leibniz*, Paderborn, mentis (2004); or see https://en.wikipedia.org/wiki/Calculus_ratiocinator.

systematically recorded. He had recognised that the integrals occurring in the calculations have a certain regular common structure and that each one can thus be traced back to certain mathematical building blocks. From this observation, Feynman derived an ingenious trick that is still intensively used today: He introduced new mathematical operators, so-called "propagators". With them, Feynman designed diagrams - now called "Feynman diagrams" - whose lines and points ("vertices") are not only graphic illustrations of particle reactions, but actually represent certain calculation rules for the propagators. He created, so to speak, directions for venturing through the impassable jungle of quantum field theory.

Of course, Feynman's propagators do not describe real paths of particles or locations of interactions and therefore must not be understood as descriptions of concrete spatiotemporal processes. The interaction between two quantum objects through exchange of particles is a process that takes place in many dimensions and cannot be fully represented by human visual comprehension. However, against this background, Feynman's diagrams work very well. The so-called g-factor in atomic theory was only explainable in quantum electrodynamics. Its theoretical value agrees with the experimental measurements to twelve(!) decimal points, i.e. the deviation between the two lies at 0.00000000001%. For no other value in physics or any other science today there is such agreement between theoretical calculation and experimental measurements. Modern quantum field theory from the USA was a triumph of very complex, but at the same time pragmatic mathematics. Again, virtuosity in dealing with variables and operators was in demand rather than the ability to think through difficult conceptual problems together with philosophical questions associated with them.

However, the calculations of interactions in the quantum equations with Feynman rules often result in infinities that could not be so easily circumvented. Finally, in the early 1970s a trick was developed that physicists today call "renormalisation". This was quite a non-elegant circumvention of the infinities, but it "somehow" worked, despite the fact that mathematics did not offer as secure a foundation as the physicists would like it.

The renormalisation is comparable to Max Planck having introduced quantised physical quantities in 1900 out of sheer desperation to somehow get ahead in his calculations and match their results with experimental ones. Planck was not happy about this, and called his quantum variable "h", which for him described the character of an "auxiliary variable" (in German "Hilfsgröße") that should be removed from the equation again as soon as possible. With renormalization the physical quantities under consideration now assumed the desired finite values, and the calculated values even agreed with the experimentally measured values. This is yet another parallel with Planck: he did not have "permission" for this intervention, but it worked. Planck's quantum introduction became the centre of a physics' revolution, while the renormalization as a tool in the computational procedures of quantum field theory to date do not exactly fit pure mathematics, so quantum field theory does not yet have a clean mathematical foundation.

Since there are no practically usable alternatives and their success proves them right, physicists and mathematicians continue to use this "dirty mathematics" to this day. However, they were and are still not entirely comfortable with it. Most of them admit that renormalisation only sweeps the problem of infinities under the carpet. Paul Dirac[14] remained a critic of renormalisation throughout his life. Even the pragmatist Feynman expressed reluctance:

> The shell game that we play ... is technically called 'renormalization'. But no matter how clever the word, it is still what I would call a dippy process! Having to resort to such hocus-pocus has prevented us from proving that the theory of quantum electrodynamics is mathematically self-consistent. It's surprising that the theory still hasn't been proved self-consistent one way or the other by now; I suspect that renormalization is not mathematically legitimate.[15]

The price for almost the entire range of focussing solely on mathematics was a lack in answering some fundamental questions.

A similar approach has today been taken with the development of neural networks and other non-linear tools to deal with statistics of data in higher dimension (see below and in Chap. 10). The goal of fitting the data well has prevailed in understanding the underlying phenomena. The success of artificial intelligence (AI) techniques in many applications ranging from pattern recognition to translation and chat bots or Go playing, as well as analysing large physical data sets (it was the physics centre CERN that was one of the first using artificial neural networks) is a testimony of its efficiency. Still, a deeper understanding of how the neural networks work is entirely missing. The question of understanding the details on what processes are involved in these activities (and how these might relate to those in the human brain, i.e. to natural intelligence) thus remains largely unexplored, and with only a few exceptions (example: French mathematicians Jean-Paul Benzécri and Stéphane Mallat) it does not really interest their developers to understand

---

[14] Side note: Dirac himself used a mathematical construct to describe quantum mechanics, now called the Dirac Delta Distribution, that did not yet have a sound mathematical foundation until a few decades later.

[15] Feynman, Richard P., *QED, The Strange Theory of Light and Matter*, Princeton, Princeton University Press (1990), page 128.

the AI black box in detail, perhaps at the cost of not being able to explore particular drifts in our brain that we are not yet able to measure.

Philosophy for Quantum Physics - Still Essential Today

Mathematics can be regarded as immaterial substance, and thus cannot really cause philosophical problems one could argue (with the exception of the notion of infinity and consistency). But what about material substances physics is dealing with? Quantum objects have no objective physical properties. Mass, charge or spin as properties are qualities that come into play solely through interaction with their environment. Philosophically, this means: There is no substance in the classical sense the properties of which are determined unambiguously and independently of the measurement, but only accidents in the form of transient interactions between quantum "objects" (in philosophy "accident" means that the considered entity contingent, i.e. depends on its environment). This results, as we saw, is a fundamental difference between the macro and quantum world:

- In the traditional metaphysics of Western philosophy, interactions come into the world as a consequence of existing things.
- In the quantum world, it is the other way round: Interactions constitute things.

Although traditional metaphysics dominated the Western philosophical thinking for 2500 years, the inverse concept of reality such as in quantum physics is not entirely new and unique in the history of Western thought:

- In the teachings of some pre-Socratics, the dualism between substance and accidence had not yet developed its philosophical dominance.
- In Kant's philosophy, the concept of reality of quantum physics already resonates: Objects only acquire their fundamental properties through our perception and thinking.
- The philosopher Edmund Husserl also speaks of two separate dimensions in the process of cognition: the "act of consciousness", i.e. the way information is processed in our brain, and the phenomenon towards which it is directed.

Fundamental questions that arose with quantum theory are still open. These correspond essentially to the open problems that today's theoretical physics has to solve on its way to the desired unifying theory of nature - from which we are still far away as it seems. We will discuss their status in detail later (Chap. 6) and only summarize them here:

1. The first question deals with the connection between general relativity and quantum physics, the two core theories of current physics, the first for the macrocosm, the second for the microworld. Can these be represented in a unified theory? This involves the question of whether all various particles and forces can be unified in a single theory that is able to explain nature as an expression of a fundamental law. The unified theory will then most likely also explain the many unexplainably free constants in the standard model of particle physics. What makes this so difficult is that quantum theory and general relativity are mathematically inconsistent against each other.
2. The second problem concerns the concept of reality in quantum mechanics, especially the transition from the quantum mechanical nanocosmos to larger systems. Here, it is particularly about the measurement problem, in popular representation by Schrödinger's cat. Can the dichotomy between observer and observed object be abolished in a universally valid quantum physics? It was this question of reality that was Einstein's fundamental problem with quantum mechanics as described before.
3. The following problem is of fundamental nature: How can we set up a rational theory for the entire cosmos, which *ipso facto*, as an unavoidable consequence, allows no external observer, no external time measurement and no experimental configurations as every observer would simultaneously be part of the entire cosmos? If the quantum theory also applies to the macrocosm the entire universe should be one huge quantum function. This is difficult to express.
4. The fourth problem is the cosmological riddle of dark energy and dark matter. Some indirect observations could be explained by postulating their existence (solving the problem why gravity seems to behave differently at the scale that describes the universe as a whole than at planetary or intragalactic distances), but there exist neither a convincing theory for them nor direct observations proving their existences (see Chaps. 6 and 12 for more details).
5. Besides the pure scientific question, we also remain with fundamental philosophical questions that are still not answered or even not treated

in philosophy any longer, like e.g. the existence of contradictions or the failure of the "Tertium non datur" principle (the law of excluded middle). Can we reasonably live without the later principle (or is it at least valid for our mesocosm, i.e. the magnitude of our lives)? How should we deal with the contradictions since we cannot resolve them? All of these issues deserve debate and philosophical investigation and should no longer be swept under the rug. Solving them would undoubtedly represent a definite advance in our understanding of our inner world and a progress for formulating a modern ethics.

Quantum physics and the loss of objectiveness in the microcosm have nevertheless not changed some fundamental beliefs of theoretical physicists today. Behind their efforts to find a unified theory lies their almost seemingly religious, but surely philosophical, basic belief that uniform principles fully assess all of nature. In other – "Platonic" - words: there is an "Idea" behind all phenomena. This creed is the ancient Greek heritage of physics. With his assertion that the unity of the fundamental laws can be grasped in the language of mathematics, Galileo transformed (some would say translated) the Greek heritage into modern times. He was the first to formulate (but not to develop, that was Newton) this fundamental conviction of physicists that is still valid today. But the last reasons why the fundament of nature is assessable in the way physicists have done it in the last 350 year remain open just as the fundamental philosophical questions do.

Scientific Revolutions Beyond Physics

We have so far predominantly looked at physics and the events in it, as its foundational change was possibly the most dramatic scientific development in the twentieth century (biology maybe beating it in the late twentieth century). However, in addition to physics, biology and medicine also contributed significantly to the development of technologies that initially determined the outcome of the Second World War and were only later - at least in the West - used for civilian purposes. They ultimately determine our modern life today by their far-reaching technological applications. With these sciences, other ones joined the constant revolutionary state equally contributing to our lives today, as we will learn in Chaps. 7–12.

Let us take a look at another decisive development in science and then technologies beyond physics: In the Second World War the US military supported with all its forces the mass production of penicillin, the antibiotic protection against bacterial infections rediscovered in the early 1940s (first seen in 1928). For D-Day, the planned invasion of Normandy on 6 June 1944, 2.3 million doses of the antibiotic were provided to the Allied troops (as we will further outline in Chap. 3). With these it was possible for the Allied invasion forces to mitigate infections in soldiers by administering penicillin and avoid major human casualties, while German soldiers still suffered strongly from bacterial infections, resulting in heavy casualties. This had a decisive impact on the outcome of World War II, of the same order as that of the atomic bomb about one year later.

The discovery and then mass production of penicillin is an example of finding fundamental scientific features by "trying things out" or even by accident, which then have a massive impact on our lives. Another fundamental breakthrough in biology was an equally largely experimentally backed discovery: finding the structure of DNA. This triggered an entire field of new biological research that is still growing today. Accordingly, while the physicists were thrilled to have finally taken a big step towards precise findings and accuracy in predicting the microcosm, the biologists were excited that they had finally discovered the fundamental structure of life (although it took the DNA model almost a decade to have it fully accepted).

The speed with which new technologies are being created has grown exponentially in the last few years. The range of technologies that shape our daily lives seems endless nowadays. It is clearly to be expected that this speed will further accelerate (with possibly dramatic consequences on our whole life).

Next to this rapid growth of science and technologies, today there exist many disciplines *across* respective fields (see Chap. 7), some examples of which are:

- Chemists and physicists assessing the complexity of molecules with the theory of *emergence* (Chap. 6),
- Biologists and chemists working together in *biochemistry* on the fundamentals of molecular genetics (Chap. 8),
- Physicists and biologists study living systems with new tools for understanding how the biology of all types of life works (Chap. 8),
- Physicists and mathematicians created the foundation for modern information theory and artificial intelligence (Chaps. 7 and 10),
- Biologists, chemists, physicists and neuroscientists developing an ever deeper understanding of the processes in our brain and thus not least of many mental procedures (Chap. 9),

- This leads to: (what we already looked at above): Deeper understanding of neural networks combined with research on artificial intelligence (Chaps. 9 and 10), and
- Physics and astrophysics now assess the entire universe (Chap. 12).

The number of research fields developed rapidly, ant their researchers exchange their ideas more and more with each other. Respective new technologies are today emerging in a space that ranges from new quantum technologies to optimisation algorithms, from nano-chemistry to reproductive genetics, from artificial intelligence to robotics, from atomic physics for imaging techniques to new brain and consciousness tools.

Next to unbelievable new technologies, pluri-disciplinary research is also key for the understanding of complex problems. It is thus worrisome that the development of the research *organization* is hampering this process by maintaining researchers in silos (see Chap. 14 for details on this point).

Replacing the historically European based approach of science to deal with the fundaments of the world and understand its structure first before applying them to technologies, i.e. applying the inverse of the ranks, 1. science - 2. technologies, has changed the world and still shapes it fundamentally today. However, besides some key question of physics, fundamental questions of *philosophical* character that people have been dealing with for millennia remain largely open (see Chap. 4):

1. How did the world itself come into existence (the very first part of the Big Bang)?
2. On the deepest level, what is everything made of (beyond 10^{-18} m)?
3. Where does life come from?
4. What constituted the human being and his self-conscious mind? Do we need knowledge beyond science to understand the essence of our mind?
5. Is there a deeper meaning of our existence (and our death)?

Sciences and philosophy have in recent years moved a little closer together again, at least in certain disciplines, but still remain much more separated from each other than before the 1940s. However, one should expect that their possible future interplay might yield further exciting insights into what builds and shapes our world as well as our subjective experience and what holds them both together. In any case, a strong interaction between philosophy and science should help us better navigate among the new possibilities offered by scientific applications and not let those applications overwhelm humanity. We shall look at the details of this in Chap. 4, after having seen in the following chapter how important (and scary) *capitalism* has become for science in the last 60 years.

3

Publicly Backed Science in Competition with Private Companies
Science as Part of Capitalism

The Transition from Fundamental Scientific Research to Technological Applications

There is joke among scientists (in some versions among the physicists only) describing the importance of financial considerations in today's science. It describes a new fundamental physical unit of measurement (next to meter, second, ampere, kelvin, mole and candela): the *US-dollar*. The function of this unit is to describe the probability of an experiment taking place.

To grasp the relationship of the sciences to money today, it is worth looking at the historical development of their funding. In the old days before the nineteenth century, rulers and rich individuals funded the work of scientists for the idealistic sake of supporting the progress of human knowledge about nature. Their driving factor was pure curiosity without any goals of direct financial profitability of the obtained new scientific knowledge. Scientists were not seeking direct financial gain from their work. It is only through the fame of it that they could hope for financial stability. Typically, the suppliers financed a particular single person or maximally a handful of scientists. Sometimes a scientist himself was rich enough to fund his work himself, Examples are Charles Darwin in the nineteenth century who was, via his parents, rich enough to finance his research that lead to his revolutionary evolution theory, or in the eighteenth century Voltaire, who actually, besides his writing, also performed a significant amount of science and mathematics work - together with and led by - his love partner, the brilliant woman Émilie du Châtelet, or last but not least the physicist Horace-Bénédict de Saussure who financed himself in August 1787 his ascension of the Mont Blanc with several servants and guides to verify the change of measured weight with altitude.

From the mid-nineteenth century onwards, science, i.e. the technologies implied by it, broadly improved the quality of life. Its need intensified tremendously as the societal demand for new useful scientific discoveries increased while the pace of scientific advancements equally grew. Consequently, governments became the major money providing entities. The money was given to the chairman at the university, who distributed it, so a professor rarely needed to recourse to extra-funding for his research. His academic chair usually came with its own funding. This continued to largely give scientists the freedom to choose their research areas. This allowed them to remain free and independent thinkers. This was called academic freedom. In the early twentieth century until about the middle of it, this led to the most revolutionary scientific progress in the history of mankind,[1] driven from Europe with the US benefitting from but not shaping the results of the new scientific knowledge.

From the early days in the sixteenth/seventeenth century onwards, modern science has often disproven existing, religiously and politically founded world views and brought more and more understanding of the natural world. At latest in the Enlightenment Era it has also become a political factor, as it triggered new views on society structures when God was no more a justifying factor for the existing ones. In the second half of the twentieth century, however, another factor added to scientific research: When the US took over the worldwide scientific leadership during and after World War II, the nature of science itself moved from the fundamental search of scientific laws to the development of technological applications primarily.

The integration of technology and science had a tradition in the US. US science already existed before WW II and came already with a certain shape, i.e. technological efforts were more directly integrated with science. One example was the work and research of Nikola Tesla, the Serbian-American physicist, inventor, electrical and mechanical engineer. He knew physics such that he even went as far as criticizing Einstein's theories of relativity. His name became even the name for magnetic flux density (one Tesla equals one Kilogram/(Ampere times second squared)). However, his life's work was eventually characterised by numerous innovations in the field of electrical engineering, such as the development of what is now known as the two-phase alternating current system for electrical power transmission which became the standard in the world. He developed all of this in the US, where he had

---

[1] For more details see: Lars Jaeger, *The new discovery of the world - How geniuses in crisis led Science and society into modernity*, Springer (2022).

moved to at the age of 28. Tesla was granted more than 280 patents in 26 countries, 112 of them in the USA. The choice of the name of Tesla for Elon Musk electric car brand is another tribute to this brilliant inventor (if one wants to call it an honour).

European scientists were rather working in ivory towers of science with not many direct links between them and engineers. The later surely existed in Europa as well, like Werner von Siemens for electrical application, Gottlieb Daimler and others in Germany for fist car production, or Gustave Eiffel, André Citroën who built one of the first front-wheel car, Conrad and Marcel Schlumberger in France, Brown Bovery and Escher Wyss in Switzerland or Isambard Brunel and Guglielmo Marconi in the UK. The latter even got the Nobel Prize in Physics in 1909 with the German physicist Karl Ferdinand Braun "in recognition of their contributions to the development of wireless telegraph".

An example of Europeans scientists creating the scientific fundament and the US engineers eventually creating its applications was, as we saw in Chap. 2, quantum physics. The entire theory came from Europe, while the applications originating from it came almost exclusively from the US, starting with the nuclear bomb, the micro-transistor and the microwave, all three of which are still shaping the world today. An example outside of physics was the large production of penicillin during the second world war: Discovered in England it was produced on a large scale in the US, see further down for the details.

In summary, the US was already a leader in technological development in the 1930s, but not yet in science itself. Most American scientists in fact received their education in Europe, particularly in Germany. With WW II the US then took over the leadership in science from Europe which resulted in the integration of science and technological advancement.

Relationship of Sciences and Technological Applications Today

Today the technological applications of scientific efforts have globally become the main role of publicly sponsored research rather than the need to further understand the natural laws themselves. Today's politicians often claim that public science centres such as universities or other publicly financed research facilities should produce new technologies and start-ups, not just new pure scientific knowledge per se. What they often do not realize is that valuable applications almost always rest on fundamental science (which is why many founded technological companies are close to top universities, such as Stanford, Harvard/MIT, in Europe Paris, Munich, Cambridge or Zurich). This does not mean that past politicians knew more about science than todays, but they simply left scientists mostly independent and thus enabled them to do their core research and to choose what were the most relevant scientific questions. Science simply did not play significant roles in their political agenda and in society in general. In contrast, since World War II scientists experience parallel to their increasing needs for funding, very strong political influences on their research by external decision makers and their choices of where the money for science is going to. Thus, the available money in favour of fundamental science has - relative to the size of scientific research overall – decreased over time.[2]

However, the overall resources accessible to do fundamental science today is also significantly higher than it has ever been before. This is drawn by the fact that a large part of today's scientific research costs a lot more than it did in the past. In fact, the costs of science have grown already exponentially even from its early times in the sixteenth/seventeenth century until the mid-1930s. This exponential growth has continued and, in many areas, even grown with a higher exponent since the 1940s.[3] We discussed the reasons that triggered this tremendous money increase already in the last chapter: At the beginning the responsibility came with the military and their need for new weapons, triggered by WW II,[4] but increased even further after the war. Then, there occurred a continuously growing civilian demands accompanying economic growth with the invention and construction of new products and the need for ever better infrastructures. In parallel, complex experiments let us discover new and more complex (emergent) laws and knowledge beyond the fundamental laws on atoms and genes in pure science, which quickly brought the chances of new technological implications.

The possibly best example for the increasing costs in pure science, is the construction of particle accelerators starting in the 1950s. Building the latest accelerator of the *European Organization for Nuclear Research* (CERN) near Geneva, the Large Hadron Collider, costed around $4.75 billion. The US

---

[2] For a discussion on the history and increase of money in the US since the 1970s as well as the financing of science in the entire world, see https://www.bu.edu/articles/2015/funding-for-scientific-research/ or https://en.wikipedia.org/wiki/Funding_of_science.

[3] Lutz Bornmann, Robin Haunschild, Rüdiger Mutz, *Growth rates of modern Science: a latent piecewise growth curve approach to model publication numbers from established and new literature databases*. Humanit Soc Sci Commun 8, 224; (2021); https://doi.org/10.1057/s41599-021-00903-w.

[4] Examples of the use of science for developing weapons are old, think of Archimedes using parabolic mirrors to burn the Roman navy defending the Greek city of Syracuse. However, the magnitude and importance of new technologies in WW II were unprecedented. And the rise after WW II was even higher.

based *Superconducting Super Collider (SSC)*, which was supposed to be multiples more powerful than all existing accelerators by the fall of 1993 would have had an estimated cost of minimally $11 billion (equivalent to $18 billion today), when its construction was cancelled by the US Congress due to these costs (while, 15 years later, they decided within days to spend 800 billion for saving banks).

With these examples, we see, that today, the availability of money has become a crucial factor for the actual undertaking of a scientific project. Since the 1990, money epitomizes the way science is moving. Today's scientific projects depend even more heavily on money.

Who Finances Modern Science?

A new economic system in the early nineteenth century enabled the marriage of scientific and technological dynamism with innovative entrepreneurship and professional production management. Starting in England, *capitalism* developed to its first peak, allowing a hitherto unimagined economic productivity and creativity to unfold, which included or was even driven by new technologies. However, in Europe these changes did for a long while not affect very much the content und structure of scientific research. Until around 1950, the main job of a scientist was to embark in the science projects that were most promising in his or her eyes.

In contrast today, most scientists have on top of their research, another important task (for many this task is even more important, or at least takes more time, than research itself): the application for the money that is necessary to undertake desired scientific projects. It is estimated that this time constitutes an average of one third of the researchers' working time! This is, in essence, quite similar to the activities of private companies asking for money to grow their ideas into money-making entities. The motto of scientists has for long been "publish or perish". However, in recent years it has become also "bring in the big bucks or perish",[5] to the extent that nowadays the CVs of scientists often contain, in addition to the list of publications, the list of grants obtained.

This "scientific enterprise" feature, however, has some major disadvantage in fundamental research entities that private companies do not share: It orientates the science practice often towards scientific conformism in order to receive the money and gives little recognition to a (often more risky) new

---

[5] Geoffrey West, Scale: *The Universal Laws of Growth, Innovation, Sustainability, and the Pace of Life in Organisms, Cities, Economies, and Companies*, Penguin Press (2017).

path that potentially leads to entirely new discoveries, just like those from 1900 to 1950 in physics, chemistry or biology. There are in fact fundamental questions for mankind still open as we saw in the last chapter. We also refer to Chap. 4 for more details.

The rapidly grown investments in technologies have been recognized by non-scientists much more compared to the recognition of funding fundamental science. So, financing science is often much more challenged than the one for new technologies. If we examine in more detail the current process by which certain scientists or groups of scientists obtain the recognition necessary to receive the financial support that is essential for their research, we quickly notice some problematic practices in the current way of judging the performance of individual scientists, with so many scientists working in so many particular areas. In order to cope with those, the scientific managers try to introduce "objective" measures for the "quantification of the quality" of individuals scientific achievements (see below the details). This is a particularly difficult task as, on top of the problem of judging the potential scientific importance of a project, personal interests play a significant role when money is being distributed. Another question is: How can the quality of a scientific project and its potential outcomes be judged before it starts? Last but not least, the problems are also aggravated by the need to rapidly take decisions by overworked peers assessing the worthiness of scientific paper drafts and their projects.

When it comes to receiving cash, scientists tend to behave and structure their research differently, often rather strangely, than if they are free from funding concerns like former scientists were. This is exactly the issue of the entirely new role of university professors. They are today often measured by the amount of funding they succeeded in obtaining rather than by the significance of their discoveries. Receiving funding (through marketing oneself) is not necessarily highly correlated with scientific quality.

Publicly financed research is largely responsible for finding new laws and structures in nature. However, researchers are increasingly interacting with the economic interests in a society. This is not a brand-new phenomenon of the second half of the twentieth century. As we saw, technologies did already grow out of fundamental new scientific discoveries and laws during the eighteenth and nineteenth century.[6] But the actual technological application was created by different people than the scientists, such as Werner Siemens for

---

[6] Railway technology and later cars have grown out of the scientific understanding of gas behaviour in the late eighteenth and first half of the nineteenth century, electricity - finally providing much more public light - was equally born out of scientific experiments (Michael Faraday) and the discovery of the laws of electrodynamics by theoretical physics (James Maxwell).

dynamo machine or Thomas Edison and Nikola Tesla for the transport and usage of electricity. Neither Faraday nor Maxwell, Pasteur or Koch in the nineteenth century, neither Albert Einstein nor Niels Bohr in the first half of the twentieth century were ever entrepreneurs.

Today most start-up companies are equally founded around new scientific entities or possible technological breakthroughs in these, however often by the scientists themselves or by employing key scientist of the field. The goal in creating such companies is in the end to make their founders as well as early investors and employees rich (even despite some scientific enthusiasm that often plays a key role among the scientists working for these companies). They attract billions of dollars in Private Equity funding and are considered to be the ultimate achievers of the new science in society. One example that is starting to gain public attention are companies that are researching and technologically designing concepts for producing energies based on nuclear fusion, or AI companies like DeepMind whose motto is "Solving intelligence to advance science and benefit humanity".[7] For decades public research has focussed only on one particular approach (the tokomak structure) for which it is still quite questionable whether it works. The public scientists are now in the process of building an even larger one (ITER) that has already cost 20 billion euros and will likely cost more than 60 billion euros in the end. Here the increasing and very interesting range of (much cheaper) different approaches that are financed privately might end up more successful than the publicly financed ITER program.

Without denying the importance and the role the private initiatives have played and will continue to play in the development of new technologies shaping the society (as seen by the nuclear fusion, or AI), for efficiency reasons in getting new important *core scientific results* we can surely not leave the future of fundamental science research to private initiatives and interests. And even new technologies developed by private companies must be publicly more controlled than they are today, because they touch - more and more often - very questionable areas both in ethical terms and in terms of risk for the survival of humanity (an historical, however government led, example for such a case was the deeply secret development of the nuclear weapon which gave birth to a danger of killing the entire humanity on this planet; public support could surely not be expected to do so).

Let us take a quick look at mathematics and how it was influenced by money: In fact, for mathematicians the role of money entered their field

---

[7] See their website: www.deepmind.com.

significantly later than in natural sciences. Thus, one can observe that mathematics kept the original nature (one may say freedom) science had in the first half of the twentieth century about twenty years longer (until the late 1960s). It however changed radically when computers entered the field of mathematics, which all of a sudden made the research in mathematics significantly more expensive, as new fields emerged that were intrinsically related to computers such as statistics (analysed in higher dimensions), dealing with non-linear equation, or finding proofs of mathematical theorems with the help of a computer (like the famous four colour theorem from 1974, the first major theorem to be verified running a computer program). Commercial interest in mathematics today is similar to that for the sciences as calculation algorithms optimized mathematically can play important role in new technologies such as artificial intelligence or quantum computing.

So, we see, the interaction between private and public support for research and technologies is often not straightforward but highly complex, and we will have to look into many difficult specific details. This is what is often lacking by politicians or the public who simply do not possess enough scientific expertise. For that reason, politicians have expert advisor, but they often do not really follow them. Thus, even public funding of science often becomes a political game instead of insisting on objective parameters. Two examples: Climate change and possible technological reactions; and research on and implementation of nuclear fusion as a new energy source.[8]

How Do We Judge the Quality of Scientific Papers Today?

It is interesting to go through the development of the structure of scientific publications dynamics which can tell us quite a bit about the change in nature of science. The evaluation of scientists' works by their peers has the purpose to judge the quality and importance of their work in order for it to be published or not in a given research journal. This process goes back to the middle to late nineteenth century. For many years and decades, the number of publications was rather contained and could be judged without many conflicts of interest, their evaluation thus happening with as little personal biases as possible. One prominent example of this was Max Planck's assessment of Albert Einstein's papers in 1905 when Einstein was still unknown, worked in a federal patent office in Bern, but published four exceptionally profound papers during that

---

[8] For details, see: Lars Jaeger, *Ways Out of the Climate Catastrophe - Ingredients for a Sustainable Energy and Climate Policy*, Springer (2021).

year. Planck was co-editor of *Annalen der Physik* and thus an important judger of what to publish. He even let Einstein's paper on light in the form of quanta go through for publication, even though he disliked the idea of physically existing "quanta" – although he had introduced the concept himself five years earlier.[9] Einstein was the first to recognise this meaning of "h". The same year *Annalen der Physik* received Einstein's paper on relativity. Here, Planck was the first scientist that noticed the importance of Einstein's relativity theory. He realized the far-reaching perspectives of it and reported favourably on it. Similarly, it happened a few years later with the papers by Niels Bohr, the not even 20-year-old Wolfgang Pauli, and equally young Werner Heisenberg on quantum physics. There was a strong and unconflicted judgement by experts on the incoming papers. Senior scientists took the time to read and understand papers produced by young researchers.

Unfortunately, today the indicators developed for measurements of scientific progress by specific works emphasize the quantitative aspect of scientists' work to the detriment of the qualitative one. Researchers thus tend to favour conformism in research that can create more papers accepted as they fit the current thinking of the majority rather than its creativity and innovation which is often linked to less papers or more controverses around the new methods. And this, although both, government and private investors, in fact, need quality measures emphasizing creativity rather than sole conformism. On top of this, the scientific competition today is significantly higher, even at very early stages of individuals' career (often already during their PhD work), than in previous times. Concretely, researchers today are being judged on quantitative criteria: number of publications, number of citations, h-index, i-index, ranking of their institutions, etc. rather than on the relevance or quality of their work. Unfortunately, these rankings favour, as we already saw, conformism in the way science is developing and rarely fully reflect the quality, significance, and uniqueness of the scientific results achieved. To have a chance to be published in a good journal, scientists need to conform to certain styles, to be linked with prestigious institutions and to look for strong support and even good relationships in the academic network. Often senior scientists dedicate more time to looking for funding than to referee papers. They rather give this work to their postdocs as it is not really rewarding for them. Furthermore, the evaluation of the potential importance of a scientific paper proposal in a particular field is more and more given by small groups of scientists, which then control via their judgements where the scientific field

---

[9] However, for him quanta were unreal and just help variables to later get rid of them again; his famous variable "h" actually referred to "help variable" – "Hilfsvariabel" - and is now universally known as the "Planck constant" playing a central role in quantum mechanics formalism.

is going. As a result, this can be very constraining for new ideas in the respective scientific field and often actually prevents the publications of innovative ideas and results. From this came the joke from the beginning of Chap. 2: "Nothing can pass the speed of light – except the velocity of creating new science papers without information."

It is clear that new solutions for more efficient criteria of scientific *quality* are important for the better development in the various fields in science. It could even be asked whether the current organisation of science, the prevailing science culture, and structure do not hinder the progress of science in some key issues. In physics for example, integrating the theory for the macrocosm (general relativity) and the microcosm (quantum field theory) into one theory is very far away from being achieved - with no major progress in many decades, which is quite unusual for a problem scientists deal with for such an amount of time. A new suggestion for the very approach to fundamental physics might move physics forward and therefore needs to be taken more seriously when evaluated for publication. How would Einstein have dealt with if the publication of physics was already structured like today? Would he even have developed as a physicist? A similarly entirely open question is the relationship between body and mind in neurology/biology and genetics. How many publications can we find today regarding this question?

Recent research on the amount of disruptiveness brought by scientific publications has shown a marked decrease.[10] It is difficult to find one reason for this. However, quoting the authors of the study just references to: "Our results suggest that slowing rates of disruption may reflect a fundamental shift in the nature of science and technology". We see that their conclusion is in line with our analysis of the situation. Perhaps, the scientific community should reflect on the conformism, which seems to invade more and more the field of research. Good science comes always by out-of-the-box thinking. The structure of science should favour this.

---

[10] Park, Michael, Erin Leahey, and Russell J. Funk. "Papers and patents are becoming less disruptive over time." Nature 613.7942 (2023) P. 138–144. There it says: "We analyse these claims at scale across six decades, using data on 45 million papers and 3.9 million patents from six large-scale datasets, together with a new quantitative metric. [...] We find that papers and patents are increasingly less likely to break with the past in ways that push science and technology in new directions. [...] Overall, our results suggest that slowing rates of disruption may reflect a fundamental shift in the nature of science and technology".

Where Should We Go in Science?

We dealt extensively with the issues of scientists themselves judging on the quality and financial worthiness of specific scientific projects. With certain persons or groups dominating a particular area and thus deciding on the publication of papers (and often already knowing their views thus having a great influence themselves), the review has become more political and conformist than it ought to be. This combined with the structure of solely quantitatively deciding on an individual scientist's publication quality and emphasizing so highly the citation numbers makes the process of deciding on the researchers' financial access today a very sub-optimal one.

These days, researchers are forced to spend 30% of their time chasing for fundings.[11] Perhaps, we should go back to some endowments or long-term fundings to relieve good scientists from the burden of constantly applying to get the necessary (often down to yearly) money for their research. Some even propose to organize - after a first screening by experts - a grant lottery that would provide a better chance to give money to innovative projects. This unconventional proposal was published in 2016 in *mBio*, a journal by the *American Society for Microbiology*.[12]

Another simple idea to solve this problem is to create a forum of a rather wide panel of experienced scientists with solid scientific judgment experience (between 20 and 30 senior scientists), not necessarily experts in the respective field, to judge for particularly large projects to be financed. This should happen with respect to its scientific value, conformity of the methodologies used, quantitative correctness (e.g. if there are mathematical structures or calculations), as well as and especially its creativity. The judgment would happen on the basis of a report by experts of the field that has itself been reviewed on the methodological side by one or two other experts. The group would then use these experts' as well as their own judgement to arrive at the conclusion. We could introduce the above mentioned controlled and random processes for financing to assessing the publication power of papers, in particular by exclusion of particular reviewers that are too close to the publishers. This could provide us with a better external control of reviewing papers and provide a new and creative paper a higher chance to be published.

---

[11] See for instance: https://www.vox.com/future-perfect/2021/12/18/22838746/biomedicine-science-grants-arc-institute, or https://www.growingproduce.com/fruits/chasing-funding-occupies-big-part-researchers-time/.

[12] Ferric C. Fang and Arturo Casadevall, Editorial: *Research Funding: the Case for a Modified Lottery*, mBio, vol. 7(2), (2016) p. 1–7.

Another issue is that currently, the work of reviewers is not sufficiently acknowledged. Perhaps, introducing some form of rewards for the reviewers would motivate senior researchers to spend more time on this fundamental step for fostering good research.

Another procedure to foster creativity in science rather than solely supporting conformism and thus to support the exploration of thoughts and oaths that the herd does not follow (which was particularly important in most important scientific achievement such as Albert Einstein's) can potentially be achieved by dedicating a certain portion of the papers in a journal to "unusual" projects – the same applies to financing science: a certain part of the (public) funding goes to such projects. The judgement of these publications should again be done by non-experts of the specific fields, however equally senior scientists with good broad judgement. In any case, rethinking the way research performance is measured for publication as well as fund allocations is a topics which we need to treat urgently for making research fit to the challenges ahead of us.

Three Past Examples from Fundamental Research to Revolutionary Technology - Penicillin, PCR and the Atomic Bomb

Let us take a closer look how science, still sponsored by government money, turned into three core technologies of the twentieth century. In all cases individual fundamental scientists started the opportunity of the technological application in their research. Plus, their implementation all costed quite some public money.

Penicillin

Until well into the twentieth century, millions of people died each year from bacterial infections such as pneumonia, meningitis, and wound sepsis. In the 1920s and 1930s, many companies were founded in Europe and America that had recognised the tremendous business opportunities of industrialising chemistry, including developing health pills. Despite all the research carried out, the mass production of the world's first antibiotic that worked reliably can be traced back to a coincidence. When the Scottish biology researcher Alexander Fleming returned to the laboratory from his summer holiday in 1928, he discovered some forgotten Petri dishes on the table, containing dangerous staphylococci whose colonies had in the meantime overgrown the entire culture medium. When he went to clean the glass dishes, he noticed that mould had grown in some of them. Such contamination was not uncommon, but there was a clearly visible zone around the mould colonies that was free of bacteria. Some substance released by the mould into the nutrient substrate must have prevented the bacteria from colonising these areas as well.

Fleming identified the mould in his Petri dishes as a species of the genus Penicillium, which was rather rare at the time. In trying to replicate he found that the biggest problem was the task of isolating the active ingredient penicillin. After many failures, Fleming began to wonder whether a remedy against bacterial infections could ever be obtained from the mould. He published the results of his experiments in 1929, pointing to the possible therapeutic application.[13]

Ten years after Fleming's discovery, in 1938, the Australian Howard Florey and Ernst Chain, who had emigrated from Germany to England, came across Fleming's work in their search for bacteria-inhibiting substances. In painstaking laboratory work, they succeeded in producing pure penicillin and proved at the same time that this substance could be used to combat, not only staphylococci, but also other aggressive strains of bacteria. They managed to produce a little amount of penicillin so that they could treat one person, a 43-year-old policeman.

The British pharmaceutical companies asked by Chain and Florey for cooperation, however, saw no possibility of bringing penicillin to the market on an industrial scale, because its production was extremely costly. However, in America penicillin finally began its triumphal march, not least because of media interest in the story of another person cured by penicillin. The military importance of penicillin was obvious, so in 1943 the American War Production Board (WPB) took control of a possible mass production of this new antibiotic. And the Americans succeeded. In 1943, the first field trials with penicillin took place in the fiercely contested region of North Africa, and its effect exceeded all expectations.

In time for D-Day, the planned invasion of Normandy on 6 June 1944, a doses of the antibiotic were available to all soldiers of the Allied troops, as we already saw in Chap. 2. This shifted the balance of power between the warring parties in favour of the Allies. For the discovery (and rediscovery) of

---

[13] As early as in 1897, the French physician Ernest Duchesne identified and described in his thesis the anti-bacterial effect of Penicillium glaucum on pigs infected by salmonella encountering the skepticism of the Institute Pasteur. The Belgian physicians André Gratia and Sara Dath observe contamination and inhibition of their culture of *Staphylococcus aureus* by a mold. They identify the latter as being of the genus Penicillium and publish their observation in 1925, which went unnoticed.

penicillin, Fleming, Chain, and Florey were jointly awarded the Nobel Prize for Medicine in 1945.

Atomic Bomb

In his special theory of relativity from 1905 the theoretical physicist Albert Einstein explicitly showed that energy and mass must be the same. He put this into his famous formula $E = mc^2$. However, an experimental proof or even application for this fundamental formula were out of sight at that time. It took more than three decades before the equivalence of energy and mass could be proven experimentally. In December 1938 the team of Otto Hahn and Lise Meitner saw that by shooting a neutron towards a Uranium nucleus the latter splits in two lighter atomic nuclei. Lise Meitner - who by then as a Jewish person had already escaped from Hitler-Germany - calculated that by splitting heavy atomic nuclei some small part of their masses (about 1/5 of the mass of a proton) was converted into quite high energy (200 MeV). This was the first explicit example of Einstein's formula above.[14] While it took 33 years from the theory to the experiment, the path towards a first technological application started almost instantly after the experiment, as Lise Meitner and other physicists quickly realized the possibility of building a new type of bomb out of the nuclear splitting, a bomb with releasing an unimaginable amount of energy.

For many years, scientists all over the world had exchanged views on the subject of radioactivity in an unbiased manner. With the discovery of nuclear fission, the international cooperation, however, suddenly came to an end. As early as spring 1939, the German military showed interest in possible applications of nuclear fission. On the other side of the ocean, the discovery led the Hungarian physicist in the US Leo Szilard to prompt Albert Einstein to write to US President Franklin D. Roosevelt a warning letter about the atomic bomb potentially being produced by Nazi-Germany. This led the US to the Manhattan Project to construct a nuclear bomb themselves. Less than six years later, the first three nuclear bombs were detonated in New Mexico (test bomb), Hiroshima, and Nagasaki. The philosopher Gunther Anders considers this as a turning point in the human history as, for the first time. mankind can destroy itself.

---

[14] For a more detailed description of Lise Meitner's actions and thinking on nuclear fission, see Lars Jaeger, *Women of Genius in Science - Whose Frequently Overlooked Contributions Changed the World*, Chap. 10, Springer, Heidelberg (2023).

Polymerase Chain Reaction PCR

The discovery of DNA in 1953 which was largely triggered by the involved scientists' curiosity about the fundamental genesis of our body structures did not mean less than finding the entity that navigates all processes in the cells of humans, animals and plants. As we saw in the last chapter, it took about ten years for the DNA structure to be broadly accepted in biology but was then the starting point of many key technologies for improving the features of plants and animals – as well as most likely in the future those of human beings.

In this process a just discovered, very powerful methodology emerged in the 1980s: the Polymerase Chain Reaction (PCR). This is a method of duplicating genetic material (DNA) in vitro, i.e. in a glass, i.e. create a chain reaction. Important for this purpose is an enzyme called *DNA polymerase* which is an enzyme that catalyse the synthesis of DNA molecules. A thus developed chain reaction refers to the products of a previous replication cycle serving as starting materials for the next cycle, thus enabling exponential growth of the genetic material. The method called PCR was developed by the biochemist Kary Mullis in 1983 - for which he received the Nobel Prize for Chemistry in 1993 (although, in the early 1970s, the Norwegian postdoc Kjell Kleppe already came up with the idea of amplifying DNA using two flanking primers in the laboratory of Nobel Prize winner Har Gobind Khorana, but the idea was largely forgotten). Mullis intention had concretely been to develop a novel DNA synthesis process that artificially duplicates DNA by repeated duplications in multiple cycles using the mentioned enzyme polymerase.

What is needed to efficiently create 100 billion DNAs out of a single one in around 37 (2^{37}) steps is a thermostable DNA polymerase which keeps its structure to temperatures of about 100 °C. Today this is easily possible and can be done in an afternoon.

PCR and thus DNA fingerprinting (taking just a few gens out of the respective person) was first used for paternity testing in 1988. Today, PCR is one of the most important methods of modern molecular biology, and many scientific advances in this field e.g. within the framework of the Human Genome Project - would not have been possible without this method. The biological process behind it is easy to execute: Next to the DNA structure to be replicated, it takes no more than a test tube, a few simple reagents, and a source of heat to get the DNA polymerase to 100°. The following provides a selection of PCR applications:

- Analyse extremely small amounts of DNA samples which often critical for forensic analysis, i.e. detecting the genetic fingerprinting which is used to compare minute traces of DNA found at crime scenes with the DNA of suspects.
- Study evolutionary relationships of organisms.
- Analysis of ancient (fossil) DNA: Since PCR can produce any amount of material from only small amounts of DNA samples, it is particularly suitable for very old DNA, which occurs in nature only in quantities that are no longer sufficient for investigations. Thus, almost all scientific knowledge about many long-extinct species is based on the method of PCR.
- Food analysis and official food monitoring to clarify and prevent unfair competition led to the technology's entry into food analysis. For example, PCR can be used to identify spices in complex food matrices, as well as to distinguish between varieties in fine cocoa.
- Detection of hereditary diseases in a given genome can be significantly shortened by the use of PCR, as any gene in question can be amplified by PCR – and then sequenced to detect mutations.
- Viral diseases can also be detected by PCR by amplifying the viral DNA or, in the case of RNA viruses, by first transcribing this RNA into DNA and then amplifying it by PCR. This analysis can be done immediately after infection, often days or weeks before symptoms appear. If the diagnosis is made this early, it makes it much easier for doctors to treat the disease.
- PCR is also used for reliable diagnosis and validation of possible false-positive antigen rapid tests for COVID-19 disease.
- Cloning a gene for research purposes is a process in which a gene is isolated from one organism and then transplanted into another. PCR is often used to amplify the gene, which is then inserted into a vector (a means by which a gene can be transplanted into an organism). The DNA can then be inserted into another organism where the gene or its product can be better studied. In practice, expressing a cloned gene can also be used to mass produce useful proteins such as drugs.
- Common methods of DNA sequencing (determination of the nucleotide sequence of DNA) are based on variants of PCR.
- A rapid and highly specific diagnosis of infectious diseases, including those caused by bacteria or viruses, can be done with PCA.

On top of all these applications, the versatility of PCR has led to a large number of particular variants of PCR.[15]

---

[15] For further details: David P. Clark, Nanette J. Pazderni, Michelle R. McGehee, *Polymerase Chain Reaction*, Molecular Biology (Third Edition), Chapter 6 (2019) P. 168–198.

Every single one of these technological applications from basic scientific discoveries significantly changed our lives on this planet.

The Most Important Technological Efforts Today - And Those in the Future

With this heavy influence of money in our minds let us have a brief look at the most important technologies today and in the future (to be elaborated in much more details in Chap. 5). What constitutes our technological potential the famous quantum physicist Richard Feynman illustrated so comprehensively. In the 1950s and 1980s he gave two lectures, which are still widely cited today and can be seen as a vision and programme for entirely new technologies in the twenty-first century. In 1959, under the title "There's Plenty of Room at the Bottom", he described how future technologies could function on a micro- and nanoscopic level. The ideas of his lecture became the basis of todays' nanotechnology. In his second visionary speech in 1981, Feynman developed the idea of a quantum computer. Such a computer, in which subatomic particles are used for data storage and processing, is based at its core on bizarre quantum properties of the particles involved. Instead of processing bit by bit like classical computers, a quantum computer computes in parallel on numerous (in fact an infinite amount of) quantum states, so-called quantum bits. This enables an unimaginably higher computing speed compared to conventional computers. And interestingly, in both potential technologies there are currently billions of dollars invested. They are the trendy research subjects of the day.

With the change of science's character due to the US leadership some wonderful technological advances have appeared in the last 80 years which ultimately created our modern age. The speed at which this has happened is amazing, as well as how much money has flown into it. Yet, technological progresses will even more increase their speed in the future, as will likely the scientific one. The world has already changed by technologies so much more dramatically between 1990 and today (2023) than between 1957 and 1990, technologies that are the results of scientific insights from many years before. It is hard to imagine what the world will look like in 2056 driven by yet again entirely new technologies.

At the same time, we will surely penetrate even deeper into the secrets of nature due to scientific insights in the coming years and recognise more and more what holds the world together at its core. Before we have even understood the scientific upheavals, we are today already exposed to, the *scientific progress* of the next years and decades will once again dramatically reshape many supposed certainties - about the world and the universe, about space and time, about matter and substance, about man and nature and not least about ourselves and our mind.

However, it is the *technological progress* - which is likely to depend significantly on tomorrow's science and to some significant degree already on today's - that will ultimately determine the future of us all. Investments as well as new ideas play the key role here next to science itself. Obviously, we will even more have to meet challenges, not only scientific and technological ones but also philosophical ones, i.e. how to socially deal with the implications of science and technology that will be raising significant ethical issues.

The current status of the different sciences and even more their future status will provide new technologies that will be even more crucial for our lives than those of today. Most likely they will even change humanist itself. The following list outlines the likely most potential key future technologies - which often have already started to appear today. They have unbelievable significances to any future human individual, but also for the entire human civilisation. The following chapter outlines them in more detail, before the details and background for each of them are then provided in Chaps. 6–11 individually.

1. Artificial Intelligence – Controlling our lives or ever improve it?
2. Quantum Computers – The future of millions of times faster computation or just a dream of physicists?
3. CO_2-Neutrality – Can we create enough alternative energies in the next few years to prevent a climate catastrophe?
4. Nuclear Fusion – The solution of our energy problems or just a subject of a century of dreaming?
5. Genetics - The victory over cancer or manipulation of mankind heredity?
6. Internet of Things – New powerful industrial technologies and smart fabrication or an entire invasion of privacy?
7. Neuro-enhancements – Improving our thinking and acting or move away from today's reality into a new and fictional world?
8. Understanding our minds – Finding our Ego or is it eventually unfindable for scientists?
9. Digital Algorithms and Big Data – New enhanced profiles for our lives or controlling humans' thinking and acting?
10. Nanotechnology – Creation of useful things from "nothing" or destruction of our bodies?

11. Stem cells – Using cells that can do everything, also for our entire body and mind?
12. Biotechnology – From frogs for pharmacists to nano-robots in our bodies, a dream or reality?
13. Food Technology – Provide food for 10 billion people or just a science dream story?
14. Synthetic Life – When humans play God, part I.
15. Life Prolongation – When humans play God, part II.

To summarise the facts regarding scientific research and technological implementation:

1. Our world is changing faster and faster by technological progress.
2. On this rapid change capitalism plays a significant role in pushing and picking up on scientific progress.
3. The difference between science and technology has narrowed. The question has moved from how does it work, to how can we use it.
4. In fact, as we saw (and will go into much more detail in Chaps. 6–12), since around 1820 and more and more since 1960 - when science became even more technologically productive (and opportunistic) - capitalism and science have played a powerful and very productive joint role together to produce new high-tech progress.
5. Scientific research (and knowledge itself) has become a crucial factor for running a company in todays' capitalism.
6. Today's tighter and tighter interaction between capitalism and science makes the changes of the technological landscape significantly faster than ever.
7. In the future this will likely move with a speed that is unimaginable today.

Will we even be able to accompany technological changes by also progressing respectively politically and socially and develop appropriate philosophical insights and doings? We will have to in order to address and control the risk and dangers of the new technologies as well as their consequences on social relationships and interactions. We simply do not want to live with all of them as we will point out in the next chapter. But can we keep up with that?

In his 1932 novel *Brave New World*, Aldous Huxley describes a society in which people are sorted into different castes already at birth by means of biotechnological manipulation, and at the same time have all their desires, cravings and appetites immediately satisfied by permanent consumption, sex and the happiness drug soma. The novel will be familiar to most readers in its basic outline. Less well known is the year in which Huxley sets his action. It is the year 2540 A.D., more than 600 years after the novel was published! Even the visionary Huxley could not have imagined that the real technological possibilities could not only reach this scenario after only one century but could far outshine it. The applications are coming at a rapid pace, the social, philosophical and ethical challenges must be met at the same speed if mankind does not want to end up in the dystopian world described by Aldous Huxley. Is the interaction of science and capitalism still hopeful for that?

4

Philosophy in Science Is Over
And Why We Need to Reinstall It

The Tradition of Science Interacting with Philosophy

Science searches for knowledge about nature. Whatever the philosophical interpretation of scientific discoveries is, it is undeniable that science provided us with an extensive knowledge about the composition of matter, the features of life, the structure of our universe, and many other features of the world. In the last 100 years, we have particularly experienced enormous scientific progresses years after years and will likely see even unimaginably more progress in the future years. This implies that our knowledge about nature has been and *will continue* to be enlarged to an astonishing degree that resulted and will further result in technological benefits that has shaped our lives and will shape it even more in the future. Unfortunately, however, scientists largely stayed away in the last 100 years from the philosophical questions their discoveries raised. Today, active scientists usually abstain to enter in philosophical or even political debates regarding their work - with the noticeable exception of climate researchers, where we have seen an extremely high involvement of scientists in the social debate about climate change more recently - philosophically raising ethical questions rather than simply epistemological ones. The IPCC, the Intergovernmental Panel on Climate Change, report every few years is a sort of warning shout of the scientific community aiming at society and politicians concerning already well measurable dangerous climate changes.

As we already saw in the second chapter, science has tremendously more interacted with philosophy in the past, such as the intense philosophical discussions among the developers of quantum physics. Two participants even became professors for philosophy later (Carl Friedrich von Weizsäcker in Hamburg, Grete Hermann in Bremen). And this was by no means the first period scientists talked about philosophy in very subtle ways being well educated in it. Since the beginning of modern science in the early seventeenth century philosophy had been a clear participating topic in discovering new scientific laws, as significant open - and often solved but not yet accepted - questions in science were regularly seen as linked to philosophical questions and discussions. One can easily say that the development of science was *intrinsically* involved with philosophical discussions, i.e. the discovery of new scientific laws were almost always only possible after a decent level of consensus about the respective philosophical implications. Three examples - besides the deep discussions between quantum physics and philosophy already introduced in Chap. 2 shall demonstrate this:

1. The creator of modern mechanics in a solid mathematical framework, Isaac Newton, believed in a strong, subtle and elastic (and also spiritually active) substance around us he referred to as "aether", which permeates every solid substance and also circulates between the centre of the earth and the celestial planets and stars. He begins by saying that it is most important to distinguish between the time and space in daily life as there they are often perceived as relative features and the absolute, true, mathematical quantities themselves which are governed by absolute laws.[1] Such absolute space and time that all movements were to be considered with respect to was quite a speculation as there existed no empirical evidence for them. Consequently, Newton was philosophically intensely debated, e.g. by Gottfried Wilhelm Leibniz who had very different views on this question. He believed in no absolute space and time, and thus for him there existed only relative movements (in some ways forecasting Einstein's relativity theory). In his manuscripts, Newton repeatedly referred to the aether, which he believed was the originator of all matter moving toward the centre of the earth or sun. It was furthermore his deep belief in alchemy and in the all-embracing work of divine, alchemical and astral forces in nature that led him to develop an idea of an invisible and universal force, i.e. gravity, which pulls all things to the ground and also keeps planets on their orbits. It is a rather paradoxical fact in the history of science that it was the secret occultist Newton who helped modern, strictly

---

[1] At the very beginning of his *Principia*'s book, inserted between the "Definitions" and the "Laws of Motion", Newton lays out his views on time, space, place, and motion.

rational science to its decisive breakthrough.[2] It is estimated that 2/3 of the words he wrote were about theology and esoteric philosophy and less than 1/3 actually about science. However, it demonstrates how the origin of physics functioned: The reference to philosophy (and here even also to esoteric believes) played a key role for Newton. Thus, at the same time he created (parallel to Leibniz) the mathematics necessary for his physical laws: modern analysis in particular differential calculus. He published his revolutionary work in physics and mathematics in the book *The Mathematical Principles of Natural Philosophy* (Original in Latin: Philosophiae Naturalis Principia Mathematica). That Newton's laws were examined in details by philosophers can be seen in Voltaire's as well as Immanuel Kant's and many other philosophers' work. Last but not least, Newton with his mechanics is considered a founder of the early Enlightenment.

A famous example of this, besides Newton and Leibniz who we discuss below, was the French mathematician and philosopher Nicolas de Condorcet who was both an active researcher in mathematics, a philosopher, a politician, and a promoter of gender equality.

2. The development of physical theories on electricity and magnetism in the first half of the nineteenth century was also strongly related to philosophical discussions. Vice versa, the scientific questions involved were also dealt with by non-physicists such as Goethe, Hegel, Humboldt and others. The basic themes of Romanticism were emotion, passion, individuality and individual experience, and soul, especially the tortured soul. They arose as a reaction to the strongly reason-oriented philosophy of the Enlightenment. As represented by Fichte, Schelling and Hegel on the one end and natural science on the other end, science and romanticism were so to say a tandem development in which the parts influenced each other. Schelling's natural philosophy, Schopenhauer's metaphysics and Fichte's thoughts - all of which defined the period of romanticism - were in fact strongly influenced by discovered electrical and magnetic phenomena. Vice versa, the influence of Romanticism on the discovery of the nature of electromagnetism by the most important experimental physicist of the first half of the nineteenth century, Michael Faraday, cannot be dismissed out of

---

[2] There are other examples of theological beliefs helping the progress of science, see for instance, the well known Moscow school of mathematics much closer to today, at the beginning of the twentieth century, that was dominated by mathematicians belonging to the orthodox sect of the "Name worshipping" (see Loren Graham; Jean-Michel Kantor *Naming Infinity: A True Story of Religious Mysticism and Mathematical Creativity*, Belknap Press, 2009). Their religious practice appears to have opened them to visions into the infinite – and led to the founding of descriptive set theory. It is in this school that great mathematicians of the twentieth century like Pavel Aleksandrov and Andreï Kolmogorov did their PhDs.

hand. Faraday did have some strong believes that were similar to those of the romantic philosophers which even helped him discovering his laws that are still famous nowadays. For example, he believed - without having empirical evidence - that for every phenomenon we observe there exists an opposite phenomenon, such as the negative electric charges to the positive ones. His insight into the nature of electromagnetism (triggered by Oersted's discovery of the action of an electric current on a magnetic needle), in which electricity and magnetism are two sides of the same coin, corresponded to his belief in holistic structures of the world each consisting of two opposing components, which most likely served him also as a guide for his physical experiments and perhaps even formed the basis of his observations. For him nature is full of complementary counterparts, "polarities" as he put it. This clearly corresponded to principles of the romantic ideas of his time. This belief may also have led him to another important basic concept of modern physics: the abstract notion of a "field". Electric and magnetic fields exert on bodies a force which covers the whole space but is not visible for our senses, i.e. they remain invisible and untouchable. This was much to the taste of the romantic thinking about nature with its bets on the functions of deeper, unifying mysteries.[3]

3. Next to physics in Romanticism the so-called "philosophy of life" equally formed as a counter-reactions to the strict rationalism that Newton's laws had carried into science and to the philosophers of Enlightenment which claimed to even explain life mechanically. Its idea that life was something very different than objects that can be dealt with in science was thus a central part of romanticism as well. Even today it is hard to believe that the entire life can be explained by science in all its features as well of its origin (and, in fact, it cannot). The ideas during Romanticism of specific animal magnetism and electricity corresponded well to the often still today upheld philosophy of life's understanding of deeper structures and functions in living beings principally unexplainable by scientific methods. Thus, until the late nineteenth century, biology was dominated by the doctrine of "vitalism": Living organisms possess a special "life force" relying on a non-physical and thus scientifically not assessable element that controls all processes within them. Without it, biological processes cannot function. Since this is lost with the death of the living body, biological

---

[3] The young James Clark Maxwell, the later integrator of electrical and magnetic fields into mutual equations, at his young age gave some contribution to romanticism and the myth of vampires, but was later clearly anti-romantic, already when he published his electromagnetic laws in 1865 (earlier but not yet complete version in 1855).

processes cannot take place outside living bodies, according to the opinion of the "vitalists". However, it is a matter of science to eventually also test such statements explicitly by empirical methods. So, in 1897, vitalism was tested empirically. And here results came to pass that most people including scientist had thought were impossible: The German chemist Eduard Buchner succeeded in recreating in inanimate matter an organic biological process that had previously only seen as occurring in the living body.[4] This was the process of fermentation, an organic process that takes place when a sugar solution such as fruit juice comes together with yeast cells. In this process, the sugar reacts to form alcohol under the action of the yeast cells. Buchner discovered that yeast cells can cause the process of fermentation in fruit juices even after they have died. Thus, the processes in living bodies obey the same chemical laws as those in the inanimate world. Vitalism lost its adherents rather quickly.

At the end of the nineteenth century, a lot of the transcendence and religious justification had disappeared in physics and chemistry. There was, in fact, now a profound philosophical consensus among the scientists that they had understood the entire range of absolute laws of the world ("natural laws") in their depth once and for all. Newton's laws were regarded as the eternally valid world formula for everything mechanical, and the findings in the fields of magnetism and electrodynamics and their laws as discovered by James Clark Maxwell in 1855 in their view rounded off the picture in the most beautiful way, equally with laws that possess external and absolute validity. Laws of nature once discovered are always absolutely valid, that was the view which finds its origin in the statements of Newton about the absolutism of his laws. At the same time, however, they always have a philosophical significance: the belief in absolute laws was a profound philosophical principle that was also reflected accordingly. One example was the physicist Ernst Mach and his "empirio-criticism". He rejects conclusions beyond the horizon of experience. In Mach's sense, theories have no ultimate truth content, but only a use in grasping nature. Truth therefore only ever exists as a temporary truth for the discussion of nature, which comes about according to evolutionary laws: there are only the strongest, i.e. most economical and empirically clear, ideas that survive. Mach's view already anticipated how the idea of an absolute truth was almost abruptly shaken with the beginning of the twentieth

---

[4] Eduard Buchner, *Alkoholische Gährung ohne Hefezellen (Vorläufige Mitteilung)*, Berichte der Deutschen Chemischen Gesellschaft, 30 (1897), p. 117–124; finale version: Eduard Buchner, Rudolf Rapp, *Alkoholische Gährung ohne Hefezellen*, Berichte der Deutschen Chemischen Gesellschaft, 32 (2) (1899), p. 2086–2094.

century and led to the profound philosophical discussions that we already saw in the second chapter.

In contrast to physics and chemistry biology at the turn of the century had not yet reached that point yet. Vitalism had just lost its foundation three years before the twentieth century, the model of evolution by Charles Darwin did not yet receive a convincing foundation.[5] Only in the early years of the twentieth century when Gregor Mendel was re-discovered did Darwin gain more and more acceptance. This started a 50 year long process of discovering the genetic foundation of life and make biology to join physics and chemistry in becoming a strict discipline of scientific research, getting rid of exotic speculations. In that process also in biology philosophy was a key element within research. Lastly, it was about the question of the basis of life.

The Breakdown of Absolute Knowledge in Science Is a Profound Philosophical Challenge

The belief of scientists that their laws were absolute led, as we already saw, to a major conflict in physics as of the early twentieth century - and in fact also in chemistry where, however, the philosophical problems about the existence of atoms were barely discussed in the nineteenth century. It was a philosopher who first concluded explicitly from the new physics that, contrary to the believes of scientists, laws about our nature never have absolute validity (even if Ernst Mach already had similar ideas): Karl Popper. In general, the path of science from dogma and superstition to rational thinking and empirical research is long and arduous, as Popper stated. He rejected the traditional *positivist-inductivist* view that the scientific method is characterised by generalisations from observations to scientific theories that will then be seen as fundamentally true. He rather followed a path based on *empirical falsifications*. According to this, scientific theories are merely uncertain speculations that empirical science should attempt to overturn by searching for contradictory observations. If they do not find one, the theory might be acceptable, but not principally true forever, as future researchers might always be able to disprove it by an empirical observation. Popper stated this on the basis of Einstein's general theory of relativity which proved Newton's laws are invalid if considered as laws in more extreme environments. For Popper, this is the realization that the search for knowledge about nature never ends.

---

[5] He could have had it, if Gregor Mendel and his experiment with peas that he did around the same time as Darwin's publication of his fundamental book of evolution in 1859 had not been completely forgotten.

This new scientific understanding was recognized explicitly for the first time in the late 1920s/early 1930s: Logically no scientific theory can ever be absolutely (as forever valid) confirmed from an experimental test, while it can be logically refuted by a single experimental counterexample.[6] This sounds quite trivial but was far outside the mindset of physicists before the discovery of Einstein's relativities in 1905 and 1915 and the structure of quantum physics that started in 1900 with Planck's (for him forced) introduction of a quantum in an equation of physics and then got momentum with Einstein's discovery of quantum properties of lights and the discovery of the contradictions in the world of atoms in 1911. All these counteracted Newton's as well as Maxwell's theory.[7]

Popper's account of the logical asymmetry between verifiability and falsifiability is at the heart of his theory of science, which is still known and largely accepted today. Popper coined his philosophy by the term "critical rationalism". De facto, according to him, a theory can only be considered scientific, if it is potentially falsifiable. As examples of pseudoscience he cited psychoanalysis and Marxism (the first still discussed today), while he states Einstein's theory of relativity as an example of science.

Many scientific laymen, however, desire immutable truths. For Popper, these views even go beyond science: At a time when populists (in Popper's life the Nazis as well as Communists) abuse this longing for their own purposes, when an irrational, absolute criticism of science was on its way to becoming socially acceptable, and when principle doubts about science were being presented ever more aggressively, we must consistently ensure that the voice of rationality is clearly heard, so Popper.

There were other philosophers that did not entirely agree with Popper and had some strong empirical points when looking at the actual history of science. Science did not quite change that quickly as soon as a falsification appear, as Popper sort of assumed. Even Planck needed some time before he could accept quantum and atomic theory. The most prominent philosopher about how science empirically develops was Thomas Kuhn with his main work "The Structure of Scientific Revolutions"[8] from 1962. For Kuhn, scientists do not adjust their view continuously as soon as certain falsification appear, as Popper suggested, but rather through periodic and often sudden "paradigm shifts", very much like quantum jumps in quantum mechanics.

---

[6] Karl Popper, *The Logic of Scientific Discovery*; German original: *Logik der Forschung* (Imprint 1935, actually 1934), English translation 1959; Abingdon-on-Thames, Routledge (2002).

[7] For more details see also here: Lars Jaeger, *The Stumbling Progress of 20th Century Science - How Crises and Great Minds Have Shaped Our Modern World*, Springer (2022).

[8] Thomas Kuhn, *The Structure of Scientific Revolutions*, University of Chicago Press (1962).

This echoes with Jean Piaget's discoveries on the knowledge acquisition from children,[9] which had inspired Kuhn. Scientific truth, as given at any point in time only relatively, is thus not just based on objective criteria but rather on a certain consensus of the scientific community which can take some time to develop. In this sense, we can view the scientific method as a way of building a rational consensus among knowledgeable people.

Popper's philosophical views were also controversially discussed by scientists, especially physicists. For many quantum physicists like Bohr and Heisenberg Popper's program was dissatisfying. In fact, Popper still believed in objective processes being decisive in science. His aim to defend objectivity seemed to contradict the point of view associated with Bohr's Copenhagen interpretation. In fact, during his entire life, Popper blamed Niels Bohr for introducing the concept of subjectivity into physics and for defending subjectivist theses about quantum mechanics in particular. He rather supported Albert Einstein's realist approach to scientific theories about the universe. Einstein did thus have a very good relationship with Popper with quite some interactions, e.g. how much he liked Popper's draft for his first book *The Logic of Scientific Discovery*, such that he wrote:

> When the book is cleansed of these drosses, it will be truly excellent. I would like to contribute to your work attracting the attention of your peers.[10]

However, Einstein also made some fun regarding Popper's views and references to philosophy. Popper's notion

> It is our inventiveness, our imagination, our intellect and especially the use of our critical faculties to argue and compare our theories which makes it possible for our knowledge to be developed.

he summarized in just a few words:

> Here there is no goal, only the opportunity to give oneself over to the pleasant task of thinking.[11]

---

[9] Saul McLeod, *Piaget's Stages of Cognitive Development*, Simply Psychology (2018, updated in 2022), available on Internet at https://www.simplypsychology.org/piaget.html.

[10] Albert Einstein: Letter to Karl Popper on 15th June 1935 (translated by LJ); German original: „Wenn das Buch von diesen Schlacken gereinigt wird, wird es wirklich vortrefflich sein. Ich würde gerne dazu beitragen, dass Ihr Werk Beachtung der Fachgenossen findet".

[11] Text by Antonio Moreno González: http://dipc.ehu.es/digitalak/orriak/english/philosophy.html.

Despite his conflicts with one of the key quantum physicists, Niels Bohr, Popper had a significant impact into the scientific thinking of the twentieth century, and his view that science can never come up with absolute truths is widely accepted nowadays. As a side remark: it is worth noting that this view has widespread consequences on what is now called "relativism" by ethicians. If there are no absolute truths in science, why would there be absolute truths in ethics? For Critical Rationalism, ethics is rather the concrete problem-solving in the social field. Do we even need absolute ethics to solve our problems today or should we adapt our ethical principles to the evolving situation rather than keep them eternal and static? So, we see that the new standard of truth, which distances itself from absoluteness, also questions eternal ethical principles.

Does Philosophy Still Have an Importance for Science?

Kant points out that science has the task of creating unity in knowledge. It does that by developing theories that summarise our knowledge in a systematic form and allows us to better understand what we know about nature. Philosophy is therefore scientific when it thinks together what belongs together. In his "Critique of Pure Reason" Kant gives what he claims to be a complete outline of the human capacity for knowledge. Whether we exceed those limits of Kant today – with experiments that he could not imagine leading to the deep insight that the defined structures of our everyday impressions such as fixed space and time or causality do not translate into areas that we cannot directly observe such as the micro- or the macrocosm – is quite questionable. Kant himself stated that space and time are not absolutely defined but rather the very conditions of our *daily* experiences. This remains intact until today: Experience of relativistic and quantum laws are not being made by us in our daily lives. As an exception one could consider certain macroscopic applications like laser, semi-conductors and magnetic resonance imaging; however, our common experiences of these effects do not counteract against our assessment of objectivity and independent experience. Technologies with only a few quantum particles that are currently being invented (such as quantum computers) in contrast might contradict Kant directly.[12] His philosophy and its relevance for quantum physics were subjects of much discussion in the 1930s among physicists.

---

[12] See also: Lars Jaeger, *The Quantum Revolution*, Springer (2018).

From a philosophical perspective upon science, looking back into the first half of the twentieth century, there seems to have been a paradise compared to today's role of philosophy in physics or other fields of science. And here Kant also had an important say (strongly represented by the young Carl Friedrich von Weizsäcker; plus, a neo-Kantianism was developing, centred in Vienna). Einstein sharpened his thinking in a philosophical circle considering that his scientific work formed part of the philosophical contributions made throughout history to understand the mysteries of nature,[13] Heisenberg was inspired by the ancient philosophers in his understanding of the fundaments of matter, and his student Carl Friedrich von Weizsäcker tried to understand the "structure of physics" in a philosophical way, for some period discussing intensely with the female philosopher and physicist Grete Hermann.[14] The endeavour to describe the world physically and the claim to critically reflect, interpret and question this undertaking philosophically for a long time appeared as two sides of the same coin. As Schopenhauer had once (a century before modern physics emerged) admonished:

> Empirical sciences, pursued purely for their own sake and without philosophical tendency, resemble a face without eyes.[15]

And already in the sixteenth century, the French author François Rabelais wrote:

> "Science sans conscience n'est que ruine de l'âme" (Science without conscience is but the ruin of the soul).[16]

As we saw, philosophy plays since the early 1940s when the US took over the lead almost no more role for understanding sciences. Many scientists believe that the philosophical discussions as they were held in the past were inappropriate for the purely empirically justified and then theoretically defined sciences of today. That is not quite logical, as empirical results and theoretical models were equally important parts of scientific research

---

[13] Einstein speaking about the reciprocity between physics and philosophy, he (together with his assistant Leopold Infeld) wrote: *The Evolution of Physics, from Early Concepts to Relativity and Quanta*, Touchstone, New York (1938).

[14] See Lars Jaeger, *Emmy Noether - Ihr steiniger Weg an die Weltspitze der Mathematik*, Kapitel 8, Südverlag (2022).

[15] Arthur Schopenhauer, *The World as Will and Representation* (German original: *Die Welt als Wille und Vorstellung*), Dover Publications (1966) (original version: 1818–1859).

[16] From Rabelais' novel *Gargantua and Pantagruel*.

before 1940. Moreover, Popper's philosophical views could serve as a counter-example here. Last but not least, the generation of scientific ideas in human minds is today still not quite as rational and clear as what scientist often claim. Even these days, the most complex sciences can rely on irrational believes. One might find an example of this in the string theory which, because of their distance from energies reached for us will barely ever be provable, and it has so many parameters today that it makes it more or less useless for science.[17]

The ideas many scientists had for their greatest theories often came from all kinds of different backgrounds other than rational thinking. Einstein himself is often quoted having once said:

I never made one of my discoveries through the process of rational thinking.[18]

Bohr, next to Einstein the other most significant physicist of the twentieth century, wrote:

Any truly deep phenomenon of nature cannot be uniquely defined using the words of our language and requires at least two mutually exclusive additional concepts to define itself.[19]

Bohr's general principle of complementarity established a rational and irrational aspect of reality, which he believed can never conflict with each other, just as Heisenberg's uncertainty principle applies: The more one aspect of reality is clarified (measured), the more uncertain (non-measurable) the

---

[17] String theory assumes the existence of six more dimensions besides the ones we know that enrolled in hyper-small dimensions (15 times smaller than protons). There are many ways of this being the case, and each one of them corresponds to a different possible universe with different collection of particles and forces. Today, string theory comes with an enormous number of versions, typically estimated to be around 10^{500}. This number, of course, is ridiculous. With it we will likely be able to accommodate almost any individual phenomenon that might ever be observed at our energy levels. Some physicists, e.g. Steven Weinberg, see a virtue in this large number of solutions as it might allow a natural anthropic explanation of the observed values of physical constants that enable life on Earth to have been developed in one particular status only.

A couple pages below we will come up with more examples of new very speculative theories in physics. For more details of the string theory, see Chap. 6.

[18] It is not clear, whether Einstein ever explicitly and exactly said these words. However, he did say things that might have shaped up this quote. As a young man, he admired the philosophical writings of Schopenhauer who said very similar things, e.g. "art and aesthetic experience not only provide escape from an otherwise miserable existence, but attain an objectivity explicitly superior to that of science or ordinary empirical knowledge", see in Dale Jacquette, Ivan Gaskell, Salim Kemal, *Schopenhauer, Philosophy and the Arts*, Cambridge University Press (1996).

[19] Niels Bohr, *Atomic Physics and Human Knowledge (collected articles)*, CUP Archive (1961).

second aspect of reality becomes. However, that only applies to the microcosm. As we saw in Chap. 2, if you extend this principle steadily into our world, the contradictions cannot be avoided any longer, as Schrödinger clearly showed: Is saying "A cat is – in reality – dead and alive at the *same time*" not irrational on our everyday scales?

Another example: For most (theoretical) physicists today, mathematics is no longer just a tool for understanding or describing the world. For them, mathematical forms are, so to speak, perfect frameworks how the world behaves - fundamental and valid everywhere and always. This would make mathematics a substance in the sense of Plato, who had seen the common origin of everything empirical in the world in absolute ideas. Thus, in their vast majority, physicists, despite rejecting philosophical topics, still follow a metaphysical belief in something fundamental in the things they study and in their independence from the images we form of them. The only difference: Instead of a substance in the form of atoms or particles, mathematical concepts such as symmetries, conservation laws, invariances, etc. have now taken this place of reference to what reality is. The mathematical structures describing these are supposed to be the carriers of absolute properties. The physicist and philosopher Carl Friedrich von Weizsäcker expresses this in the following way (interestingly quite similar to Galilei's famous expression: "Philosophy [...] is written in the language of mathematics"):

> And if one therefore asks why do mathematical laws apply in nature, the answer is: because these are its essence, because mathematics expresses the essence of nature.[20]

These mathematical believes by physicists that are still being held up today are not necessarily more rational than e.g. the philosophy of Plato. Another important feature is that the mathematical form of laws in physics today are simply not followable by 99.9% of the people, so there cannot be a socially broad discussion about the interpretation of it. The physicists (or even only a small part of them) remain among themselves which prevents to let flow relatively easily new and very different ideas – as it was rather typical for philosophy (as well as sciences in the past).

Today psychologists have a much deeper knowledge about our thinking, and they can make a clear statement: *No human thinking is always entirely rational*. However, if a non-empirical statement comes to the wrong conclusion – as it often does – science has strong techniques, much stronger than

---

[20] Carl Friedrich von Weizsäcker, *Ein Blick auf Platon – Ideenlehre, Logik und Physik*, Stuttgart (1981); translated by LJ.

those in (non-empirical) philosophy, to sort it out. In a way, the scientific method could be seen as the best path to reach an agreement among rational individuals.

Let us look a little deeper into the claims of scientists about philosophy and how reasonable this is, given the nature of their research. There are three arguments by the scientists today why philosophy is rather useless for science[21]:

- The most drastic one: "Philosophy is dead. No more relevant things come out of it." This comes from Stephen Hawking:
  - *"Philosophy has not kept up with modern developments in science, particularly physics. Scientists have become the bearers of the torch of discovery in our quest for knowledge."*
- The lack of contribution to science: "There are simply no subjects where philosophy has been useful for science".
- Different topics: "Philosophy and science have different subject matters, therefore philosophy is useless for science."

All three contradict the fundamental role philosophy had historically played in science for 250 years. Until around 1940 it had been especially the philosophical discipline of *epistemology* (the philosophical branch reflecting about the nature, origin, and scope our knowledge) that guided much of the discussion among scientists. If scientists claim today philosophy is useless for them since their discipline became more complex and mathematical, one can easily state a counter example out of another philosophical discipline: Since WW II *ethics* has become an important philosophical reference to technological applications of science starting most prominently with the use of the nuclear bomb by the very country having taken over the leadership among democracies. Unfortunately (and dangerously), until today the refusal of ethical evaluation of technological developments has been widespread, especially in the US, where most of the development of technological applications has so far taken place (see Chap. 3). And is this perhaps one of the reasons why the establishment of new technologies created significant risks in the last 75 years - from extracting energy without consideration of the climate to potentially suppressing human beings by big data or artificial intelligence? While *ethics* has become well important and applicable – and should be applied - to technological applications given their dangers in various fields,

---

[21] Sebastian De Haro, *Science and Philosophy: A Love–Hate Relationship*, Foundations of Science, Volume 25 (2020), p. 297–314.

and scientists indeed accept ethics somewhat, but rather in theoretical than in practical form -, *epistemology* remains as important for science as it was until 1940.

Open Questions in Science Today Are as Well Open Philosophical Questions

Is perhaps the refusal of scientists to employ philosophy in form of epistemology (and ethics) as reflection opportunities for today's scientific laws (as well as the scientific undertaking) the reason why they simply have not made any progress in some key areas for an unusually long time (without even any solutions in sight)? Might philosophy perhaps provide a new perspective in a field where science has been stuck into the corner for too long? The potential significance of dealing with a philosophical perspective in modern science shall be illustrated by the following examples:

1. Physics' standard theory of particles is by the views of physicists and mathematicians a terribly ugly and inelegant theory (see Chap. 6 for more details). Moreover, they have been dealing with the same fundamental problems in it for over fifty years now! Although the physicists know that the standard theory *cannot* hold up at higher energies, there seems to exist no way to get out of it and create a new and better theory at energy levels we can reach today. Even worse, we have no idea about the level of energy where the functionality of a new theory prevails. With almost any new particle experiment in the last 30 years (with ever higher energies) the physicists have hoped for results that contradict the standard model, but each times the results matched the ugly model to 100% which makes the standard model to be the most successful scientific model in history.

    As a last aspect: The phenomenon that affects our daily life, gravity, is not even considered in the possible extension of the standard theory. Here it is surely worth to step back and wonder, if there are possibly fundamentally different model types (one can argue, physics has already done that, e.g. with the string theory, but maybe that is a little far-fetched). Here, physics does not seem to advance any time soon. Rejecting philosophy altogether is not worthwhile here.
2. Around 95% of the matter in the universe consists of two kinds of matter – briefly mentioned in Chap. 2 - that no one has ever observed or been able to perform a measurement of yet: dark matter and dark energy. The first one, dark matter is said to account for almost 27% of the energy

density in the universe, while "baryonic matter", which makes up all the matter we know, only contributes five percent. Yet, Scientists actually have no idea about the precise nature of dark matter. They also have no direct empirical evidence for them. There are some possibly indirect ones going back to the 1930s when the Swiss astronomer Fritz Zwicky studied the movements of galaxies in the Coma galaxy cluster. These were too fast for the visible matter - stars, gas and dust - to hold them moving in the galaxy cluster in line with their gravity only. Scientists today are desperately trying to find out what exactly lies behind dark matter. Clear is only that this must be a completely new form of matter. The second type, the dark energy, is even more exotic and speculative and takes up, according to the theory, by far the largest proportion of the total density of matter and energy in the universe - around 68%. Dark energy is an energy field that is supposed to accelerate the largest cosmic elements being driven apart, as cosmologist have been able to measure (for more details, see Chap. 6). This hitherto mysterious form of energy counteracts the gravity of the matter in the universe, which would be characterized by a contraction if only the visible energy and matter would exist.

We see: Physicists have little or no idea about 95% of the universe's substances! Should they here perhaps also rethink their models? In history, physicists have already spent quite an amount of time and energy chasing Newton's aether that did not exist! Here, too, physics does not advance so quickly. Thus, here, too, philosophical reflection is in demand.

3. Physics is surely not the only science with open key problems. Biology equally has core questions left unanswered, despite its many successes in the last 75 years. The German physiologist Emil du Bois-Reymond was one of the most influential representatives of the natural science in the nineteenth century. In a speech from 1880, entitled "The Seven World Riddles",[22] he named the open questions he considered could never be answered by science. Two of them – those for biology - are[23]

---

[22] English version: Gabriel Finkelstein, *Emil du Bois-Reymond: neuroscience, self, and society in nineteenth-century Germany*. Cambridge, The MIT Press (2018). pp. 272–273. German original: *Über die Grenzen des Naturerkennens*. Ein Vortrag in der zweiten öffentlichen Sitzung der 45. Versammlung deutscher Naturforscher und Ärzte zu Leipzig am 14. August 1872 gehalten. Leipzig: Veit & Co. (1872).

[23] The others are: What are matter and force?, What is the origin of motion?, Where does the first life come from?, Where does the purpose ("the deliberate purposeful arrangement") of nature come from?, Where do rational thoughts, and "closely related" languages, come from?, Where does the "free" will that feels committed to the good come from? Here du Bois-Reymond saw the first two and the last as "transcendent".

(1) How did life first come into being?
(2) How does consciousness ("a subjective conscious sensation") arise from unconscious nerves?

These are precisely questions that have been debated in the philosophical context since antiquity and in almost all cultures. The second question du Bois-Reymond considered as "transcendent". The fact that they have all lost none of their mysteries to this day (see Chaps. 8 and 10, respectively) strengthens du Bois-Reymond's position that science might not be enough to deal with questions like this. His position was hotly contested at du Bois-Reymond's time, as well as it is today with quite a few scientists believing they are quite close to providing scientific answers to these questions.[24]

4. These three examples have been central problems of epistemological nature since antiquity. In addition, there are the ethical questions that also stem from philosophy. Today's technologies, the result of many new scientific insights, have changed human life tremendously, and so far, much into favourable terms. However, all new technologies demand strongly ethical discussions, judgements, and then corresponding actions. Already the nuclear bomb from more than 75 years ago, as saw a direct result from quantum physics, shows us that new technologies can have some very dangerous consequences. But other than nuclear bombs, that have always been controlled by governments, today's risks from new technologies are much more diverse, often not controllable and thus often more dangerous, as different from the development of nuclear bombs, as, the developers in firms with billions of dollars made available to them by private equity investor are, other than governments, not able (or willing) to consider the risks of their technology. Examples are potential genetic manipulations of humans in a company, a tracking of all our actions that will eventually take away our freedom as China is already doing today, but equally Facebook, Google and Amazon in the West, creating artificial intelligence that ends up ruling us, and many more which we will discuss in the next chapter. And all this does not even mention the climate change due to various emissions being part of our daily lives as well as the activities of companies, which might end up making the Earth as unhabitable as a nuclear war.

---

[24] Du Bois-Reymond was anything but an opponent of science. In his field, physiology, he advocated a consistent research programme, for he was convinced that natural science can, within certain limits, achieve considerable gains in knowledge; for him, the history of natural science was even "the real history of humanity". And yet, among those who were not dazzled by the successes of science, he was the first to publicly question its omnipotence with his *Ignorabimus*.

When scientists blame philosophers to be speculative and far away from empirical results, they should consider the first three points above. Science clearly does not have all the answers to fundamental question we ask ourselves broadly and will most likely not have them in the near future - even if some scientists claim that (well, scientific answers can hardly be predicted beyond 20, 30 or 80 years from now). When scientists furthermore claim that philosophy has led to strange and useless ethical theories, the last point, point 4 above, should make them reflect on their own ethical issues. Clear is with these open questions and critical developments that philosophy should be considered a potential key player in science again, as it is essentially the very foundation of critical thinking in epistemological as well as (newer) ethical challenges. This is necessary regarding the mentioned points and many more, where key philosophical thoughts help scientists to take a wider look on their current problems and, potentially much more important, provide the basis for the society's *ethical* judgments regarding consequences of technological developments. Even outside of science, but directly related to technologies, e.g. in the global economy, philosophy can play a key role.

Generally speaking, philosophy helps us

- to find out, where life can potentially come from, what the fundamental forces and movements in nature could be, what possibly created the universe, what our place in the universe is, what our consciousness exactly means, or what living a "worthwhile" life actually involves (there are, in principle, the claimed points by du Bois-Reymond 1880 science principally does not cover).
- to evolve, justify und communicate appropriately our proper morals and ethical values in light of our current lifestyle being largely determined by technologies.

Summarizing, we can go as far as claiming that it might be fundamental for the future of mankind that philosophers and scientists put their forces (back) together in order to more profoundly understand how new scientific discoveries, respectively technologies, affect our way of understanding the World (epistemology) as well as how to judge on the many ethical problems (ethics). In that process philosophy should not direct the search for the truth per se (as it had in the past for so many years claimed to be its efforts) but should help decide on the best interpretations of scientific results as well as safe on ethically conform applications of technologies thereof. We see already the need of ethical committees supervising independently the work of scientists. But it is also time for the scientists themselves to supervise their own work and to ask internally as well as society for guidance in terms of choosing the right research subject avoiding certain fields with the potential of developing dangerous applications.

Why the Influence of Philosophy into Science, Besides All Its Needs, Is Still so Low Today

As we saw in Chap. 3, the current structure of science is to a large degree plagued by (besides money and capitalism) conformism and individual scientists being self-limited towards quit narrow research topics. This leads to philosophical perspectives lying far away for most scientists. It seems like philosophical reflections, if at all, occur rather by older researchers that are not any longer busy applying for funding (it is estimated that they spend a third of their time on this) or in trying to understand some particular details of a scientific field in order to progress their career within fierce competition, and who thus start enlarging their perspective. However, at this point their philosophical reflections are largely shaped by the many decades of fundamental research they had performed.

There is yet another reason for this separation that lies within philosophy itself and its relationship to mathematics. Modern science in the twentieth century has become, as we already saw, quite abstract (such as physics being governed by mathematics) and multi-dimensional (e.g. biology has split up in so many different areas) that most philosophers are quite far away from the details of the complex knowledge of modern science. Already with the appearance of quantum physics and general relativity it was difficult for philosophers to follow these abstract theories. Nevertheless, there were quite a few philosophers understanding this, e.g. Grete Hermann and Karl Popper. But the sciences all have become more and more abstract since the 1950s. This complexity and the mathematics required to understand today's scientific theories can easily impose philosophers believing that these are unreachable for non-scientists. Despite this, there are currently quite a few philosophers who are also well trained in science and complex thinking. The problem remain the seemingly disinterest for researchers in both fields to cooperate.

A response to philosophers with the perception of them having no chance to understand science today is to realize that most of the *broad* problems, trends and determinations of science are actually not that difficult for philosophers to follow and to deliver important views on them (as we also claim to do in this book). The open problems in physics, the nature of mind and many other fields in science can be assessed without much mathematics or complex structures as we already saw at various points in this book so far. The research by a philosopher does not have to include the details of the research analysis in papers like the following abstruse formulation (most prominent paper by Lisa Randal, who is today also a member of the *American Philosophical Association*). Here is just the abstract (summarizing), not even the mathematical part:

> [...] The Weak scale is generated from a large scale of order the Planck scale through an exponential hierarchy. However, this exponential arises not from gauge interactions but from the background metric (which is a slice of AdS_5 spacetime). [...] There are fundamental spin-2 excitations with mass of weak scale order, which are coupled with weak scale as opposed to gravitational strength to the standard model particles. [...][25]

(Note: The "Planck scale" is an energy scale of particles of around 1.22×10^{19} GeV, i.e. around 10^{15} times stronger than what we can achieve in an experiment today.) Who could escape the impression that today's standards of writing about physics are quite different from those in Europe until about 1940? Let us have a look at Einstein's language in 1905:

> Examples of this sort [*moving magnet and an electrical conductor*], together with the unsuccessful attempts to discover any motion of the earth relatively to the "light medium," suggest that the phenomena of electrodynamics as well as of mechanics possess no properties corresponding to the idea of absolute rest.[26]

There are also easier ways to access scientific efforts today than the one by Lisa Randall above:

> Dark Matter: This phenomenon cannot be explained with the standard model of particle physics either. "Normal" matter can emit electromagnetic radiation, for example with wavelengths in the visible range. Dark matter, on the other hand, is invisible. It is assumed that it cannot emit electromagnetic radiation in any waveband. Only gravity has an effect on it. It is estimated that there is about five times as much dark matter in the universe as normal matter.[27]

---

[25] Lisa Randall, Raman Sundrum, *A Large mass hierarchy from a small extra dimension*, Phys. Rev. Lett. 83 (1999) p. 3370–3373; https://inspirehep.net/literature/499284.

[26] Albert Einstein, *On the Electrodynamics of Moving Bodies* (original title in German: *Zur Elektrodynamik bewegter Körper*), Annalen der Physik (Juni 1905).

[27] Lars Jaeger, Women of Genius in Science - Whose Frequently Overlooked Contributions Changed the World, Springer, Heidelberg (2023).

We saw that the spectrum of possible discussion between philosophers and physicists can be quite broad. In doing so, philosophers can avoid texts like the two above without missing out very much on the necessary general knowledge about modern science for discussing about it. However, modern scientists nevertheless have lost interest in the discussion with philosophers, the reason of which we want to summarize shortly as follows:

- The tight subject each of the researcher is covering today prevents discussion about philosophical views and implications.
- The strict and by itself stressful measurement of a researcher's performance prevents him or her from more philosophical reflections that takes time away from competition for publishing in a narrow field.
- Philosophical reflections bring no money into the budget.

A Key Mathematician Steps Out of Science Protesting Against Its Nature of Leaving Out Important Topics

Let us consider a particular scientist who introduced the ethical dimensions of his discipline (mathematics) into his life in a very direct and practical way. Alexander Grothendieck was one of the most important mathematicians of the second half of the twentieth century, to whom mathematics needs to be grateful for a complete reconstruction of algebraic geometry he did in the 1960s. In 1966 he was awarded with the highest award in mathematics, the Fields Medal. However, influenced by political ideas of May 1968 in France, he largely withdrew from his central position in the mathematical life of Paris in 1970. The particular reason was the lack of consideration of mathematicians regarding broader questions like ecological and political ones (and the sponsoring by the military of the Bures-sur-Yvette center he was directing). In the same year, he founded, together with his two friends, the mathematicians Claude Chevalley and Pierre Samuel, the group *Survivre et vivre* (survive and live), which, with the backdrop of the 1968 movement, advocated pacifist and ecological ideas. Grothendieck saw amazingly early an ecological catastrophe approaching humanity that would make it impossible for him to deal with mathematics in the future, and consistently turned to non-mathematical questions like ones of philosophy, metaphysics and religion (in particular Buddhism). However, he still dealt with quite a few mathematical questions, but later dealt further with all kinds of ecological as well as religious and philosophical questions, even with Christian mythology and esotericism. He is a good example of a remarkable mathematician/scientists who saw a much larger framework of problems of his time (which have reached an even higher intensity today), disagreed fundamentally with society and its philosophy (or lack thereof), in particular with scientists not thinking about the philosophical implications of their work. His frustration and removing himself from the norm of mathematics and then his philosophical undertakings, all of which was discredited by most of his colleagues, illustrates the dichotomy between science and philosophy today.

Another example for a high level protesting scientists is the (theoretical) physicist Jean-Marc Lévy-Leblond, a younger friend of Alexander Grothendieck, who was a professor himself teaching in the departments of physics philosophy and communication. He published numerous more general articles on epistemology, quite exceptional for an active researcher (who also published article on theoretical physics and mathematics). His goal was to more generally "put science into culture". He has been advocating with alarm the need for a public intelligence of the sciences, where knowledge, research, culture and politics would be linked, as there exists a severe gap in understanding between scientific specialists and the general public. He also cultivated the need for a history of science, against the illusion of the universality of scientific knowledge born in well-defined steps and thus, against the global standardisation that the domination of technoscience is what dominates the world development.

Besides most scientists' scepsis with respect to interactions with philosophical thoughts, there is a parallel movement on the side of philosophy itself. For example, according to the Austrian philosopher Paul Feyerabend, no universal and ahistorical scientific methods can be formulated concretely. Further, defined methodological rules do not generally contribute to scientific success and there is thus no generic "scientific method". Productive science must rather be allowed to change, introduce and abandon methods at will (as they have done in some ways in the past with changing mathematics and experimental devices). Moreover, according to Feyerabend there are no general standards by which different scientific methods or traditions can be evaluated. The lack of general standards of evaluation leads him to a philosophical relativism according to which no theory (scientific or any other kind) is universally true or false (the statement of "true" was supported by Popper, however, surely not the statement "false"; for him "false" is "false"). Feyerabend was critical of any guideline that aimed to judge the quality of scientific theories by comparing them to known facts. He read through the history of science against Popper's approach ("grain" as he called it) and provides many examples that scientists in reality often do not adhere to fixed rules and nevertheless, or precisely because of this, achieve success. Before an existing paradigm in science disappears often the scientists representing it themselves need to be gone as they hold on to the old models.

Feyerabend was, however, subject to various critics, especially for his misrepresentation of the practices, methods and goals of science itself. Defenders of the enlightenment view of rationalism, are next to Popper, Steven Pinker,[28] or Jacques Bouveresse in France the later referring strongly to Poincaré.[29] It is worthwhile to quickly have a look at the philosophy of Poincaré as an example how mathematicians at the turn of the nineteenth century operated in philosophy. Poincaré is often considered as one of the last universal genius of mathematics. After him, mathematics blossomed in so many fields that it was not possible anymore for one individual to grasp all its concepts. In 1880, he worked jointly with Émile Boutroux on an edition of Leibniz's *Monadology* and wrote an article comparing the physics of Descartes and Leibniz. In the 1890 he then published about 20 articles in *Revue de métaphysique et de morale* in which he worked on philosophical ways of interpretating the new mathematics working broadly against the founders of modern logic and set theory such as Cantor, Russell, Zermelo, and last but not least Hilbert. With respect to logics and foundations of mathematics, Poincaré's position was essentially governed by these two theses:

- Logical judgments alone are epistemologically not adequate to express the structure of a valid mathematical reasoning in view of its comprehensibility. There are logical antinomies.
- Thus, one should avoid any impredicative (only self-referencing; a term introduced by Russell into mathematics in 1907) concept.[30]

Movements that contested science from inside as by Grothendieck and Levi-Leblond as well as by philosophers like Feyerabend accompanied the raise of science to a growing power in society. However, whether any significant impact of such thinkers into science remains is quite questionable. Any quest for integration of philosophy into science will depend on general philosophical views and necessarily must include personal commitments to science. What we observe is that both remain rather contained within the science or philosophy community. However, as we saw in the last chapter, we should act

---

[28] Steven Pinker, *Enlightenment Now: The Case for Reason, Science, Humanism, and Progress*, Viking (2018).

[29] Jacques Bouveresse, *Une épistémologie réaliste est-elle possible? Réflexions sur le réalisme structurel de Poincaré*, Publications du Collège de France, 2015. https://books.openedition.org/cdf/4017 [archive].

[30] For details on Poincaré's philosophy: https://plato.stanford.edu/entries/poincare/#LogFouIntPre.

against the capitalist today's power to reduce the freedom of science and then exploit it. Science should primarily help mankind to evolve more and more to knowledge, responsibility and freedom, as the combination has been the most powerful one in making us freer, happier and more respectful of nature. And for this, we only repeat ourselves, involves a much larger community than scientists and technologists alone. Furthermore, this request does not primarily come from theoretical philosophy but from "practical" philosophy as Kant called it: the philosophical branch of ethics.

A Good Example for an Interaction of Philosophy and Science: Research on the Nature of the Human Ego-Consciousness

An important open question for today's life sciences is (as we already saw in the words of Emil du Bois-Reymond in 1872) around the origin and nature of our Ego-consciousness, what modern philosophers call the "self-model". Most biologists today see this less as a problem of philosophical principles than one of science. However, until the late twentieth century Descartes' view that mind and body are two different substances had remained rather untouched by natural scientists. This was somewhat surprising given that biology for quite a few years had been a pure science, even if this dualism, the division between body and soul, matter and mind into two different spheres, has shaped large parts of Western philosophical since the ancient Greeks.

After having remained in the background for much of the twentieth century, it was not until the 1990s that fundamental brain research returned back on the agenda of scientific research and discussions, and this with rapid advances in the researchers' understanding of the human brain and consciousness - though not yet of the Ego-consciousness. Here, it quickly met the philosophical perspectives on our Ego. Many of the underlying physical processes that manifest in us subjectively as memory, emotions or consciousness can now be systematically investigated empirically. For example, physicians can localise those regions in the brain that are damaged in patients with disorders of consciousness. And modern imaging techniques, e.g. MRI, make it possible to monitor the brain at work directly. With the insights resulting from the new technological possibilities of the last 30 years, brain researchers believe they are now on the way to nothing less than an explanation of our Ego-consciousness on a purely physical (materialistic or better naturalistic) basis. Thus, a major goal of neuroscientific research has become the finding of its concrete neural correlates, thus penetrating into the thinking of many philosophers. This has resulted in lively discussions between natural science and philosophy, which is a good example for an interaction of the two fields today.

Let us dive a little deeper into the current status quo of scientific research on consciousness and the philosophical discussion on it: Clear is that the available empirical results of our materialistic search for the fundament of Ego-consciousness are still very far from being conclusive. Current imaging techniques are simply still too fuzzy in their spatial and temporal resolution to measure our brain activities with sufficient accuracy. Furthermore, brain researchers do not even know yet how information is precisely stored in the enormously complex architecture of the brain. The statements and theories of brain research on the problem of Ego-consciousness are therefore still at a rather speculative stage. Nevertheless, neuroscientists are already finding first neurological correlations of certain states of Ego-consciousness (see Chap. 9 for details). With this they can already say one thing: According to all they have learnt so far in the last 30 years, Ego-consciousness is not a particular material item, not a at one location in the brain located, well-defined substance, nor a single entity. Thus, there is no particular place where Ego-consciousness directly reveals itself. It is rather a global occurrence in the brain that cannot be broken down into its individual part, a typical characteristic of the new scientific discipline: a phenomenon of *emergence*. Yet, and perhaps precisely because of this, neuroscientists are still far from having a clear grasp of ego-consciousness. On the one hand, they have thus not been able to avoid some serious philosophical questions, on the other hand, the turn of the brain researchers to deal with our Ego-consciousness has vice versa brought movement into philosophy. The "new philosophy of mind" is the first with an empirical foundation.

In addition to the empirical research data, there is an important second aspect to the recording of our Ego-consciousness: the phenomenological level, i.e. the subjective inner experience in us as such with all its multi-layered facets and variations. How can a scientist objectively capture a person's subjective experiences? Both researchers and philosophers therefore agree (by a majority) that the problem of Ego-consciousness represents a completely new type of scientific questions. This is already evident from the fact that, on closer inspection, it is not even clear what exactly the riddle of Ego-consciousness consists of. What could a scientific explanation for this look like at all? What would we even accept as a convincing solution to the puzzle? How can a multitude of subjective universes (those of the respective Ego-consciousness of each of us), that arise constantly and pass away rather quickly again in each of our mind's global universe, be described assuming the *objective existence* of such in the search for universally valid laws for them? This is why, so many begin with the dualism between subject and object that we are all so familiar with. In addition, subjective consciousness, with its evolutionary-historical development, must have an objectively accessible history of origin. In the long process from single-celled organisms to Homo sapiens, mind and Ego-consciousness must have emerged at some point (likely 70,000 to 100,000 years ago). The characteristics and dynamic structures of self-organisation in the brain that underlie Ego-consciousness must have emerged evolutionarily from brains that did not possess Ego-consciousness - or ones at a lower level, as this was probably a gradual rather than a spontaneous process.

The emergence view is opposed by the thesis that Ego-consciousness has its place in very specific regions in the brain that become active whenever I become aware of an impression, i.e. become aware that it is I who am having this experience. Its proponents postulate a direct identity of brain states and (subjective) mental states. They argue that many structures of human experiences and thoughts today described with the attribute of emergence will be explained over time by a more precise knowledge of the functioning of corresponding components of the brain (a similar viewpoint of Einstein regarding quantum physics).

In addition to the question of whether Ego-consciousness is holistic-emergent or correlational-atomistic in nature, consciousness researchers struggle with two other fundamental questions.

- The first is the question of determining characters of the human Ego-consciousness itself (for example, in comparison to possible states of consciousness in animals): Is it essentially "phenomenological" in nature, i.e., related to perception and feelings, or does it only have a specific effect in mental reflections, i.e., does it only articulate itself on the conceptual level of thought?
- The second - epistemological - question is what we perceive in our experiences: Do we experience the external world "as it is in itself" or does the world of our perception only emerge in our brain? And what is the "out there" that we experience?

We will dive into these questions more deeply scientifically as well as philosophically in Chap. 9 after having discussed in detail the modern brain research. There we also present the details of the fundamental problems in naturalistic views on the mind (1. qualia, 2. intentionality) and discuss some various position about the Ego-consciousness held by today's philosophers like Thomas Metzinger and Daniel Dennett.

More and More Important: The Relationship Between Science and Ethics

So far, we have mostly contained our discussion to the relationship between the philosophical study of nature, the origins and limits of human knowledge on the one hand and science and technologies on the other hand, i.e. mostly referred to epistemology. The today equally important *relationship* between philosophy and technology, we only referred to on less than a page, when we defined ethics as the relevant philosophical branch for how to deal with the technologies coming out of science. The necessity of integrating ethics into their development or even seeing it as the central philosophical involvement has, however, become quite clear after many risk issues have become obvious in the last 60 years. Here are a just a few examples, some of which we saw already briefly before:

- Nuclear risks: The most prominent and often cited one are nuclear bombs (not just the US atomic bomb form 1945 which we discussed in details in Chap. 2), but the development of various types and sizes of nuclear bombs is also a risk that has gained attraction (again) with the Ukraine war. Another nuclear risk comes out of nuclear reactors used globally today. Besides possible accidents, they can become blackmailing targets of a conventional war, as we saw in Zaporizhzhia, Ukraine.
- Risk of Artificial Intelligence: AI might take over significant or possibly all parts of our society or could help only very few individuals to control it (end of democracy). An example already existing today is the monitoring of an entire society with the help of AI in China and some similar attempts in other Asian countries like Singapore. Airports in their tracking of passengers are implementing more and more AI technologies. And big companies' HR departments already do a first selection of job applications using AI. The general risk is that governments or even private companies can track and easily interpret us in whatever we do (and say) easily. In Western democracies the GAFAM firms (Alphabet (Google), Amazon, Apple, and Meta (Facebook), Microsoft) already do this to a large extent (more details in Chap. 10).

- Risk of recognition everywhere: Facial recognitions are today able to simply spot and follow every possible pedestrians and see where they are going (as this already happens in London). Privacy on the streets is thus simply impossible. But further steps are to monitor completely ethnical groups, like Uighur Muslims in China; where the coverage now extends across the entire country in everybody's doing, scan the streets for "people of interest" in Russia, or track enemies within the country (like the Israeli following the Palestinians inside the West Bank) or the US killing of heads with drones.
- Risk related to our environment: It has become clear that the applied technologies of conventional production of energy has hit its limits today. They are likely to change the globally environment to levels which threatened human survival.
- Risks related to digitalisation: We depend more and more on IT systems that depend themselves on electricity. Think for example of the enormous amount of energy needed to mine bitcoins whose function and use remain questionable until today.

For centuries the core discussions between science and philosophy have focussed on epistemology and got eventually, as we saw stopped by the US. Is this the end of the relationship between philosophy and science? Surely not, as we saw in this chapter. And even *if* the discussions between science and epistemology stay non-existent, ethics, often called the "prima philosophia", remains of central importance (before 1950 ethics barely played a role in the philosophical discussions among scientists). Given the range of technologies that has emerged out of science with some severe risks, ethics has likely become the most important discussion with science overall. But epistemology is also still of great importance for the sciences, as we have pointed out in this chapter.

In the next chapter, we will concretely outline the various key technologies which constitute an impressive list, each one with high potentials for changes as well as high risks. Then, in the following chapters, we will get to the individual sciences. Their respective philosophies will remain part of the discussions.

5

Promising and Scary Developments in Future Technologies
An Overview

In light of the dramatic shifts in the world due to the development of a whole variety of new technologies in as little time as the last 30 years it is highly likely that in the *next* 30 years the speed of changes will grow even significantly further which will ultimately make it the most important phase in mankind's history.

A significant reason for this is that from the 2020s to the 2050s there will be a growing and most possibly mankind shaping race which we can already follow -and should shape - as of today: It is on the one side the climate change due to mankind changing the composition of our atmosphere by emitting large amounts of CO_2 and other climate changing gazes like methane, versus on the other side new energy technologies with the goal of having a sufficient supply of energies without CO_2 emission. Examples of the latter are wind-, solar, geothermal- or a new type of nuclear energy all of which combined with improved efficiency of energy consumption will hopefully save the world for us from too heavy climate changes. At the same time, the set of new technologies in the next 30 years will, even if we manage to resolve today's environmental problems, likely by itself create new potential battles that will decide on how we want to live our lives. The last thirty years have already seen significant life changers (just look at the internet and cell phones), much more than the period from 1960 to 1990. However in 2050, the world and our own individual lives will have changed much further having restructured them in ways still unimaginable today.

In order to prepare for the individual details of the last 60 years' and today's science and technology features as well as their future potentials, this chapter shall first provide an intense, exciting as well as scary walk through fifteen developments of technologies which are about to come out of sciences and will likely shape mankind's future. We shall also (equally briefly) mention the according risks which we will equally deepen in the following chapters.

The Future Technological Application Changing the World and Human Beings

Artificial Intelligence - Improving or Controlling Our Lives?

The idea, that human intelligence or more generally the processes of human thought could possibly be automated or mechanised, i.e. that we could design and build a machine which exhibits man-type intelligent behaviour, is quite old. The earliest source is usually referred to Julien Offray de La Mettrie and his work *L'Homme Machine*, published in 1748. Today, this Enlightenment notion of "man as a machine" is called the *strong AI*: an intelligence that replicates any human thought, a machine that responds intelligently or behaves like a human. A strong AI could thus independently recognise and define tasks, rather than just deal with specific problems, and acquire and build up independently knowledge of the corresponding application domain – or potentially of everything humans can think of. The goal of building strong AI, however, remains visionary after decades of research. In contrast, the *Weak AI* has no spontaneous and independent creativity and no explicit abilities to learn independently in a universal sense. Its learning capabilities are mostly reduced to training recognition patterns (machine learning) or matching and searching large amounts of data. The American philosopher John Searle's said the following about it: Weak intelligence "would be useful for testing hypotheses about minds but would not actually be minds".[1]

While strong AI is still far from being reached, what in contrast weak AI has reached today stands at rather high levels. It is sometimes no longer limited to the specific purpose for which it was created, such as playing chess, searching databases or recognizing faces. The learning and optimization methods underlying today's AI, so-called "deep learning", enable a massive machine intelligence increase across the board. This also affects more and more areas that most people today still regard as unchallengeable domains of human ability: intuition, creativity, or sensing other people's emotions. The

---

[1] John R. Searle, *Minds, Brains, and Programs*. Behavioral and Brain Sciences, 3(3), (1980), p. 417–424.

latter in particular will probably already be a standard capability of AI systems in the next few years. AI programmers speak also of "affective computing." Will machines soon be able to recognize our emotions even better than other humans do? Movie makers have already anticipated this movement like in the film: "Ich bin Dein Mensch" (I'm Your Man) from Maria Schrader where a robot is built to match the expectations of a woman and be her "ideal man". Also book writers have found that topic such as Ian McEwan with his novel "Machines Like Me" from 2019, in which Alan Turing (who survived the 1950s) achieves a breakthrough in artificial intelligence.

In socio-political and global political terms, the further development of AI is of great significance: the country with the most advanced AI is very likely to become the dominant economic and military power on this planet. Currently, two countries are vying for global AI supremacy: the US and China. China has caught up strongly in the last few years and is now even aiming to leapfrog to first place. The Europeans have long since been left behind in this race and relegated to the status of extras. The basis of the Americans' and Chinese' lead is not smarter researchers, better AI algorithms or better computer programmers, but simply the availability of data. Data is thus often considered "the oil of the twenty-first century."

Thus, an increasingly connected world is developing, in which our privacy threatens to disappear. With the appropriate software for face and image recognition and a dense network of cameras, the creation of movement profiles of individual people in real time has long since ceased to be a problem. What will be in store for us if there is no democratic control by the state, seeing how far completely controlled tracking data are used and the feeding of AI algorithms can go, is shown by the example of China, where people are already being steadily monitored individually by AI.

There are many questions society will have to find answers to, if we do not want to end up in an Orwellian world. How can we keep our personal freedom? Can democracy survive an age where machines would take political decisions? Moreover, what if the machines become so powerful that they acquire forms of self-consciousness (thus become strong AI), are we prepared to deal with this? (for more details on AI, see Chap. 10, for self-consciousness, Chap. 9).

Quantum Computers – Millions of Times Faster Computation or Just a Dream by Physicists?

For a long time, quantum computers were the stuff of science fiction. The term alone seems, to most people still today, as eerily bizarre as excitingly futuristic, combining the technological omnipotence of digital computing with the awe-inspiring complexity and abstractness of the most important physical theory of the twentieth century.

As we saw in Chap. 3, in 1981 the famous quantum physicist Richard Feynman gave a lecture on the possibility of quantum computers which is still widely cited today. It was the first vision and programme of possible quantum computers. Such a computer, in which subatomic particles are used for data storage and processing, is based at its core on quantum properties of the particles involved. Instead of processing bit by bit like classical computers, a quantum computer calculates in parallel on numerous quantum states, so-called quantum bits, qubits (the fundaments of quantum computer, entangled states of quantum objects, were still not fully developed in 1981; in fact, only a year later it was proven by experiment that such entanglements are even possible). This enables an unimaginably higher computing speed compared to conventional computers.[2]

In today's reality, the development of quantum computing advances rather quickly. It promises a new technological revolution that could shape the twenty-first century in much the same way that the development of digital circuits did for the twentieth century. Back in the fall of 2019, Google announced that its engineers had succeeded in constructing a quantum computer (relying on a superconducting type that uses electricity that faces no resistance) that could, for the first time, solve a problem that any conventional computer would cut its teeth on.[3] This was still more of a symbolic milestone, as the mathematical problem was still of a highly academic nature. Then, in October 2021, a Chinese team reported that its Jiuzhang 2 processor (a quantum computer that relies on photons, i.e. is light-based) could complete a task in one millisecond that a conventional computer would require 30 trillion years to finish. This marked a new top speed for a quantum processor whose qubits are light-based, not superconducting.[4] Future developments in quantum computing could generally impact different technological fields dramatically. This shall be further developed in Chap. 7.

Ethically, quantum computers can potentially threaten existing protections of data for individuals as well as companies, or even governments and militaries. Furthermore, with its fast computation it could essentially increase AI capacities incredibly, exacerbating the above-mentioned risks of AI.

---

[2] For more details see Lars Jaeger, *The Second Quantum Revolution*, Springer (2018); See also in Chap. 7 for more details.

[3] Google (many authors), *Quantum supremacy using a programmable superconducting processor*, Nature, 574 (2019), p. 505–510.

[4] The *Zuchongzhi 2* quantum processor reportedly completed the Google task one million times faster.

CO_2-Neutrality – Can We Create Enough Alternative Energies in the Next Few Years to Prevent a Climate Catastrophe

After three decades of deep sleep and criminally missed opportunities, the resistance - political as well as economic - to the global energy turnaround for saving our climate is finally falling (although there remain powerful economic forces that still do not want to act accordingly as well as setbacks due to the war in Ukraine). Some momentum has been established for our drive towards CO_2-Neutrality and thus protecting our climate. However, it remains open whether we are taking the right steps *fast enough* to prevent the worst scenario for the global climate.[5]

After all, with the exception of some parts of agriculture, all human influences on the climate can be traced back to energy production and consumption. On the positive side we can state that year after year even the most optimistic forecasts about the advancement of *new technologies* in these areas have been caught up with and often even surpassed by actual developments (though their *implementations* are not always following them as quickly). There is thus no shortage of ideas, technological possibilities and concrete initiatives to deal with and lower the CO_2 emissions. Driven by amazing advances in the fields of photovoltaics, wind power, geothermic energy extraction and battery energy storage (and possibly the most amazing one: nuclear fusion energy, see next point), as well as in nanotechnology and artificial intelligence for optimal energy use, we are on the cusp of the fastest and most profound shift in the energy sector in the last 150 years. In fact, today's professional investors' money is flowing much more into alternative energies than into coal, oil or gas, as they have finally recognised the potential of renewable energies und thus seeing it as an increasingly attractive capital investment. At the same time, fossil suppliers have lost massively in investors' favour.[6]

On the negative side, we might not get to the important CO_2 neutral production of energy fast enough. Reasons for that are:

---

[5] See Lars Jaeger, *Ways Out of the Climate Catastrophe - Ingredients for a Sustainable Energy and Climate Policy*, Springer (2021).

[6] For details in the reference above, see: Lars Jaeger, *The Comeback of Sustainable Energy - Why Fossil Energy Sources Are Only a Footnote in Human History*, Chap. 12: The marketability of renewable energies, Springer (2021).

- A look at *political* actions is quite alarming: The pressure from political institutions to change the energy towards CO_2-neutrality is very slow – partly also due to the intense lobbying of energy producers.
- Can we really expect the market reaction to be fast enough into this direction and implement quickly new CO_2 neutral technologies? The past of the energy market does not quite support a"yes" to this questions.
- There are still many people who oppose the views on climate development by our energy use, which are shared by almost all scientists (100% chance as the IPCC stated in August 2021).[7] Moreover, the free-riders, people refusing to pay for the transition hoping the others will do, should be added to those who frontally oppose to be charged for it.
- For the first time of human history dealing with the climate is a process that has to be developed *globally*, i.e. in consensus with all countries in the world. However, the interests of the developed countries who historically have emitted most of the CO_2 or related gazes into the atmosphere which has put us in into the situation today might be quite different than those of developing countries that are willing to grow economically to increase their standard of living for which they need the energy of today.
- The developments of the climate conditions are likely not to be linear in their dependence to the CO_2 amount in the sky for too much longer. However, many people assume a general linearity in their assessment of the temperature rise, as it has actually been so far. At certain ranges in the CO_2 amount that linearity of the global temperature increase is likely to disappear leading to a much faster rise of it. Unfortunately, we do not exactly know when, but the researchers estimate that as of about a two degrees increase global temperatures can start building up non-linearly. I.e. increase quite abruptly with further CO_2 (we already have reached 1.3 degrees today). This will then be a process which *cannot be reversed* anymore. In that case, significantly higher average temperatures on Earth for a long time are unavoidable.

It cannot really be the money that is missing for avoiding the climate change. The countries have come up with amounts of money for dealing with the Covid-19 pandemic issues rather quickly which is roughly the amount that will be needed for the global use of alternative energies.

---

[7] CIMP6 report of the Intergovernmental Panel on Climate Change (IPCC); https://www.ipcc.ch/assessment-report/ar6/.

Nuclear Fusion – The Solution of Our Energy Problems or Just a Topic of a Century of Dreaming?

Meanwhile, without much public attention, scientists are making some significant progress in an area that could solve the problems of global energy supply once and for all: the peaceful use of nuclear fusion. At stake is nothing less than the dream of capturing unlimited, clean and safe energy from the thermonuclear fusion of atomic nuclei, the same energy that power our sun and the stars.

In addition to the mammoth ITER project in Cadarache, France, which is being funded with massive amounts of *public* money (so far 20 billion Euro, which will end up most likely north of 60 billion Euro) and is expected to produce its first results in the 2030s (and actual broader electricity production rather as of the 2050s), a number of *privately* financed companies have also devoted themselves to fusion research. However, they are taking different approaches than the ITER researchers. With alternative and much smaller reactor technologies, they want to generate electricity from fusion already in the next few years, and thus much earlier than ITER. A public–private race is brewing to find the best solution for fusion technology.[8]

If we were actually able one day to produce energy like the sun and thus give ourselves access to the most efficient, safest and environmentally friendly form of energy that nature has to offer, this would certainly not only be another major technological advance, but rather a leap forward in civilization that would be on a par with the invention of the steam engine, which 250 years ago gave us the energy to completely transform our society. So, it is surely worth keeping an eye on the interim results from this race for the ultimate type of energy.

Genetics - The Victory Over Cancer or Manipulation of Mankind?

The success in developing vaccines against the Corona virus that quickly was based on immense progress in genetic engineering of recent years. "Genetic vaccines" contain the genetic information of the pathogen, which after administration is translated by the body's own cells into corresponding proteins, whereupon as in a real viral infection a defence reaction of the immune system is triggered.

---

[8] See also: Lars Jaeger, *An Old Promise of Physics – Are We Moving Closer Toward Controlled Nuclear Fusion?* ATW International Journal for Nuclear Power (December 2020).

However, genetic engineering processes are not only being developed for vaccines against infectious diseases, but also in the fight against cancer. Here, too, encouraging results are emerging. For example, researches are today being conducted on cell vaccines in which the mRNA sequence in the vaccine is designed to encode cancer-specific antigens. There are already more than 50 clinical trials for mRNA vaccines for a range of cancers, including blood cancers, melanoma, glioblastoma (brain tumour) and prostate cancer. So, it could well be that the Covid-19 pandemic will be the initial step in a broad breakthrough in the treatment of cancer and infectious diseases through genetic vaccines and thus patient-specific medications.

The most important bio-medical and genetic engineering breakthrough of this century remains CRISPR, a powerful new gene technology discovered in 2012, largely unnoticed by the public. Since then, it has become the most important tool in genetic engineering. Its name will soon be as well-known as DNA or AIDS. It describes sections of repeated DNA in a genome.[9] The specific CRISPR/Cas9 complex enables extraordinarily precise, rapid and inexpensive intervention in the genetic material of living organisms. It thus allows genetic engineers to directly access and manipulate individual genes. This genetic engineering process could make scenarios such as the definitive cure of cancers a reality much more quickly than even the greatest optimists among genetic engineers thought just 10 years ago.

The technology has long since been used in practice, particularly in the modification of the genetic makeup of plants. But CRISPR is also entering a new phase in the use in animals and humans. Direct interventions in the human genome are no longer a problem today, technically speaking. For some medical applications, the technology has already progressed to the stage of clinical trials. This will revolutionize, on the one hand, the treatment of many hereditary diseases previously considered incurable, caused by genetic defects, as well as plagues of humanity besides cancer such as HIV, malaria or even diabetes, and other age-related diseases.

On the other hand, the enormous potential of CRISPR has also triggered new ethical debates (and exacerbated old ones). Will we see the birth of genetically enhanced babies or the artificial creation of enhanced brains? What would be the consequences of playing the sorcerer's apprentice on our heredity? The potential of CRISPR is immense in almost any direction. In November 2018, for example, a previously unknown Chinese scientist announced that he had created the first genetically edited humans. He had altered their genome to make them immune to HIV (some negative side

---

[9] CRISPR stands for "Clustered Regularly Interspaced Short Palindromic Repeats".

effects became obvious later). It had not taken more than a moderately equipped lab and basic knowledge of genetic engineering. With CRISPR children, we have finally arrived in the age of human experimentation and designer babies. The weighing of these inacceptable goals against the immense opportunities of CRISPR must go through the filter of reflection and, as we outlined in the previous chapter, philosophical (ethical) assessment of what can and should be done and what should be forbidden and abandoned.

Internet of Things – New Industrial Technologies and Smart Fabrications or a Full Invasion of Privacy?

All of us have likely already become accustomed to numerous everyday digital helpers, from the app that informs us in real-time about train delays, the electronic measurement of the steps we have walked or run, to Tinder, which displays contemporaries in the immediate vicinity who are willing to mate. We are also familiar with the refrigerator that automatically reorders groceries. But what about (just two almost random examples).

> … an umbrella that flashes when rain is forecasted, attracting the attention of the person who wants to go out of the house, or
>
> … a wallet that becomes increasingly difficult to open the lower the bank account or credit card balance.

These tools are also no longer a technological problem today. With increasing computing capacity, faster networking through ultra-fast mobile Internet and ever smarter data processing, the development of such "smart things" will continue apace. Already in 2019, 5G was switched on, enabling breath-taking speeds of up to 10 gigabits per second on our cell phones. This network continues to be expanded globally. To achieve what we want, we no longer need to use a computer; everyday things will take care of themselves without our direct intervention.

With the increase of IoT ("Internet of Things") objects, however, the risk of cyber-attacks and intrusion in privacy are growing fast. Linked to this, the collection of massive amounts of private data by the internet providers could become another threat, especially when coupled with the AI developments and the processing speed of quantum computers. Are we prepared for such a huge invasion of privacy? We see already how pressure groups can use these data to forbid women to have access to abortion pill in the US. This is only the tip of the iceberg of what could be ahead of us. With a clearer understanding of the issue, we need a better regulation and legislation here.

Sure, the boarder between privacy and social sharing is moving fast, but we need to be aware of the difficulties of fixing a good frontier and move on our reflection regarding what is essential for preserving (commercial as well as individual) freedom to let data out for using internet devices while keeping social coherence.

Neuro-Enhancements – Improving Our Thinking and Acting or Move Away From Today's Reality?

In the last twenty years, knowledge about the structure and dynamics of our brain has multiplied, as we already saw in Chap. 4. The more we understand how it works, the more precisely we can influence our feelings, thoughts and experiences. Scientists are even working on microchips that can be implanted in the brain and *permanently* improve our state of mind, raise our sense of well-being, increase our intelligence, our memory and our ability to concentrate, or even provide lasting happiness.

People are doing in general an enormous amount of research into the genetic, chemical, and neurological backgrounds of emotions such as trust, compassion, forbearance, generosity, love, and faith. The more we grasp these, the more we can (and probably will) use this knowledge to manipulate ourselves and others. Thus, the more deeply neuroscientists grasp the workings and processes of decision-making in our brains, the more precisely they can influence how we feel, think, and experience:

- Today, the injection of neurotransmitters already alleviates mental illness, but our performance as healthy individuals can also be manipulated in this way. Think about the nudge theory that has flourished in applications even in our men's pissoirs.[10]
- Direct stimulation of the appropriate sites in our brain by neuro-electrical impulses also changes or controls mood, attention, memory, self-control, willpower, comprehension, sexual desire and much more.

Brain researchers are already allowing brains and machines to interact. Using brain-computer interfaces, for example, they are transferring content from a person's brain to a machine so that the machine, in turn, can assist the person with various tasks. This technology will probably be first widely used in video games to make them more experiential with sensory input projected

---

[10] See R. H. Thaler and C. R. Sunstein, *Nudge: Improving Decisions about Health, Wealth, and Happiness*, Yale University Press (2008).

directly into the brain. It has already been used to help paralyzed patients to drive their computers or seats. But consciousness technology can go much further, e.g., create the impression of climbing Mount Everest, along with the real and emotionally breath-taking feeling of the challenge of climbing the highest mountain on earth, as the Icelandic producer of the game "Everest VR" (Virtual Reality) advertises.[11] When the sensory impressions are no longer just about the visuals, but also literally bring wind, cold and exhaustion into play, the game experience will be indistinguishable from "real" reality. The very concept of "reality" is then put in question, and we already see manifestations of that in another way: with the fake news phenomenon. People are rendered suspicious by the fact that reality can be manipulated. This pushes them to be over-suspicious and then again believing in strange objectively unreal circumstances.

It is clear that these developments of manipulating our very consciousness comes with severe ethical problems. But can we really stop this from happening – with all the commercial interests? (see Chap. 9 for more details).

Understanding Our Minds Through VR-Technologies – Finding Our Ego Or Is It Unfindable for Scientists?

Until now, our perception and how we see and experience ourselves (our "self-model", as philosophers call it,[12] or Ego-consciousness in Chap. 4 regarding our self-experience) depended solely on our connection with the reality around us. What we experience from outside and how we experience ourselves on it has directly been given by the stimuli of the external world. With the new technologies referring directly to our consciousness, our connection to this reality is faltering. By playing new realities to our brain, our perception and self-model will be changed almost at will. VR technologies therefore not only alter our external living world, but increasingly also our subjective "inner space. That has some severe philosophical consequences, as standard philosophy has largely separated the inner perspective of our being (possibly also the "Ego-consciousness") and the external world. A significant philosophical insight from our experience with virtual reality so far is that

---

[11] Colin Campbell, *New VR tour will take you on a climb up Mount Everest*, https://www.polygon.com/2015/11/10/9707306/new-vr-tour-will-take-you-on-a-climb-up-mount-everest.
[12] Also called the *self-model theory of subjectivity*. This concept comprises experiences of first person perspective of ownership (including) experiences of ownership and of a long-term unity of beliefs and attitudes. (Source: https://en.wikipedia.org/wiki/Self_model).

the mental self-image of humans is anything but stable and can be manipulated relatively easily (against much of the philosophical belief in the last centuries and millennia). Incidentally, experiences with hallucinogenic drugs in the 1960s and 1970s also lead to this insight. It is thus hardly surprising that we can with suitable setups very easily identify ourselves with an artificial body image instead of our biological body, with a so-called "avatar". We see already the premises of this in the development of virtual reality system like Metaverse. Already, some modern philosophers, like Niklas Boström,[13] claim that it is not possible to be sure that we are not ourselves "simulations". Will we, with the respective manipulation technologies, reach an entirely new interaction of ourselves with the environment, and how does this have to be evaluated ethically?

Digital Algorithms and Big Data – New Profiles for Our Lives or Controlling Humans' Thinking and Acting?

With all the collection and sharing of data about us and our possessions on the internet we have, just like our contacts and connections, long moved beyond computers or phones. For example, our smartphones are often directly connected to the heating system and other everyday appliances in our homes. Also popular is the idea of so-called "wearables," which are pieces of clothing that incorporate various sensors directly and connect to apps that permanently measure (and store) our pulse and other body values. It no longer needs a computer at all for all this. The objects of our everyday life regulate their needs directly with each other in the above already described Internet of Things. How simple this makes our lives!

But there is a catch. The data we leave behind everywhere, like bacteria after a sneezing fit, is collected, processed with ever more powerful algorithms and ever smarter AI, and used for ever more comprehensive purposes. From them, our behaviours, preferences, and character traits can be specifically read out and the patterns of our lives calculated - and then manipulated.

In a world of total networking, our privacy will ultimately entirely disappear. With the appropriate software for facial and image recognition and a dense network of cameras, the creation of movement profiles of individual people in real time is no longer a problem. Banks, too, are increasingly determining the granting of loans through AI that is well able to capture the essence of our finances with data they can access. Life insurances compute the

---

[13] Niklas Boström, *Are You Living in a Computer Simulation?*, Philosophical Quarterly. 53, 211 (April 2003), p. 243–255.

premium using access to health data of the client. So soon we will hardly be able to do anything without someone finding out about it. As Eric Schmidt, the head of Google, put it, "If there's something you don't want anyone to know about, maybe you shouldn't do it anyway."

Reactions are already happening to this development. The EU has put in place new legislations like the RGDP to protect our data. The US is thinking about it but the lobbies there have been very effective so far to prevent new regulation put in place. But it goes much deeper to profound ethical questions: Is this the world we really want to live in?

Blockchain Technology - Is It a Groundbreaking Innovation or Just a Passing Trend?

Blockchain technology is best known for cryptocurrencies such as bitcoin, but it has great potential in other areas. The decentralised nature of the blockchain enables secure and transparent transactions, data storage and identity management. Blockchain technology is expected to gain prominence in finance, supply chain management, healthcare, energy and more. This technology by expanding the scope of its applications will stay on the agenda.

Cybersecurity – Is It a Consistently Significant Concern or Merely an Occasional Problem?

With the growing reliance on technology, the risk of cyber-attacks is on the rise. Establishing robust security measures and protocols is essential to safeguard our digital systems and sensitive data. Additionally, artificial intelligence and machine learning are increasingly employed in cybersecurity to identify and counteract these threats. Building the resilience of society against cyber attacks will stay on the agenda for a long time. Thus, cybersecurity must keep pace with the growing importance of information technology in our societies.

Nanotechnology – Creation of Things from "Nothing" or Just a Dream?

Remember the title of a lecture Feynman gave in 1959 carrying the title "There's Plenty of Room at the Bottom". As early as in 1986, the researcher

Eric Drexler also foresaw the advent of molecular machines.[14] In fact, today, next to quantum computing, the most ground-breaking vision of future quantum technologies is the construction of ultra-small machines that can perform work at the level of atoms and molecules. Such nanomachines could assemble individual atoms as if using a building-block principle, and use them to synthesize any chemically possible (i.e. energetically stable) compound. They could even manufacture or replicate themselves. Nature has long since shown us that this is possible: DNA is nothing else than a self-replicating nanomachine capable of producing the most amazing structures - just think of the multiform synthesis of proteins, the foundation of any form of life.

Thus, the respective "second generation of quantum technologies", which contains quantum computers, will change our lives at least as much as the first generation with computers, lasers, atomic energy and imaging techniques in medicine. However, also nanotechnologies with all their promising future come with some ethical concerns, mostly related the possibility of their military applications- (by using them in autonomous lethal weapons. Moreover, there are also dangers posed by self-replicant nanomachines that are not controllable any longer. Are such possible? – an open question.

Stem Cells – Using Cells That Can Do Everything. Also for Our Entire Body and Mind?

The next revolution in reproductive medicine - after in vitro fertilisations, the first of which was in 1978 - is already on the horizon. It is conceivable that in the future hundreds of embryos can be conceived in the Petri dish from parental cells. After DNA analyses have determined in detail the predispositions of each individual embryo, parents will be able to choose their desired child. Perhaps an egg can even be created from a man's skin cell so that homosexual couples can have children with their own genes. Or a person may even be both biological father and biological mother.

All this will be possible through the use of stem cells. These are cells that have not yet differentiated into special cell types such as muscle, skin or fat cells. Stem cells are also important in medicine, as they can be used to create any part of the body. For a long time, it was considered as certain that once cells were established they could not be re-developed into stem cells. But in 2006, Shinya Yamanaka, who later won the Nobel Prize, showed that stem cells can be artificially created from a person's body tissue (so-called "induced pluripotent stem cells"). They are able to renew damaged cells and

---

[14] Eric Drexler, *Engines of Creation: The Coming Era of Nanotechnology,* Doubleday (1986).

thus massively support the regeneration of cartilage and bone, for example. Stem cells have been used successfully for decades to treat leukaemia and other types of cancer. In 2011, it was even possible to grow human heart tissue from stem cells. In the same year, significant improvements in vision were achieved in the USA with stem cell therapies in patients suffering from retinal disease. Even the cure of serious diseases such as Parkinson's, diabetes or paraplegia is within reach with stem cells. Last but not least, as cellular all-rounders, stem cells are particularly suitable for genetic engineering. Here the same ethical issues come up that we have already raised with genetic engineering above.

Biotechnology – From Frogs for Pharmacists to Nano-Robots in Our Bodies, a Medical Dream or Future Reality?

Until the 1960s, every pharmacy still had an aquarium with warm water in which clawed South African frogs were kept. The reason: They were needed as indicators of pregnancy in humans. The morning urine of the woman suspected of being pregnant was injected into a female frog. If the frog spawned within 18 h, the urine contained a certain hormone - proof of the woman's pregnancy. Today, that detection is much faster and less complex, and the large analytical devices in laboratories have shrunk to tiny systems. Through chemical processes biomarkers cause colour reactions in extremely small amounts of liquid. Every pregnancy test and every alcohol tube into which a driver has to blow is such a mini-laboratory. A "lab-on-a-chip" can diagnose numerous disease symptoms in minutes. Respective nano-level biosensors can also work electronically using microprocessor chips. Since September 2016, the police have been testing test strips that check saliva samples for cannabis and other drugs right at the roadside.

More and more worked with in biotechnology are nanorobots. They are already being used to transport medication specifically into diseased tissues, so that pathogens or mutated cells can be attacked directly with active substances. In this context, one speaks of "medical miracle bullets". However, the future promises of using nano-machines in medicine are even more exciting. They range from ultra-small nano-robots (so-called "nanobots") that permanently move through our bodies looking for pathogens, from automatic nano-checks every morning while brushing our teeth through, biomarkers that indicate serious diseases in their early stages, to the replacement of defective body parts with implants made of corresponding nanoparticles. Nano-machines could also be inserted into the human body with the task of independently searching for and destroying cancer cells. They would therefore move through the bloodstream to diseased organs and release drugs there or even perform surgical procedures microscopically, thus molecular machines could on the long-run replace today's large scale surgery.

Risks of biotechnologies are not quite so obvious but involve largely the chance to change medicine as well as agriculture in non-desired forms (forever). Furthermore, bioterror attack in a densely populated area could be massive.

New Food Technologies - How We Will Provide Food to 10 Billion People or Just a Science Dream Story?

The twentieth century is rich in significant technological achievements and formative technologies. However, if one had to name the most significant technical invention of the last century for mankind, the choice would probably go to the Haber–Bosch process of 1908, which made the large-scale industrial production of ammonia from hydrogen and atmospheric nitrogen possible. Without the Haber–Bosch process, it would be impossible to feed even half of today's world population. However, even artificial fertilisers and modern agricultural technology will not be able to cope with the task of feeding 10 billion humans. Thus, the question arises: How can the growing world population be supplied with food? The farmland can only be increased worldwide at the price of massive environmental damage; water is becoming already increasingly scarce. In addition, more and more people want to eat meat, which aggravates the situation even further because large areas of cultivated land are not cultivated with food for humans but with animal feed for meat production. In addition to the (controversial) use of genetic engineering, which increases the yields of rice, maize and co., more food must inevitably be produced industrially. In other words, chocolate from a printer and artificial meat will probably become the norm.

This is exactly where massive changes could soon occur as a result of technological advances, namely through meat that comes like from 3D printers. Such "printers" use muscle stem cells from cattle that are artificially grown and multiplied and then mixed with nutrients, salts, pH buffers, and so on. The result very likely tastes more delicious and is at the same time healthier than any animal meat to date, and … will come with virtually no CO_2 emissions! Anyone who doubts that such artificially produced in-vitro meat is more appetizing or that a diet based on it is healthier should spend a few hours in a large slaughterhouse or visit a reasonably sizable agricultural producer. Then he or she will likely quickly lose his or her appetite for today's animal meat.

There are already today a large number of plant-based substitute for animal protein that taste (almost) like meat and other animal products. Furthermore, there is an increased use of "green genetic engineering": Recent genetic engineering applications include the production of plants with higher nutrient content. The best-known example is "golden rice", a genetically engineered rice variety with additional beta-carotene. With CRISPR, genetic engineers have been given a new, powerful tool for plant breeding.

It is not easy to find ethical problems that comes with health-wise better, safer and environmentally more friendly sorts of food for humans than those we eat today. However, some are generally and importantly mentioned around the genetic manipulation of food such as potential harm to human health or drastically damaging the environment (if one includes the genes of animals).

Synthetic Life– When Humans Play God: Part I

In addition to gene optimisation, life extension through cancer treatments, enhanced crop breeding, stem cell therapy, genetic engineering opens up yet another exciting and equally scary possibility of human technological intervention: the creation of completely artificial life tailored to specific purposes. Creating life from scratch is what genetic researchers today call "synthetic biology". This new field of research in biology is aiming at creating life forms that have never existed on our planet before.

A first breakthrough in synthetic biology was announced by gene pioneer Craig Venter in spring 2010. His team had succeeded in building an entire artificial genome in the laboratory. They then implanted this in a bacterium called *Mycoplasma capricolum*, which had previously been freed from its natural DNA. Their artificial bacterium was thus a reproducing creature with a new, completely artificially produced DNA (in a traditional cell framework, however). The plans of the artificial bio-research guild go further and have already made some significant advances since such as approaching the goal of a complete synthesised genome of the many times more complex yeast with 16 chromosomes. This is a much more difficult undertaking than the creation of synthetic biology DNA announced by Craig Venter (see Chap. 8 for more details, also on how the building of new life is actually done).

When it comes to the future development of human creation of synthetic life, which is likely to become more and more sophisticated, the natural question that arises is: Is it an egregious example of scientists "playing God"?[15] And should we do that? Think of the "frankenviruses"[16] built in laboratories and the risk of pandemics they pose to us.

Life Prolongation – When Humans Play God: Part II

The average human lifespan in the most highly developed countries has risen from about 65 years 100 years ago to more than 80 years today. And at 2.5 years per decade, it continues to rise briskly. Its doubling in the last 150 years is almost entirely due to advances in our medical and scientific knowledge, including better standards of hygiene, more adequate nutrition and more comprehensive emergency medical care. But why do we get older in the first place - and eventually die? Science still cannot give an exact answer to this question (while sociology, religion and psychology can). None of the various theories of ageing is broadly accepted.

Simplified, it could be said that over time our cells and organs simply lose their ability to function. But most genetic researchers today assume that this process could be stopped or even reversed. Will this makes a primal human dream come true: the fountain of youth of eternal life? Animal cells, for example those of the threadworm *Caenorhabditis* elegans (a popular model organism in biology and genetics), can already be completely reprogrammed today through targeted editing, i.e. changing, rewriting or deleting individual nucleotides in the DNA strand. For example, after a certain gene that forms an enzyme called "SGK-1" was switched off, the manipulated worm lived 60 percent longer than its untreated conspecifics. Even more spectacular is another small animal which lives in the Mediterranean Sea and bears the name *Turritopsis nutricula*. It is a small type of "hydrozoan", not much more than a floating disc of goo (a slimy substance) in the water. But it has an amazing property: It is immortal (as long as it is not being eaten). This is because this special jellyfish has a cell programme that reverses the usual transformation of young cells into differentiated cells. It thus "rejuvenates" its adult cells permanently (see Chap. 8 for more details). Why should this not somehow be possible with human cells?

---

[15] For more details on the ethical discussion about synthetic life, see: Gregory Kaebnick, Thomas Murray (eds.), *Synthetic Biology and Morality: Artificial Life and the Bounds of Nature*, The MIT Press (2013).

[16] "Frankenviruses" are viruses built artificially by combining parts of or entire viruses. The etymology comes from "Frankenstein" and "virus".

However, is this really desirable? Would we really want to live in a respective society? How about giving birth to kids in light of the fact that thus the population just grows indefinitely? We would have to critically ask the question of what effects a significantly extended life expectancy of humans would have on our society which would surely change dramatically.

Are These All the Technologies that Will Shape Our Future?

We see from the list above how many technological developments are today shaping up mankind *at the same time*. Before the 1950s, in each of the 200 years before people also faced particular technological upheavals. However, technological progress has simply been slower and tended to take place over a few generations. This is entirely different today. We are not dealing with just one "technological sorcerer's apprentice", but with a whole series of them. The associated complexity and speed of social change overwhelms most of us – and drives many of us into a phase of ignorance, together with greed. And this will even accelerate further, possibly getting to a point where humans can generally by no means oversee the changes of the society any longer. It should be clear that this is a very critical situation, as we see already today the spreading of new technologies, some of them being quite critical. Many are global, and all of them without any significant governmental control, not in democracies, nor - which is even more worrisome - in non-democratic states.

The spectrum of future technologies is enormous. Each of the fifteen ones and the two concerns mentioned in this chapter can - and likely will, possibly even in some ways not yet known today or even being accompanied by yet unknown technologies - change our life such that *taking them altogether* the future developments of the world and human life itself is indescribable. Many of the future technologies come with severe risk for our lives and our freedom taking directions that we cannot - or better do not want to support. We will now in the following chapters work out the details of these threats *and* excitements, *as well as* the properties of the modern sciences themselves.

6

Physics from 1960 to Today
And What We Do Not Know Yet

The New Quantum World – How to Deal With An Uncountable Number of New Particles

By the late 1940s Americans had entirely taken over the lead in physics. One of the first among those was Richard Feynman, who we already met in Chap. 2. He was born in 1918 and, as Heisenberg and Pauli, was below 30 years old when he reached his peak in physics. In 1945, he had succeeded in developing a first theory of quantum electrodynamics which made it possible to calculate the *electromagnetic* interactions between quantum objects, and thus for the first time to see what *quantum fields* are like. However, in this process Feynman did not care any longer about the physical concreteness of the quantum objects, a topic that was so hotly discussed by quantum physicists before 1940. He was the first theoretical physicists that left the area of trying to imagine the quantum objects and interactions entirely and instead focussed on a proper *mathematical foundation for describing their functioning*. It was this giving up of concreteness that made his approach in quantizing the electromagnetic fields so successful. Unlike generations of physicists before, Feynman thereby did not focus on any philosophical question or interpretation at all. These appeared even further difficult as his results emerged to be even more complicated than the quantum physics for atoms.

Mathematically his theory worked very well: He had recognised a certain regular structure in the integrals occurring in tracking the interactions of the wave functions with the quantum electromagnetic fields. Each one could be traced back to certain core mathematical building blocks. And from this observation Feynman derived an ingenious trick still used intensively today, the so-called "propagators". With them, one can well describe the behaviour of the wave function of a quantum object during its quantum electrodynamical interaction with other quantum objects.

As Feynman entered a new range of mathematical complexity, an important development took place in experimental physics: Already at the early stage of quantum physics theoretical physicists predicted the existence of new particles, like Dirac in 1928 with his antielectron (positron) thesis or the neutron. Experimental physicists were a few years later able to prove their existence. However, this sequence reversed rapidly as of the later 1940s: From now on, experimental physicists found a higher and higher number of unknown particles, at a rate that theoretical physicists were not able to keep up with. In other words, the world consisted of much more than protons, neutrons and electrons, which called for a new theoretical foundation.

- Many new particles appeared in the measurement of cosmic radiation, for example pions and muons.
- In the early 1950s an important experiment which is still a central activity of experimental physicists today was developed: the particle acceleration (and then crashing to get new particles). Small physical particles are brought to very high speeds and then collide with others, using electromagnetic fields. The higher the energies, the more "fragments" of the involved particles are created and the deeper the physicists' insight into what they call the "fundamental structure of matter".

In the 1950s the list of discovered particles (in the newly established field of particle physics) quickly grew to a few dozen without any explanations for their existence. The questions the theorists were confronted with were:

- Why are there so many different particles in nature in the first place?
- Is there possibly a liaison to relate them to each other (which would require a new physics)?
- Is there a fundamental theory that lets us put *all physics* into the same framework of equations?

While the theorists vigorously searched for an explanation, the number of experimentally proven particle types continued to grow. Every attempt to grasp the characteristics of the particles already identified and develop a unifying particle theory was accompanied by the discovery of more particles with new features and thus to even more confusion. Soon the physicists ran out of names for the new discovered particles, so they gave them only ancient Greek (capital) letters as names: Σ-, Λ-, Ξ-, Ω-hyperons are just a few examples still in use.

Another important factor was the assumed presence of two new forces (next to the long-known electrodynamics and gravity) that act only at the level of those elementary particles:

- The strong force had to be postulated to explain how the atomic nucleus was bound together despite the protons' mutual electromagnetic repulsion. It was believed to be a fundamental force that acted on charged protons as well as neutral neutrons by pulling them together so nuclei can consist of protons kept together. It thus had to be significantly stronger than the electromagnetic force.
- The weak nuclear force was an equally hypothesized force proposed already in 1933 by the Italian Enrico Fermi explaining certain aspects of the beta decay, which however had not been understood very much back then. It is a thousand times weaker than the electromagnetic force and exceedingly weaker than the strong nuclear force.

However, it was yet unknown how these forces actually act, i.e. which laws there are following. It was also not clear how to order them into a structured and possibly unified quantum system taking electrodynamics into account as these two forces are only acting at such small distances (the gravity force is way too small as a force to be considered on atomic or nuclear levels). Dimitri Mendeleev had solved a similar problem about a century earlier when he arranged the chemical elements known at the time into the periodic table. Only one thing was clear to particle physicists: the classification of quantum particles would be a lot more difficult than Mendeleev's classification and, in particular, require much more complex mathematics. In the following, the reader will be introduced as clearly as possible to the structure of today's physics' particle theory. If he or she feels overwhelmed, he or she can turn to the section thereafter (chaos theory: back to normal life) without any further comprehension problems for the rest of the book.

A First Theory Integrating Various Particles and the Strong Force

The mathematics of particle physics became more and more complex. Nevertheless, we want to present here the physics as precisely as possible for making it accessible especially to non-physicists. In the mid-1960s, American theoretical physicist Murray Gell-Mann, another genius of twentieth century physics, succeeded in creating a first classification within the particle zoo. Born in 1929, Gell-Mann studied at Yale University from the age of 15 and completed his doctorate at 21. Being 26, he became a full professor at the renowned *California Institute of Technology* in Pasadena, Los Angeles. Even before becoming a professor at "Caltech" (in 1955), Gell-Mann had set himself the task of shedding light onto the jungle of particle physics. His attention was initially focused on particle called "K-mesons" (also called kaons) which were discovered in cosmic rays in 1947 and later produced by the strong nuclear force acting in high-energy collisions between atomic nuclei and other particles. Gell-Mann believed these particles should also decay through the strong force. However, it turned out that they decay under the influence of the weak force. It seemed strange that particles created by means of the strong nuclear force do not decay as a result of this force. That is why physicists today call kaons and their related particles "strange particles".

Gell-Mann approached this complex problem just as pragmatically as Mendeleev had done. He turned to hadrons (the name had just been created[1]), which are characterised by the very fact that they are subject to the strong nuclear force. He quickly saw that there are two types of hadrons:

- *Baryons* (the name was introduced by the Dutch-American physicist Abraham Pais, a close collaborator of Gell-Mann, in the late 1950s and comes from the Greek word for "heavy" (βαρύς)[2]): These include the heavier particles like proton and neutron.
- *Mesons*: They include less heavy particles such as the "pi meson" (or pion), a particle also detected in 1947. The Japanese physicist Yukawa called his

---

[1] The term *hadron* comes from the ancient Greek word hadrós, for stout or thick. It was introduced by L.B. Okun right in a plenary talk at the 1962 International Conference on High Energy Physics at CERN in Geneva.
[2] Robert Crease, *Abraham Pais 1918–2000*, National Academy of Science (2011).

predicted particle "meson", from the Greek word μέσος, mesos, for "intermediate", because its mass was between that of the electron and that of the proton.[3]

At first, the sorting seemed to him like a puzzle. Protons and neutrons hardly vary except for the fact that they are differently charged. In particular, they have almost identical masses. Some other particles are also very similar. Gell-Mann noticed that, in addition to their charge and mass, the members of a group can be characterised by a so-called "Strangeness Number". This is zero for most conventional particles and only non-zero for the strange particles like the K-mesons described above. One property of the strangeness number fascinated Gell-Mann in particular: In all reactions, its total sum remained constant across all particles. In physics, such invariances have always corresponded to physical conservation laws, i.e. particular physical variables that do not change under any physical transformations, with the inverse also applying, as the (female) German mathematician Emmy Noether already pointed out in 1918.[4] Does the invariance of total strangeness for any reaction maybe correspond to a not yet known physical variable that remains invariant under the equally not yet known equations applied to the process?

From the "puzzle" of particles reacting to the strong force, Gell-Mann developed a basic scheme for the classification of particles in the early 1960s. He called it the "Eightfold Path" (an expression that comes from Buddhism where it describes the path to the highest knowledge). But why eight? Well, what came to help him were particular mathematical sets of variables, the so called "Lie groups". Mathematicians knew this abstract concept from the Norwegian Marius Sophus Lie in the nineteenth century, with which elements of a set can be very nicely connected. Physicists had used Lie-groups before for describing the invariance of equations under certain transformations, which were directly related to the conservation of physical variables. Gell-Mann thus followed the footsteps of Emmy Noether, with a very new, much more complex group as the ones she had considered.

A particular Lie group proved to be perfectly suited to Gell-Mann's purpose: the eight-dimensional SU(3) group. Latest here, the mathematics of modern particle physics are at risk of ending all comprehension by non-mathematician/non-theoretical physicists. But one should not worry too

---

[3] He had originally named the particle the "mesotron", but he was corrected by Werner Heisenberg who pointed out that there is no "tr" in the Greek word "mesos".
[4] For more details on Emmy Noether and her theory, see Lars Jaeger, *Emmy Noether - Ihr steiniger Weg an die Weltspitze der Mathematik*, Springer (2022) only in German; for an English discussion on Emmy Noether see Lars Jaeger, *Women in Science – Hidden but key contributions in science in history*, Chap. 12, Springer (2023).

much about that as this shall not be decisive to understand the following in this book and will only be described shortly here. Within Gell-Mann's scheme, eight members fitted into one group corresponding to the dimension of SU(3). It was still rather a guesswork by him. But it was perfectly applicable to a particular group of known baryons, which could be distinguished exactly as eight particles with certain properties. However, for mesons this did not work, as only seven such mesons were known. So Gell-Mann did something that Mendeleev had already done 100 years before him with chemical elements: he postulated the existence of an eighth meson. And indeed, in 1964, physicists running a particle accelerator discovered a new particle they called η' ("eta prime meson") which turned out just to be the particle predicted by Gell-Mann. Finally, the theorists had predicted a particle again. A first structure had appeared in the jungle of particles.[5]

But Gell-Mann now suspected even more behind his new scheme. It could not be a coincidence, he thought, that the SU(3) group gave the order that brought a group of particles together, and that the eightfold path matched the particle order so exactly. The members of a group differ in mass, charge and strangeness, but have equal quantum numbers of spin and parity (where the mirror image remains the same). The differences in their masses are so small that it can be assumed that the real differences only exist in charge and strangeness. Gell-Mann also knew that the strong nuclear force makes no distinction whatsoever about the charge of a particle. It works the same for protons as for neutrons, as well as antiprotons or antineutrons (which were postulated in the early 1930s by Fermi and Dirac and experimentally discovered in the mid-1950s). He therefore asked himself: Is there possibly more behind the grouping into eight particles? He had the idea that the baryons could consist of even smaller particles that no longer carry integer charges. He assumed that all the particles of the baryon group were composed of three individual smaller particles, which he called "quarks", again following his penchant for strange names. He had taken this name from a line in James Joyce's (indeed "strange") novel *Finnegans Wake*.[6] Gell-Mann then called them "up quark", "down quark" and "strange quark". The names "up" and "down" refer to one of the physical quantities that must be attributed to quarks: the so-called isospin, an inner symmetry under the strong interaction. For the up-quark this points upwards, for the down-quark downwards. Strange quarks were contained in all particles that possessed the "strange" properties introduced above. According to his considerations, protons and

---

[5] Murray Gell-Mann, *A Schematic of Baryons and Mesons*. Physics Letters. 8, 3 (1964), p. 214–215.
[6] It seemed a good fit for particles that come in threes, as the line in *Finnegans Wake* reads: "*Three quarks for Muster Mark!*".

neutrons should thus each consist only of up and down quarks, the proton of two up quarks and one down quark, the neutron of two down quarks and one up quark. For reasons of symmetry, the strange quark was to be joined by another quark, named the charm quark. But it was not until 1974 that a particle consisting of this new type of quark was discovered in two independent experiments which is why it got the double name J/ ("J/Psi"). In parallel with Gell-Mann (and his colleagues, German Harald Fritsch and the Swiss Heinrich Leutwyler), a young researcher named George Zweig at CERN developed a very similar model in Europe. Zweig assumed that the quarks were actually physical building blocks of particles, an idea that Gell-Mann did not follow in his abstract scheme, for reasons described below. Feynman called the quarks "partons" but was overruled in the naming.

Gell-Mann's idea created the much longed-for possibility of sorting and classifying the ever-growing particle zoo and finally tracing it back to even more elementary particles. The following scheme emerged: quarks combine in triplets to form baryons and in duplets to form mesons. There should be four different kinds of quarks with their complementary "antiquarks" (for the respective anti-particles). All known baryons and mesons including their anti-particles could be represented by the various combinations of quarks. As time went on, two more quarks were added that enabled the physicists to classify the continuously growing number of particles: the "bottom" and the "top" quark (the latter was only discovered in 1995, the former in 1975).

But why can individual quarks not be explicitly observed until today? Gell-Mann found an explanation for this, too. With the SU(3) symmetry property in the abstract eight-dimensional space, another classification pattern emerged for the quarks. Each of them can be assigned a "colour", which, analogous to the charge in the electron, represents the type of interaction with the strong nuclear force: red, blue and green (this is, of course, not to be taken literally, but only serves as an illustration), as well as "anti-red", "anti-blue" and "anti-green".[7] All particles subject to the strong nuclear force thus obey a "colour theory": A combination occurring in nature can only exist as a "colour-neutral particle", either as a combination of red, green, blue, where the three colours just cancel each other out[8] (which create baryons), or as a combination of a colour with the corresponding anti-colour which also neutralises the colour (creating mesons). These colour properties also gave the theory of strong nuclear forces its name: "quantum chromodynamics" (*chromos* is the Greek word for colour). The deeper reason for the colour neutrality is that within baryons and mesons, the strong nuclear force

---

[7] They were first named that way in 1964 by Oscar Greenberg.
[8] See the RGB colour model used in modern video displays.

between the quarks acts more strongly the further away the individual quarks are from each other. This is comparable to a rubber band or a spring whose restoring force acts more strongly the further it is pulled apart. As a result, each quark is confined with the other quarks in the baryons or mesons. If you apply brute force (or a lot of energy) you can eventually rip them apart - just to find that instantly a new quark/antiquark-pair will emerge to take the place of the ripped away partners of the original quarks. Quarks cannot be observed as individual particles.

Besides the classification of particles, the interaction of particles with the strong force needs to be described mathematically as well. Especially important for that are those transformations that do not change the physical properties, i.e. those that are invariance transformations à la Noether. These invariances altogether form a so called "gauge symmetry", the invariances themselves "gauge invariances", and the respective field theories "gauge field theories". The latter are physical field theories that satisfy a - local or global - gauge symmetry. In a *local* gauge symmetry, the interactions that leave the equations invariant can be freely chosen locally. In contrast, in a *global* gauge symmetry the equations invariants are acting the same way globally.[9] The SU(3) in the theory for the strong force provides a locally invariant symmetry.

But how do these forces express themselves physically? By particles that make the quarks and thus baryons and mesons interact with each other. The eight exchange particles for the eight-dimensional theory of strong interaction are the so called "gluons". They are massless, electrically neutral, but "colour-charged" particles, while the photon, which transmits the electromagnetic force, has equally no electrical charge, but also neither any colour charge. Gluons therefore participate in the interaction of their own force with themselves which keeps the quarks together in a cloud of short reaching gluons.[10]

To non-physicists, concepts like colour properties and quark theory may seem a bit playful. But the underlying mathematics is highly complex and has at the same time beautiful consistencies and symmetries. However, it was in the end the great fit to the experimentally obtained particle features that gave physicists a deep confidence in their new theory. Quantum chromodynamics became part of today's "standard model" of elementary particle physics, in

---

[9] The transformation $\overline{\psi} = e^{i\theta}\psi$, where θ is a real number, is a global gauge transformation. In contrast $\overline{\psi} = e^{i\theta(x)}\psi$ (with complex numbers) is a local gauge transformation. Here, x describes the set of relevant physical variables.

[10] They work in such a way that there is no meaningful propagation which can be the speed of light, although they are massless (in principle all massless particles have the speed of light) - as they interact with each other all the time.

which Gell-Mann's eightfold path had created a first structure in the particle jungle - for which he received the Nobel Prize in 1969.

The Standard Model of Elementary Particles

The next step towards unifying the three atomic forces (strong, weak and electromagnetic) was to come up with a combined theory for all those particles that do *not* undergo strong interactions, but only the weak and/or electromagnetic force, such as electrons. These are called "leptons" (from the Greek *leptós,* "thin, small, fine"). This was successfully done by the three physicists Steven Weinberg, Abdus Salam and Sheldon Glashow. In the late 1960s, they succeeded in describing the electromagnetic force and the weak nuclear force in the atom as different manifestations of a single force, so they can be combined into the "electroweak force" (for which they received the Nobel Prize in 1979).

While the Feynman interactions in quantum electrodynamics are described by the exchange of *massless* photons, which have the speed of light and are being far reaching, the electroweak force theory explains the small range of the weak interactions. The reason why the weak interaction is only acting in such small spaces is that their exchange particles are very *heavy* The charged W^+- und W^-- bosons and the neutral Z-boson have masses in the range of 100 giga-electronvolts (the mass of protons and neutrals are only 1% of it, around one GeV).

The integrated electroweak theory is also an example of a (local) gauge field theory with a gauge group corresponding to the product $SU(2) \times U(1)$. Here SU(2) stands for the three-dimensional invariant transformation group for the weak interaction as provided by W_+-, W_- and the Z- bosons and U(1) for the one-dimensional unitary group that describes the invariances of the electrodynamical interaction linked to the conservation of charge.

Today's standard model for all elementary particle physics is summarized by the following image:

We see the differences between the fundamental quarks (strong force) and the leptons (electroweak force) as well as the particles for the different interaction (all in colour red):

- Gluons for the strong force (eight of them in total, here represented in only one square)
- W- and Z-bosons as well as the photon for the electroweak forces

The interacting particles are all bosons, i.e. particles that have an integer spin. Also all mesons are bosons (made of a quark and an antiquark, which leads to a spin of zero or one). Baryons (made of three quarks, each with a spin of ½ or − ½) and leptons, on the other hand, are all fermions, i.e. particles with a half-integer spin: + ½ or − ½ for all leptons and ½ or 3/2 for all (known) baryons.[11]

A last question stayed open to particle physicists is how all the masses for particle are being created. With the given symmetries expressed by the Lie groups their masses should in fact all be zero. This question can now also be answered by modern physics, for a very particular particle has experimentally been proved: the Higgs particle, named after the British physicist Peter Higgs. It is electrically neutral, has spin 0 and decays after a very short time. The Higgs particle reflects a theory proposed as early as the 1960s by Peter Higgs according to which all elementary particles except the Higgs boson itself acquire their mass through interaction with the corresponding ubiquitous Higgs field. Higgs's first paper on the theoretical model was however rejected. The editors judged it "of no obvious relevance to physics".[12] It was only his second attempt that was published by the prestigious *Physical Review Letters*.[13] However, for almost 50 years the Higgs particle remained unobserved, until finally, on July 4th 2012, it was discovered. This discovery ultimately rounded up the standard theory.

Now we have all parameters for the standard model. It is characterized by the SU(3) × SU(2) × U(1) symmetry model. This has now been the foundation of it for more than 50 years. Its experimental support stands at the peak of all scientific history: No discipline in overall science has been confirmed experimentally more exactly than the standard model in physics. Should physicists not be proud and happy about this? Well, they are not. The SU(3) × S(2) × U(1) field lacks any elegance, i.e. it is not *one* unified theory with *one* symmetry structure for all particles and forces. In fact, all these symmetry structures in today's physics represent overall a rather abstruse and anything but elegant structure and thus contradict the ideas of the physicists that their discipline must at last be elegant and uniform. Furthermore, numerous important parameters, 17 in total, do not receive their values from the model itself but have to be determined experimentally.

---

[11] The distinction between fermions and bosons is important as they interact differently with each other. In short: Bosons can stick together in constrained spaces in almost unlimited numbers, whereas fermions only like other fermions that are different in at least one quantum feature. So, fermions form things with characteristic structures, like atoms.

[12] Cited in Higgs Wikipedia page en.wikipedia.org/wiki/Peter_Higg (Dec. 16th 2022).

[13] Peter W. Higgs, *Broken Symmetries and the Masses of Gauge Bosons*, Phys. Rev. Lett. 13, 508 (19 October 1964).

Physicists do already know, that the standard model will have to fail at higher energies than those achieved so far and be replaced by another, hopefully more elegant one. But they have no idea at which levels this will be the case, and physicists have until now observed no deviation from the standard model. The highest energy reached so far is 7 TeV (10^{12} eV) – at the CERN accelerator near Geneva – which corresponds to the energy of the motion of a flying mosquito. However, what makes this value so extraordinary is that it squeezes energy into a space about a million of a million times smaller than that of a mosquito.

There are many alternative theories that the theoretical physicists have come up with for overcoming the lack of elegance in today's theory. For long the most popular of these was the 24-dimensional SU(5)-theory combining SU(3) × S(2) × U(1) that we already saw in Chap. 4. Today SO(10) has become quite popular (we will avoid any further details of it[14]). However, there has been not a single piece of experimental evidence for either - or for any other suggested theories integrating the different forces of today's particle theory. And how the gravitational force is to finally be integrated remains even more unknown. But here too, there exist theories, which are even some more orders of magnitude of energy away from us. Thus, we can possibly barely ever validate - or falsify - them, which even questions their belonging into science.

"Chaos" Theory and Emergence Patterns in Today's Physics

From the smallest scales let us return to the sphere of our daily life, where actions appear to be characterised by processes that are easily explainable with a simple law of causality: Same causes have the same effects. This physical view of the world goes back to Isaac Newton, whose laws of the relationship between force and matter and the theory of gravity elegantly described the behaviour of bodies and made the world appear as a gigantic mechanism that works like a clockwork. The paradigm of determinism was born, according to which all events could be predicted as long as there was sufficient information about the location and speed (or momentum) of all matter particles in the universe. In 1814, the famous French poly-mathematician, astronomists and philosopher Pierre Simon de Laplace thus said:

---

[14] For experts: The SO(10) together with chiral fermions in the 16-dimensional spinor representations, defined on non-spin manifolds.

Given for one instant an intelligence which could comprehend all the forces by which nature is animated and the respective situation of the beings who compose it an intelligence sufficiently vast to submit these data to analysis it would embrace in the same formula the movements of the greatest bodies of the universe and those of the lightest atoms for it, nothing would be uncertain and the future, as the past, would be present to its eyes.[15]

Since then, this statement became contradicted by two physical theories: quantum physics and non-linear dynamics. As we have dealt with the first topic and the Heisenberg uncertainty quite extensively in the previous chapters let us take a look at the other one which arose later, i.e. in the second half of the twentieth century. The statement that *if* we know the value of all relevant physical parameters exactly the movement can be predicted precisely remains correct. However, it is only a theoretical statement. Do we ever know all parameters in a system 100% exactly? We do not. At last, when we get to the atomic level, Heisenberg's principle makes this ultimately impossible. But if we do not know the system's parameters exactly, how predictable is the behaviour of the system over time compared to the one with exact parameter knowledge. If we only know the parameters of the system with a variation within 0.001%, how does the system develop compared to the exactly specified one? For many years the physicist thought that the behaviour is likely to be very, very similar to the one starting with the exact location.

The strictly deterministic way of thinking about nature was broken as late as in the twentieth century, first at its very beginning by the French mathematician Henri Poincaré, who came across systems whose behaviour are unpredictable, even though all the underlying physical laws were known. The most well-known one is the three-body system (e.g. three planets): Poincaré succeeded in showing that this system can exhibit strange behaviour. Except for very particular initial conditions, even minimally small changes in the parameters at start lead to great differences in its motion sequences. However, such systems were seen as very theoretical and not really representative for nature. It was not regarded as yet convincing that nature is not evolving linearly, although physicists were already familiar with non-linear equations for nature. Non-linear means that variables in the equation appear as variables in a polynomial of degree higher than one or, more generally, the argument of the function is simply not a polynomial of degree one.

---

[15] Pierre Simon de Laplace, *Philosophical Essay on Probability*, ed. by Richard von Mises, Leipzig (1932); Original edition: Paris (1814). Sentence taken on page 4 of an electronic copy of the text, given under: https://archive.org/details/philosophicaless00lapliala/page/n5/mode/2up.

However, in the winter of 1961, the American meteorologist Edward Lorenz rediscovered the phenomenon of the sensitive dependence of a system on its initial conditions. To simulate the weather, he had set up a simple (but non-linear) model of convection (the rising heat flow in a gravitational field) of liquids and gases. When he entered this into a computer and then used rounded values from an earlier calculation as intermediate results in order to save time (computers were not as fast and could not store as much data as today's), Lorenz made a surprising observation. Tiny changes by rounding numbers from the system's previous values led to completely different results in the state variables after only a short time.

The realization of Lorenz's results led to what is today often called the "chaos theory". Lorenz himself came up with the now well-known statement of chaos theory: Even such minimally small changes as the flap of a butterfly on one side of the earth can lead to a completely different temporal development of the global weather, which can mean a hurricane on the other side of the earth.[16]

"Chaos theory" eventually found a much stronger impact in physics (and later other scientific areas) when computers became efficient enough to calculate various non-linear equations. These illustrated that even relatively simple physical systems can spontaneously exhibit very complex and unpredictable behaviours.

A rather straightforward one (but one with complex numbers) goes back to the French-American mathematician Benoît Mandelbrot. It resulted in the first narrative representation of chaos, the so-called "Mandelbrot set", also known colloquially as the "apple man" (as its visualisations take a corresponding shape). The colourful illustrations of this set caused a great stir in the media and in the public in the mid-1980s. Suddenly, chaos theory was on everyone's lips. Through it, natural scientists hoped to gain insights into many still not entirely understood phenomena of everyday life, such as weather or complex ecosystems. And indeed, it actually started a process that very much improved weather and climate forecasts using increasingly powerful computers. Today, non-linear systems and their properties are key for many research areas. Some researchers even saw chaos theory as the third great theory of twentieth century physics, alongside relativity and quantum theory, and spoke of a "paradigm shift in physics". Unlike the "other great theories", however, chaos theory was not about new laws of nature and basic equations, but about the exploration of phenomena within an existing and established theoretical framework.

---

[16] Or, quoting the title of a speech of Lorenz in 1972: "*Does the flap of a butterfly's wings in Brazil set off a tornado in Texas?*".

A major result of chaos research over the last 40 years is the discovery that chaotic systems, despite their seemingly irregular, long-term unpredictable behaviour, exhibit certain general, universal patterns of behaviour, which enable a systematic study of their properties. An essential feature of chaotic systems is that their motion takes place in subspaces of their actual state space that do not have integer dimensions. For example, if we plot the motion of a two-dimensional chaotic system on a piece of paper, we get a pattern that occupies only a certain part of the paper. Mathematicians call such a structure an "attractor", because it seems to "attract" the trajectories of the motion. Since attractors do not have integer dimensions, they are also called "fractal spaces", a space of fractal dimension (the term came from Mandelbrot). With chaos theory, a new mathematical discipline arose that investigates such spaces with non-integer dimensions. The fact that these spaces can be visualised in beautiful images by computers helped chaos theory further to become so popular. Nowadays, many mountain landscapes or large concentrations of armies in cartoons or in films like the "Lord of the Rings"[17] are designed using such fractal structures.

Chaos theory has undoubtedly brought new insights to physics and beyond. Besides simple physical systems like double pendulums or non-linear oscillating circuits, chaotic behaviour can also be found in technical and other non-physical systems: in the weather, in pinball machines and billiard, as well as in the heart rhythm. We find it in turbulent flows (which obey the non-linear "Navier-Stokes equation" established as early as 1822 by the French physicist Claude Navier), in chemical reactions, in neuronal currents in the brain, in biological systems and population dynamics and in other parts of the human body. Chaos theory has thus developed into an important interdisciplinary research field far beyond the boundaries of physics.

However, today, the word "chaos" is barely used any longer by mathematicians or physicists (or scientist from related applied fields). The proper expression is "non-linear dynamics" (or "deterministic non-linear systems" in examples like the afore-mentioned butterfly effect). The features of currently well understood non-linear systems with known types of behaviour apply to systems with relatively few state variables. Physicists therefore also speak of "low-dimensional non-linear systems". The dimension (free variables) in such a system is less than five or seven, often only two or three. Unfortunately, many phenomena in nature cannot be described with such few variables. The best example is the weather, which depends on very many different factors, such as pressure and temperature, precipitation, humidity,

---

[17] The software "Massive" was actually developed to render realistic battle scenes for "The Lord of the Rings".

and solar radiation, and this at numerous locations. Physicists are still unable to treat them very well with the methods of non-linear dynamics because they hardly differ in their behaviour from random systems. The differentiation of "high-dimensional non-linear dynamics" from stochastic processes is a great challenge for today's non-linear-dynamics research.

However, if one looks at the quality of weather forecasts today, which are going quite reliably up to a week or even 10 days, compared to those from 30 years ago, for just a couple of days, one sees great improvements. These are due to the combination of faster computer handling of complex systems with a tighter grid of data collection as well as an improved (non-linear) modelling of the weather. Thus, also the modelling of the climate change - equally highly non-linear - has become better and better over the last few years, although the effects of climate change are challenging existing weather models.[18]

However, despite all their efforts and their initial excitement forty years ago physicists and mathematicians have still not yet succeeded in getting non-linear dynamics to an independent research field. Although chaotic systems possess numerous universal properties, the absence of a unified chaos theory is a big impediment. Instead, current chaos research encompasses a long list of applications across many different fields.

A growing subfield of non-linear dynamics involves the study of "self-organising systems". These are multi-component systems in which the individual parts are interconnected through non-linear reciprocals as well as permanently connected in changing relationships to each other such that complex forms and structures can spontaneously emerge from within the system itself. These structures can usually no longer be traced back to just the properties of the individual parts of the system. In this case, one speaks of "emergence" (from the Latin *emergere*, meaning "to appear"). So, the characteristics of the larger system cannot be concluded from the properties or behaviours of the individual parts of the system. They only emerge when the parts interact in a wider entity. In this field of "self-organising systems", there are many overlaps with disciplines such as biology, geology, computer science (artificial intelligence[19]) and, as some economists have examined, even capital markets. Emergence has meanwhile actually become a term in philosophy, describing an entity that have properties its parts do not have on their own, i.e. properties or behaviours that emerge only when the parts - up to human beings - interact in a wider whole. In biology, in turn, the phenomenon of life can be seen as an emergent property of chemistry - if one leaves out

---

[18] For more of climate and its modelling, see Lars Jaeger, *Ways Out of the Climate Catastrophe Ingredients for a Sustainable Energy and Climate Policy*, Springer (2021).
[19] For more on AI: see Chap. 10.

complex molecular levels that are themselves results of emergences, as they are operating on different levels with their own emergence "laws".

According to Nobel prize winner Robert Laughlin emergence can even be the key feature of physics. He is convinced that all - and not just some - of the laws of nature we know arise from collective actions of particles through emergence, down to the smallest levels we know. And those are emergent from again smaller levels. For him, emergence is a physical ordering principle. Even gravity as well as space and time are not to be considered fundamental according to him, but emergent at larger length scales. He thus only wants to deal with theories that are empirically founded. The Big Bang theory, known to a broad public far beyond astrophysics (see Chap. 12 for more details), for example, is "nothing but marketing". Laughlin calls such theories "quasi-religious", especially those concerning a potential "world formula". Future physics should increasingly deal with macroscopic phenomena such as the self-organisation of matter, as particles cannot be explained by single atomic or subatomic processes.[20] In this context, the concept of unity of matter and energy seems for him to be like the search for a Holy Grail - that does not exist.

This view is pushing the discoveries of modern physics philosophically to the extreme. Laughlin proposes to abandon all the philosophical principles that have built the occidental philosophy. However, he has not gained the support of the majority of physicists.

Today's Situation in the Macrocosm – Will We Soon Get Answers to the Fundamental Questions About the Universe?

We stated in this chapter that physics today is still quite far away from a unified quantum theory for the smallest particles. But how is it on the other side of the size scale, in galaxies, possibly even in the entire universe? *Cosmology* is concerned with the study of the large-scale structures of the universe and attempts to understand the properties of the universe. Other than in particle physics astrophysicists have seen some impressive - but equally riddling - discoveries in cosmology in recent years. Just as in particle physics today cosmology also comes with a standard model, the so called *cosmological standard model*. It is based on the general theory of relativity

---

[20] Robert Laughlin, *A Different Universe – Reinventing Physics from the Bottom Down*, Basics Books, New York (2005).

in combination with astronomical observations. That standard model is much younger than the one in particle physics, and still subject to changes according to observations. The current ΛCDM or *Lambda-CDM model* describes the evolution of the universe since the Big Bang with a few - in its fundamental form six - parameters. It is currently the simplest model that is in good agreement with cosmological observations and measurements (see Chap. 12 for more details).

Just as gravitation has no role in the microcosm, we could assume that quantum physics has no role in astronomical models today. But eventually it has to for understanding specifically the (very) early universe (very) shortly after the Big Bang, when density and temperature were unimaginably high, high enough such that quantum physics applies. The same applies to the features inside black holes. But so far, we know nothing about how the quantum effects work in those states. We only think that we understand the (relatively straightforward) particle and nuclear physics inside stars. Plus, as we already saw, quantum theory and theory of general relativity have features contradicting each other. An expanded (and possibly final) understanding of the universe will thus only be achieved when general relativity and quantum physics are integrated in one common overall theory that explains everything from the smallest to the largest scale. For that the physicists already have a name: "theory of everything" (TOE) albeit not having any understanding of it. We are in fact likely still very far away from it.

Other than the standard model for particle physics, recent cosmological observation have changed the standard model for cosmology's characteristics over the last few years. There are in fact some more recently (only indirectly) observed two key features which we already discussed in detail in Chap. 4, that had to be included into the model: First, the apparent presence of invisible matter in space, so-called *dark matter*. It has however not yet been proven to exist, and there exist alternative models for these phenomena, e.g. the MOND-Hypothesis (Modified Newton Dynamics). Second, even more abstract, the *dark energy*, supposed to explain the repulsive force against gravity what makes matter (or better: the space) in the universe fly apart at an accelerated rate. In both cases, we clearly see the effects, but can only use the term "black" to demonstrate that we do not yet understand the underlying principles.

Bringing the two great theories that are the foundation of modern physics together in a single theory with a scope over all scales proved with increasing efforts to be theoretically impossible. From the perspective of quantum theory, the general theory of relativity is still a "classic theory", which requires no quantum leaps or probability waves, while from the perspective of general relativity, quantum theory remains a "background independent theory", i.e. it knows no influence of matter on the structure of space and time. This creates a physical inconsistency: As gravity is linked to the matter, which in turn is described by quantum theory, space–time cannot be both static and dynamic.

Black holes now fall into the scope of general relativity and quantum theory at the same time and therefore represent a test case for both. However, we have found no way yet of examining the features inside them. The debate around black holes is thus by no means finished by the first picture of one in 2019. We can even state that it has only started. At its end, coinciding with a complete, physically consistent description of black holes, must stand the last and "final theory" of nature providing the deeply desired link between quantum field theory and general relativity theory, the TOE mentioned above. If it exists at all, that is.

How Realistic Is a Unifying Theory for Physics?

Such a TOE has already been suggested - theoretical physicists would not be theoretical physicists, had they not come up with answers even for that. However, the mathematics involved is much more abstract and complicated even than that of general relativity and quantum field theory. With the help of the string theory, we already saw in Chap. 4 that physicists believed in the 1980s that the solution of the TOE was found (originally the strings were considered as a core particle for the strong force in the early 1970s, but that did not bring the theory about the strong force anywhere).

String theory indeed does have some powerful benefits and is a serious candidate for a unifying quantum field theory and general relativity in a final theory of quantum gravity. However, despite more than a quarter of a century of intense research, the string theorists have not yet delivered any predictions that could have been verified in experiments. The reason is: The unification of the forces of nature is only expected on energy scales that are not achievable in the foreseeable future. As we saw in Chap. 4: This requires energy in the range of the Planck scale from which we are still about 10^{15} times away. A further problem with string theory, according to some physicists, is that it comes up with too many possible universes. It predicts not one but some 10^{500} versions of spacetime structures, each with their own laws of physics (just to compare: there are "only" around 10^{81} atoms in the entire universe). This model is thus sufficiently diverse to accommodate any phenomenon that might be observed at the low energies we reach,[21] without any feasible explanation why our word has developed with the properties that it has (and brought up a universe that can support life). The question critical physicists ask "Does it thus make sense that a part of theoretical research continues to deal with such a theory".[22] Last but not least, it was Wolfgang Pauli, who basically characterised experimentally unverifiable (and also non-falsifiable) theories for physical phenomena and called such model "not even wrong".[23]

A competing theory of quantum gravity is the so called "loop quantum gravity" which was developed around the same time as string theory. This theory describes space as a dynamic quantum mechanical spin network that can be graphically represented by diagrams of lines and nodes. A consequence of this theory would be the quantisation of space and time in the range of the Planck length (approx. 10^{-35} m) and Planck time (approx. 10^{-43} s). On scales of these order, *all* phenomena of physics, including gravity and space time, are no longer described as a continuum, but quantised. Loop quantum gravity is able to correctly describe some already known or suspected phenomena such as long-wave gravitational waves on a flat background spacetime, the radiation emitted by black holes or a positive cosmological constant (the last is a term in Einstein's equation of his general theory of relativity that describes the accelerated expansion of the universe[24]). However, one of the criticized features (among others) is that loop quantum gravity has not yet been able to derive - the in most cases very well functioning - general relativity as a low energy limiting case.

Philosophically, ideas such as that reality is mostly composed of dark matter and energy that are not reachable by humans seem quite strange. The complete disinterest of modern physicists for philosophy is perhaps also one of the reasons why physics remains blocked on fundamental questions of cosmology, as well as of quantum particles, both for which experimental evidences remain awaited for.

---

[21] This only happens if supersymmetry is broken. For a valid supersymmetry model, there are not as many options, in fact ultimately only one unique model as the various existing superstring theories can be regarded as different limits of a single unique theory, today called "M-theory".

[22] See in particular (for a broader spectrum of people): Lee Smolin, *The Trouble with Physics*, Houghton Mifflin Harcourt (2006).

[23] See also Peter Woit, *Not Even Wrong: The Failure of String Theory and the Search for Unity in Physical Law*, Basic Books (2006).

[24] Einstein's original idea was that of a static universe, though.

Research in Physics Today

How do the big open questions of physics influence the research work of physicists today? Well, these are in fact barely the topics individual scientists deal with today. If one asks a scientist what he or she is working on, the person will most likely not say "I am trying to understand the universe" or "I want to come up with a solution for the models beyond the standard model in particle physics". His or her response is very likely something very narrow and specific in some niche area. They realise that the big questions are important but claim they must be tackled in very small "bite sized steps" with the hope to make incremental progress towards a solution for the big problems. Still, most often the steps they take are not directly relevant for the big questions. Or they are trying to obtain a useful application of their work in the industry. In other, more direct words: Many scientists can simply not afford to work on more global questions. Thus, only very few scientists try to solve a big question in one go. 100 years ago, it was the standard for physicists trying to find the "final" theory for their problems. Once in a while a particular person had a new major insight - for example, the theory of relativity or in quantum physics - and shared this with the others who understood at least most of it, if not all.

In fact, most scientists today might have forgotten how it is to deal with some bigger question. The reason for this is obvious and has already been discussed in previous chapters: the expansion of the areas in physics in the last 70 years. Today there are about a million physicists working worldwide. The number of publications has exploded exponentially as well, from a few hundred, maximally a thousand (mostly single-authored) approved papers to more than 100,000 to 200,000 (mostly multi-authored) approved ones per year, reaching even one million if one considers the publications on web sites (non-approved). The high number of authors on physics papers is also due to the experimental setups that have become more and more complex and involve many people. In theoretical parts, teamwork has also become more common already at the research level - while before, communication often happened after the publication. Today, being part of a network that involves as many publications as possible is essential already for a young scientist to start his or her career, as well as to continue it.

Enrico Fermi wanted to encourage individual creativity and innovation by his PhD students and required them to select their problem themselves, solve them, and then submit the results for publication under their name alone. That has changed fundamentally. And with the need that scientists today have to apply for government grants for their specific research goals, an applicant has to publish as many as possible publications in order to ensure that the submitted proposal had significant cachet for continuing support. The more papers they publish the more chances they have - while the quality are less important. A vicious cycle begins. So here is a final Fermi-inspired question:

> How many of today's tenured faculty members or research directors have never written a single-author paper?[25]

Philosophy of Physics Today

With this knowledge of the state of physics today, the role of philosophy can be considered more distinctively than in Chap. 4. As we saw, the development of particle physics as well as the understanding of nonlinear systems described above are examples for the new culture in physics pushing the field of philosophical reflection completely to the side. This even implies the important question of how well we can even predict the future of physical states with quantum physics in the microcosm and chaos theory at the macrocosm. Richard Feynman, Murray Gell-Mann and Steven Weinberg, Edward Lorenz and Benoit Mandelbrot were all brilliant theoretical physicists (or mathematicians in the case of Mandelbrot) with no particular interest in philosophy at all.[26] *Against Philosophy* is even the title of a chapter of a book by Steven Weinberg,[27] who argues quite eloquently that philosophy is more damaging than helpful for physics (while forgetting or ignoring, however, the historical development of physics closely related to philosophy).

Did perhaps only this neglection of philosophy give physicists a free path to the highly complex and abstract theories? Does their science have no link to anything that can be discussed philosophically, as the mathematical part requires a level of abstraction that disables any philosophical insights and that no outsider like a philosopher can ever understand in the first place? It was, in fact, exactly the path to such complex theories that made them succeed at describing the fundament of all natural laws, and this at precisions never reached by other sciences until today. But do we really need to exclude philosophy *after* the theory has been established? If we had the unique and

---

[25] Philip Wyatt, *Commentary: Too many authors, too few creators*, Physics Today 65, 4, 9 (2012); also under https://physicstoday.scitation.org/doi/10.1063/PT.3.1499.

[26] A good example is Gell-Mann's popular book *The Quark and the Jaguar: Adventures in the Simple and the Complex*, Freeman and Company, New York (1994): There is not a single word or reference given to philosophy.

[27] Steven Weinberg, *Dreams of a Final Theory - The Scientist's Search for the Ultimate Laws of Nature*, Vintage Books (1994).

satisfying theory for everything (including an integration of gravity) we might be able to argue like that. However today - and most likely in the foreseeable future - there are still way too many fundamental principles open as we saw in Chaps. 2 and 4 and here.

There seems to be a twofold structure in modern physics and philosophy. A less abstract theory such as Newton's physics or Maxwell's electrodynamics one can still apply and refer to philosophical discussions (see the philosophy of Immanuel Kant[28] or the discussion about the nature of physics of the first half of the twentieth century). However, when we move to modern, higher complex physics, philosophical discussions seem to be indeed more and more difficult. While in quantum mechanics the philosophical discussions of the founders were still very intense and filled with subtle knowledge about philosophy, it was becoming more and more difficult with the emergence of quantum field theories. However, Einstein, living in America already and possibly observing the new American way of leaving out philosophy from science, said about this:

> Independence created by philosophical insight is - in my opinion - the mark of distinction between a mere artisan or specialist and a real seeker after truth.[29]

In fact, the core ideas of the standard model, and thus the essential elements that we need in a philosophical debate about it, can today be explained also to non-physicists reasonably well. The mathematical details of the structure of SU(3) and SU(2) symmetries are barely needed for a philosophical debate. It is likely to be enough to understand the number and importance of the new particles, the value of symmetry laws (Noether theorem, not in its mathematical form, though) and the open problems of the standard model (as well as its lack of elegance). For the first two points a picture like Fig. 6.1 above are quite helpful. And Noether's theorem can also be explained to non-physicists, if they make a quite doable respective intellectual effort.[30] The open questions in physics are equally open question in philosophy, and the inability of physics to solve them at any time soon should lead physicists back to philosophy. Thus, far from being immune to philosophy, due to its lack of solving fundamental questions, physics does remain deeply affected by it today. However, the inverse applies, too.

---

[28] See Immanuel Kant, *The Critique of Pure Reason* (1781).

[29] Albert Einstein, Letter to Robert A. Thornton, 7 December 1944, University of Indianapolis Press (1994), p. 111–138.

[30] See e.g. Lars Jaeger, *Emmy Noether. Ihr steiniger Weg an die Weltspitze der Mathematik*, Springer (2022).

Philosophers simply not wanting to understand the implications of modern physics on philosophical questions are equally ignorant with their attitude, and unfortunately, one can find this among (especially older) philosophers quite often. Just as the best way of doing science listens keenly to philosophy, so does the best philosophy listen carefully to science. This has certainly been the case in the past, from Aristotle and Plato to Descartes, Hume, Kant, Husserl and Popper. The best philosophy has always closely interacted with science. Aristotle himself wrote a book called "Physics" attempting to establish general principles of change that govern all natural bodies. Even if his approach is outdated, and the term "physics" nowadays has a different and more precise meaning, it shows that for him, and most of the philosophers of his time, philosophy and the understanding of the behaviour of nature were inextricably linked.

We should, however, note here: The philosophical consequences of quantum mechanics, such as the disappearance of the principle of excluded

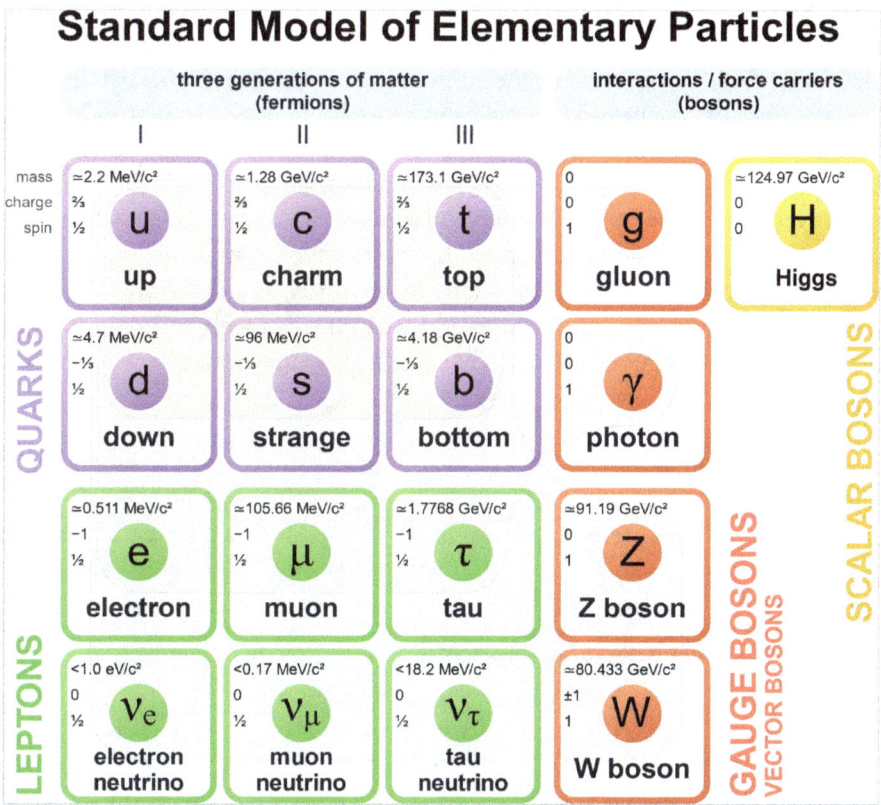

Fig. 6.1 Elements of the standard model of elementary particles

middle, have not really been taken up by the philosophers. Plus, the existence of contradictions and the need to live with them would have profound consequences for constructing a new ethics more in line with the discoveries of the twentieth century. Furthermore, the disappearing of absolute laws in physics should also be reflected upon when thinking of values in ethics. How can we reconcile all this with a modern approach to ethics and ontology? (Ontology deals with the definition of being; Quantum mechanics poses serious questions on the definition of what is and what is not.) These questions should open a vast field of reflections in philosophy. However, we have hardly seen any of this in modern philosophy.

At the same time, creating new technologies (see next chapter) based on modern physics has reached a level that creates highly relevant and severe ethical questions that need to be significantly more integrated in the planning of future technological developments in physics and science in general. And this has to happen rather quickly. The *American Physical Society* describes their Guidelines on Ethics through the following:

> As citizens of the global community of science, physicists share responsibility for its welfare. The success of the scientific enterprise rests upon two ethical pillars. The first of them is the obligation to tell the truth. The second is the obligation to treat people well.[31]

This relates to the ethics of each individual: Truth and respecting others. However, new technologies created by physicists[32] can have other ethical issues that do not immediately relate to an individual's action, but rather concern the society as a whole. Creating weapons can well be integrated into the two guidelines above and can still be ethically questionable. The physicist Lawrence Krauss thus says:

> To our great peril, the scientific community has had little success in recent years influencing policy on global security. Perhaps this is because the best scientists today are not directly responsible for the very weapons that threaten our safety, and are therefore no longer the high priests of destruction, consulted as oracles as they were after World War II.[33]

(one can probably replace "scientists" with "physicists" here.)

---

[31] Accessible under https://www.aps.org/programs/ethics/.
[32] And let us not forget that in many other fields of applied sciences (computer science, biology, chemistry, and others) physicists are actually behind the theoretical and applied developments.
[33] Lawrence Krauss, *A Universe From Nothing*, New York Times, (Jan 17th 2013).

Another condition today makes treatment of science, in particular physics, ethically difficult to assess: The velocity at which science is changing the world. In the even faster changing dynamics, technological and scientific revolutions and their consequences will be far more profound than those that arose from the physics of the twentieth century, which was already incomprehensible for most people and neglected by most philosophers.

7

Computers, Nanotechnology, Internet and Many Other Technologies

What Benefits and Challenges Physics Brought Us and Will Bring Us in the Future?

What New Technologies Physics Has Brought Us: A Tremendous Amount of Life Improvements and a Few Important Open Questions

If the phenomena and properties of the microworld seemed like magic to the physicists at first, over time they learned how to deal with the features of this enchanting world using mathematical means and tricks. And finally, they were able to tame it, even if they could not understand it completely. With this theoretical basis, the physicists went from being magicians to scientists again and then later to engineers (unfortunately losing on the way all characteristics of a philosopher), for the new theory made possible numerous technologies that were as astonishing as they were tremendous. The first of these arose when the physicists applied quantum physics to the atomic nucleus because in the process they realised: It contains a huge amount of energy at this scale. This had some severe implications soon to be realized, both on the military side (bad consequences) as well as on the civilian side (positive – but later less and less accepted – consequences).

Yet, energy is by far not the only important application of modern physics shaping our life. Besides all the issues raised in the previous chapter (like US physicists' objections against philosophy, an ugly standard model for all known particles in physics, the lack of a unified theory in physics, a lack of ethical considerations, research in so many different fields with little unification efforts) physicists have in fact through technologies enabled a drastic and mostly positive change in scientifically minded societies. Even if nobody outside of physics really understands quantum field theories or the general theory of relativity, most people in the West have benefitted tremendously from today's technologies which are often based on quantum theory and even general theory of relativity. Perhaps you know: Einstein's equations are needed for an accurate positioning on Earth by satellite navigation. Thus, Einstein's theories are also in our cell phones.

Let us take a look at the various scientific applications of physics that mostly changed the world to the better, and which have been essentially developed in the USA. In doing so, we will also look into their future, which looks as exciting as scary.

Nuclear Technologies

There are many ways of using nuclear energy. In its destructive form, it is the nuclear weapons which are based on the development of explosive devices based on nuclear fission (classical atomic bomb) or nuclear fusion (hydrogen bomb, luckily never used so far in war). The hydrogen bomb which requires very high energies to be ignited is based on an idea of Enrico Fermi to fire it with a fission bomb. He proposed it in 1941 to Edward Teller, who participated in the Manhattan project for the atomic bomb. Later, Teller was put in charge of developing this bomb despite the opposition of Oppenheimer and many other colleagues from the Manhattan project. The first US H-bomb exploded the 1st of November 1952 with much higher energy than those of atomic bombs. The main idea for it is due to Stanislaw Ulam who was later denied by Teller. Nevertheless. It is still today called the "Teller-Ulam configuration". Ulam (together with John von Neumann) also in this context developed the "Monte Carlo Method" for computing probabilities of outcomes. A method that has become ubiquitous today in many fields from theoretical physics to genetics, economics, finance and risk management.

The field of nuclear energy that currently receives much more attention than the dangerous nuclear weapons (in the discussion about a possible decarbonated future energy production) is the controlled use of the released energy from nuclear fission processes in nuclear power plants for civilian energy supply. First nuclear power plants were developed in the 1950s. There are currently around 400 of them worldwide. However, nuclear fission power plants still are very controversially discussed because of the emission of nuclear waste and the potential danger of accident like in Chernobyl in 1986 or in Fukushima in 2011. (However, with the 2022 war in Ukraine and the European dependence on Russian natural gas, the debate has restarted in

Europe on the subject.) There could in the future be a way out of this controversy by using *nuclear fusion* processes which are much cleaner and safer, but also much more difficult to set up, and still mostly in research state.[1]

Furthermore, there are important applications of nuclear processes in medicine, in particular in imaging devices and radiation treatments of the human body, i.e. ionizing and magnetically influenced radiation to make images of the inside of the body and in radiation therapy against cancer. Following the pioneering work of Marie Curie,[2] there are many research centres developing medical applications of new radiation therapy like the Institut Curie in Paris or the Maria Sklodowska-Curie National Research Institute of Oncology in Warsaw. In fact, many medical examinations and therapies would be impossible without developments from nuclear physics.

Electronics, Digital Technologies and the Miniaturisation of Processors

The combination of proper digital data processing and smaller and smaller processors has led to the development of today's computers the speed and calculation power of which we have today all grown accustomed to. Their computing power has in fact become so strong that even things like the world-wide hyper-complex climate models have become possible to calculate reaching a very high coincidence with the real development of the climate in recent years.

Two geniuses stood at the beginning of the computer architecture: John von Neumann and Alan Turing. In 1945, von Neumann published the first draft of a report on the EDVAC (Electronic Discrete Variable Automatic Computer) in which he described a computing architecture that is now named after him. His central concept of the modern computer was based on an article by Alan Turing which von Neumann openly acknowledged. In fact, Turing's work became compulsory reading for von Neumann's co-workers in the EDVAC project. In short, one could say Turing invented the Turing machine (mathematical model that defines an abstract computer model, in which manipulation of signs are carried out according to defined rules), and von Neumann built it. In 1949 von Neuman developed the

---

[1] For more details see: Lars Jaeger, *An Old Promise of Physics – Are We Moving Closer Toward Controlled Nuclear Fusion?* International Journal for Nuclear Power, Vol. 65, Issue 11/12 (2020).

[2] In 1931, Marie Curie was awarded the Cameron Prize for Therapeutics of the University of Edinburgh. It is paradoxical that she died only three years later from her long-term exposure to radiation, causing damage to her bone marrow. Unfortunately, the damaging effects of ionising radiation were not known at the time of her work.

first universal computer, i.e. a computer that can process any computational program with one and the same hardware (the original purpose of von Neumann's computer and their algorithms was to perform calculations on the explosion behaviour of hydrogen bombs). His programming concepts and architecture have become the basis of all modern computers until today. He also immediately saw the importance of Ulam's insight to estimate statistically the probability of a successful outcome and asked him to work together. Neumann and Ulam thus developed the first simulations for the hydrodynamic computations on von Neumann's digital computers.

Another important development towards our modern computers was the integration of semi-conductors. In 1947, the physicists John Bardeen, William Bradford Shockley and Walter Houser Brattain working at Bell Laboratories plugged two metal wire tips onto a small germanium (semi-conductor) plate and were thus able to control the p-conducting zone (lower number of electrons, i.e. positively charged ones) with the second wire tip with an electrical voltage. They thus built the tip transistor (bipolar transistor) with about 1/50 of the space that had been used so far. This was the beginning of designing smaller and smaller electronic processing tools, until today when billions of transistors are taking place on a one square-centimetre large device. Making electronic tools faster in their calculation processes, thus more and more efficient and powerful, is the very foundation of our modern computers and cell phones that today we all use on a daily basis.

Digital Revolution (also Known as the "Third Industrial Revolution" or "Microelectronic Revolution")

The term "digital revolution" refers to the upheaval triggered by digital technologies and faster and faster computers, which since the end of the twentieth century has brought to most countries a change in almost all areas of life and thus led more and more to a digitally networked lifestyle (see internet below). In significance this is comparable to the way the industrial revolution changed society 150 years earlier.

The digital revolution is essentially the shift from mechanical and analogue electronic technology to digital - and thus much faster and more efficient - electronics. The changes in the world of business and work, as well as in in the public and private life have been taking place at great speed wherever the material prerequisites for application and use possibilities of progressive digitalisation exist. In the professional lifetime of one of us authors, computers moved from the basement of his university (where the Univac mainframe was sitting in a refrigerated location) to the smart phone in the palm of his hand today. In the same time span, documents and memos used to be painstakingly produced on typewriters by secretaries (mostly females) and are now easily put on any laptop using the most sophisticated word editors and then sent electronically via email to various (often too many!) addressees. And this continues to go on. New media are increasingly influencing communication behaviour, socialisation processes, language culture, and political as well as intellectual and religious belief systems.

Lasers

As early as in 1916 Albert Einstein described the possibility of stimulated emission of photons as a reversal of absorption, which is the basis of today's lasers.[3] Lasers are different from regular light, as their frequencies fall into very tight ranges. In 1928, Rudolf Ladenburg succeeded in providing an experimental proof of Einstein's theoretical assessment. And in 1950, Alfred Kastler proposed the method of optical pumping, which was experimentally demonstrated two years later by Brossel, Kastler, and Winter. The first working laser - a ruby laser - was completed by Theodore Maiman in May 1960 based on the theoretical work of Charles Townes and Arthur Schawlow at Bell Labs. Starting in the 1970s, the first semiconductor lasers (laser diodes) were introduced. Their smaller footprint, energy efficiency, and lower price point opened the door for further uses, even though the spectrum of light colours were limited.

Today, lasers have numerous applications in technology and research as well as in our everyday life, from simple light pointers (e.g. laser pointers in presentations) to distance measuring devices, cutting and welding tools, reading optical storage media such as in CDs, DVDs and Blu-ray discs, message transmission, laser scalpels and devices using laser light in everyday medicine.

In the 1990s, new pump geometries for high laser powers were realised, such as the disk laser and the fibre laser. At the end of the 1990s, blue and ultraviolet laser diodes reached market maturity. And at the beginning of the twenty-first century, non-linear effects were exploited for the first time to generate attosecond (1×10^{-18} s)[4] pulses in the X-ray range. This made it possible to follow live processes inside an atom, very much like the stroboscopic camera of Georges Demenÿ allowed a century before to analyse

---

[3] The word laser is an acronym for "Light Amplification by Stimulated Emission of Radiation".

[4] For comparison, an attosecond is to a second what a second is to about 31.71 billion years, more than twice the age of the universe!

human movements. Today, the laser is an important instrument that cannot be imagined to be missing in industry, medicine, communication, science and consumer electronics. Lasers also constitute one of the ways explored today to produce energy via nuclear fusion.[5]

Mobile Phones

The first mobile phone calls became possible via terminals mounted in motor vehicles - car telephones - in 1946. It was the US Bell Telephone Company that offered its Mobile Telephone Service, through which the first calls were made on June 17th, 1946. It was the communication through low frequency (radio) electromagnetic waves and sophisticated algorithms for treating the signals and jumping from one antenna to the next that enabled cell phones to be processed until today. The respective networks for cell phones are distributed over land areas called "cells", each served by at least one fixed-location transceiver.

In 1983, the *Motorola DynaTAC 8000 ×* was the first commercially available handheld mobile phone, although, at 790 g and quite bulky, it would not appear particularly portable or mobile to today's smartphone users. It was the first-generation (1G) systems which still used analogue cellular technology. Digital cellular networks appeared in the 1990s, enabled by the adoption of new (MOSFET-based RF) power amplifiers. In 1991, the second-generation (2G), now based on digital cellular technology, was launched, in 2001 the third generation (3G), around 2010 the fourth (4G), and as of 2019 the current fifth generation (5G) which can now communicate with high efficiency, i.e. processing speed and throughput at 1 Gigabyte/s. Despite the fears for health, the 5G actually exposes people to less radiations than the 3G or 4G as the density of antennas is higher and they each emit less energy. Used in combination with the internet, it allows the managing of industrial processes from distances ("Industry 4.0") or live videos in high quality. For instance, Nokia has designed its factory in Oulu (Finland) around the 5G technology and claims that they achieved productivity gains of 30% and reached 50% savings in time of product delivery to market.[6] This technology also promises to improve a lot the communication between cars allowing to potentially manage large fleets of 100% autonomous vehicles, thus bearing the potential to revolutionize our transportation style.

---

[5] For more details see: Lars Jaeger, *An Old Promise of Physics – Are We Moving Closer Toward Controlled Nuclear Fusion?*, International Journal for Nuclear Power, Vol. 65, Issue 11/12 (2020).
[6] https://www.nokia.com/about-us/news/releases/2019/07/03/nokias-digitalization-of-its-5g-oulu-factory-recognized-by-the-world-economic-forum-as-an-advanced-4th-industrial-revolution-lighthouse/.

Cell phones have become one of the most important daily used technological devices worldwide. At the same time, they have actually turned into pocket computers combining the computation power and the communication. Cell phones in the early 2020s are faster than the fastest computers in the early 1990s!

Internet

Internet can be considered the most important, i.e. most influential technological device of the last 30 years. The cultural significance of the internet is often equated with the invention (at least to the western world) of movable-type printing by Johannes Gutenberg in the fifteenth century.[7] The spread of it has led to significant and comprehensive upheavals in many areas of our life. It contributed to a modernisation push in many economic sectors as well as to the emergence of new economic branches and has led to a fundamental change in communication behaviour and media use in the professional and private spheres. The role of physics for building the internet was essential. Computers, fibre optics and networks do not find their origin in the minds of engineers attempting to build an internet, but rather those of physicists having ideas about calculating equations, how to interact with colleagues or distribute information.[8]

The internet began on 29 October 1969 as *Arpanet*. It was used to network the mainframe computers of universities and research institutions. The initial aim was to use the computing power of these mainframes more efficiently, first only in the USA, later worldwide. The first nodes were financed by DARPA (Defence Advanced Research Projects Agency; an agency of the United States Department of Defence that conducts research projects for the United States Armed Forces).

The most important application in the early days were e-mails. In the year 1971 the amount of data transmitted by e-mail was more than the amount of data transmitted by the other protocols of the Arpanet, namely Telnet and FTP. This fell short of the goal of outsourcing computational work according to supply and demand. In 1981, the basis of most connections on the internet today were specified (RFC 790-793 IPv4, ICMP and TCP). These were active on all hosts on 1 January 1983 after an announcement period of almost two years. With the changeover from the Arpanet protocols to the Internet

---

[7] See (only in German): Lars Jaeger, *Sternstunden der Wissenschaft* - Eine Erfolgsgeschichte des Denken, Süd Verlag (2019).
[8] For more details see: Michael Raymer, *The Silicon Web - Physics for the Internet Age*, CRC Press, Boca Raton, Florida (2022).

Protocol, the name "Internet" also began to gain acceptance. This represents the first protocol changeover in the history of the Internet. Its initial spread was closely linked to the development of the Unix operating system.

In 1990, the US National Science Foundation decided to make the internet commercially viable, which made it publicly accessible beyond universities. Tim Berners-Lee developed the basics of the World Wide Web (WWW) at CERN in 1989. On 6 August 1991, he made this project of a hypertext service publicly and globally available via "Usenet". The internet subsequently spread beyond the borders of the USA. Usenet for some time became the dominant application of the internet. The first rules of conduct (netiquette) were formed and thus the first signs of a "net culture".

The internet received a rapid boost from 1993 onwards when the first graphics-capable web browser called *Mosaic* was released for free via download, enabling the display of WWW content. Particularly through AOL and its software suite, there was a strongly growing number of users and many commercial offerings on the Internet. Since the web browser of Google displaced almost everything else in the last almost 20 years, it is also referred to as the "killer application" of the internet. It became a major catalyst for the acceleration of the digital revolution. Nowadays, Internet access for everyone is considered essential. There is virtually no hotel in the developed countries as well as touristic places all over the World that do not offer Internet access. Internet and social media are the main information source of youngsters.

The internet is also playing an increasingly important role in the military sector. One of the first cyber-attack by the Russians while invading Ukraine was in February 2022 against the satellite network Ka-Sat. This network had been used for internet access by the Ukrainian military. Moreover, by accessing satellite pictures over the internet, Ukrainians could spot Russian military convoys and follow the progression of troops.

Superconductivity and Superfluidity

Superconductive materials, whose electrical resistance tends (abruptly) towards zero when the temperature falls below the so-called transition temperature, have already been discovered in 1911 by Heike Kamerlingh Onnes, a pioneer of low-temperature physics. The temperature required for this was only a few degrees above absolute zero ($-$ 273.15 degrees). It is in fact the first known macroscopic quantum effect (before quantum theory was even created). The A15 phases discovered in the 1950s, especially Nb_3Sn, became particularly significant for applications that require strong magnetic fields. Important technical applications of superconductivity are indeed the generation of strong magnetic fields - for particle accelerators, but also for magnetic resonance tomography, levitation, potential nuclear fusion reactors - as well as measurement and energy technology. In 1986 high-temperature superconductors were discovered by IBM researchers in Rüschlikon, Switzerland, Bednorz and Müller (Nobel Prize 1987). Surprisingly, they usually are ceramic materials. While Nb_3Sn features a transition temperature of below − 250 degrees, the newly developed materials gain superconductivity at much higher temperatures above minus 196.2 degrees (77 K), and thus can be cooled by liquid nitrogen, which is produced industrially at relatively low price. In fact, the question of how and why superconductivity arises in high-temperature superconductors is one of the major unsolved problems of theoretical solid-state physics today.

Another macroscopic quantum effect is a state of a fluid in which it has lost all internal friction. It was simultaneously discovered in 1937 by Pyotr Kapitsa on one side, John F. Allen and Don Misener on the other side. Just as higher temperature superconductivity superfluidity cannot yet be fully explained theoretically. However, there are various approaches that describe the properties of e.g. superfluid helium at least qualitatively. In physics and chemistry, superfluid ^4He is used in spectroscopy as a quantum solvent, e.g. in the study of gas molecules: A single molecule solvated in a superfluid medium gives a molecule an effective freedom of rotation and enables it to behave similarly to how it would in the "gas" phase. It also has application in more effective cooling devices close to absolute zero.

Satellites in and Beyond the Atmosphere

In 1955, US President Eisenhower ordered the development of an American Earth satellite, whereupon the Soviet Union announced a similar project four days later for propaganda reasons. The successful launch of the Soviet satellite Sputnik 1 on October 5th, 1957 surprised the world public and caused a veritable Sputnik shock in the West. The first US satellite, Explorer 1, followed on February 1st 1958 and provided evidence of the Van Allen radiation belt which made research into the ionosphere possible. The subsequent development of satellites was for a long time severely under the influence of the Cold War.

In 1969, the year in which the US went to the moon for the first time, there were already about 1950 artificial objects in space. In 1996, according to ESA data, there were said to be around 8500 pieces of "space debris" (larger than about 10 cm in size). And in 2009, the Joint Space Operations Center of the United States Strategic Command knows of more than 18,500 man-made celestial objects.

Especially, the importance of satellites is multifield today:

- Earth observation satellites can provide images and measurements for different purposes, such as weather and spying-activities, and this through different techniques such as radar, infrared waves, scanning methods, sensor technology or photography/firm recording.
- Communication satellites fulfil all kinds of commercial tasks.
- Broadcasting satellites transmit radio and television programmes directly to viewers, eliminating the need for earthbound broadcasting and cable networks.
- Navigation satellites enable precise worldwide position and time determinations and even the automatic steering of vehicles.
- Geodetic satellites are used to measure the earth's shape and gravity field achieving accuracies in the mm range today.
- Astrometry satellites measure the position and proper motion of stars for scientific purposes, e.g. microgravity experiments. The recently launched James Webb Space Telescope, a technological prowess, with its 6.5-m-diameter gold-coated beryllium primary mirror made up of 18 separate hexagonal mirrors and its giant sunshield protection, is able to explore in the near-infrared the very beginning of the universe or to analyse the atmosphere of exo-planets (see Chap. 12 for more details).
- Military satellites are used for surveillance, defence and other military purposes, such as those of spy satellites or killer satellites (for the destruction of enemy missiles).

New Materials That Do Not Exist in Nature

Artificial substances are all around us today. The best example is plastic which is (unfortunately) nowadays distributed (by ending up as waste in the oceans) to almost any place on this planet, even to uninhabited regions. Artificial materials are today used in many areas such as (just two perhaps less obvious examples):

- An implant, i.e. an artificial material implanted in the body to remain there permanently or at least for a longer period of time for health treatment. That goes from a prosthesis used to substitute a particular part of the body

such as artificial bones to small parts dealing with the heart or other soft features of the body.
- Artificial stones used to mineral-, cement- or resin-bound materials that are manufactured with aggregates of gravel, sand and rock flour.

Solid-State Physics

The most influential physical theory of the twentieth (and so far twenty-first) century was without doubt quantum theory. It made, as we saw, semiconductors and laser technologies possible. Without it, there would be no microchips, no computers, no mobile phones, no satellite navigation, no microwave ovens, no nuclear technology, no laser, no imaging procedures in medicine and many other technologies that have become a matter in our everyday lives. It is estimated that between a quarter and half of the gross national product of industrialised nations today is based directly or indirectly on technologies with a quantum theoretical basis.

One area where quantum physics played a particularly significant role is solid-state physics, i.e. physics in the solid state of aggregation. The development of computer and information technology with its needs for fast logic and high storage media devices would be unthinkable without solid-state physics. New phenomena are still being discovered in solid states and used in ever smaller solid-state structures. Examples are the "one-electron transistor" or the "giant magnetoresistance" in multilayers. Even as late as in 2004, new useful and astonishing materials were discovered by solid state physicists (Russians Andre Geim and Konstantin Novoselov, who received the 2010 Nobel Prize in physics): "wonder material" graphene. This material has become a valuable and useful nanomaterial due to its exceptionally high tensile strength, electrical conductivity, transparency, and being the thinnest two-dimensional material in the world. Other examples of applications of solid-state physics in the framework of quantum physics are (list anything but complete, and more applications have been suggested that are not yet available):

- *Spin glass*: A magnetic system that is disordered (like the particles in a glass) with respect to its spin structure and the position of those spins. The have a very high number of meta-stable conditions.
- *M-Glass*: Innovative coatings for solar control glass based on the theory of optimised spectral transmittance.

- *Quasicrystals*: In quasicrystals, the atoms or molecules are arranged in an ordered but aperiodic structure (discovered in 1982).
- *High-temperature superconductors* (see above): Lithium doped, single layer graphene has been found to be superconductive (2015). While the transition temperature is still low (around 6 K), research with multi layered graphenes could lead to significantly enhanced superconductors.

The following two topics are results of solid-state physics, which are producing what we call "the second generation of quantum technologies". Here, unlike for the technologies above, which are all dealing with large numbers of quantum particle, we are talking about controlling and processing a v*ery small number of quantum particles*. These technologies will change our lives at least as much as the first generation of quantum technologies with computers, lasers, nuclear energy and medical imaging.[9]

Quantum Computer

A computer, in which subatomic particles are used for data storage and processing, is based on so-called quantum bits (qubits), which were introduced in Chap. 5 and which we want to elaborate on in some more details here. Such a computer features potentially an unimaginably higher computing speed compared to conventional computers. The basis of quantum computers is a specific quantum effect: Different quantum particles can be in so-called *entangled states*. It is as if the qubits are coupled to each other by an invisible spring, in direct contact with each other, yet without any explicit acting force. Each qubit "knows" at any time over any distance what the others are doing. With an ensemble of entangled qubits, physicists hope to be able to operate simultaneously on all possible states as such a structure could enable a parallel calculation process impossible in classical physics. While a normal computer has to process all the bits one after the other in many, many steps, i.e. flip from 0 to 1 or from 1 to 0 respectively, a quantum computer can process all these steps at once and store the information in a combination of 0 and 1 (also with complex weight numbers) not only in binary form. This high-level parallelisation of operations increases the computer's computing power *exponentially* with the number of qubits: With a given number of entangled particles that have a certain computer power, adding one *doubles* the computer power. This is in contrast to a classical,

---

[9] For more details see: Lars Jaeger, *The Second Quantum Revolution - From Entanglement to Quantum Computing and Other Super-Technologies*, Springer (2019).

sequential computer, whose computing line only increases linearly with the size and number of available computing components (processors). Therefore, even quantum computers with only a few dozen qubits have an unimaginably higher computing speed compared to common computers.

The idea of a quantum computer came, as we saw in Chap. 5, from a visionary speech by Richard Feynman in 1981, who developed the concept of yet another possible exciting future technology out of quantum theory (besides nanotechnologies, see below). With its help, problems could be solved that are still far too complex even for the "supercomputers" used today in physics, biology, weather research and elsewhere. Here are the main areas for that:

- *Cryptography*: Today's common ciphers are based on the re-factorization of the products of two very large prime numbers. Above a certain number size, this task can no longer be solved by a classical computer. A quantum computer could do this in minutes - thus threatening global data security. However, at the same time quantum computers allow advanced techniques for security which are absolute secure.
- *Finding new chemical compounds*: Complex optimization problems arise repeatedly in the simulation of quantum systems themselves (quantum chemistry). The elucidation of the electron structure of large molecules or crystals could let us find entirely new molecules. While for conventional computers shifting through the many alternatives for determining the best possible, i.e. energetically most favourable chemical compound, the configuration of electrons in complex molecules is too difficult, quantum computers could directly map the behaviour of the electrons involved, i.e. calculate their status much, much faster, since they themselves behave like a quantum system. This would be of great importance in materials research, quantum chemistry or drug development. Example for this type of construction of synthetic molecular compounds and solids are:

    Active ingredients in pharmaceuticals in medicine,
    Efficient catalysts for chemical reactions (chemical production),
    Tools for finding materials with completely brand-new properties. Researchers hope, for example that batteries or more generally, components of molecular electronics could be precisely calculated and thus optimised. Potentially, this would give us significantly more efficient batteries.
    New tools in biotechnology (to improve our lives).

- *Search in, or sorting of large databases*: When searching unsorted data sets, a classical computer must consider each data point individually. The search duration therefore grows linearly with the number of data points and thus quickly becomes too large for classical computers in increasingly large data sets. With a quantum computer, the search duration would only follow a root law. Instead of taking a thousand times as long for a billion data entries as it does for a million, a quantum computer would only take slightly less than 32 (square root of 1024) times as long – a breath-taking improvement in the case of very large numbers. Searching fast in a database is not merely a question of speed. It also brings a qualitative jump in the use of data, e.g. by AI algorithms (see below). When sorting large data sets, quantum computers could be even bigger time savers, as time complexity with traditional computers follows quadratic or higher grade functions.

- *Solving complex optimization tasks*: The task of finding the optimal solution from many variants of a system is considered particularly tricky among mathematicians. These tasks apply to various complex areas, e.g. finance, cryptography, industrial logistics, street logistics (such as organising traffic), the design of microchips, and in finding the most efficient use of energy. Even with a small number of variants, classical computers drop out when asked to calculate optimal solutions. The best you can hope for is a "relatively good" solution. Quantum computers, on the other hand, could solve such optimization problems even for larger systems in a comparatively short time.
- *Artificial intelligence*: The "deep neural networks" used here involve hard combinatorial optimization problems that, as we saw above, could be solved far faster and better by quantum computers than by classical computers. This elevates "deep learning" to a new dimension – and possibly up to a sentient or even self-conscious machine?

For a long time, 50 qubits were considered the magic limit. Above this size, the superiority of quantum computers over the most powerful conventional computers is supposed to become noticeable. In this context, Google speaks of *quantum supremacy*. Their corresponding computer chip *Sycamore* (53 qubits), which they announced in 2019 and we already saw in Chap. 5, needed just 200 s for the (very) special computing task that - according to Google's estimation - would have taken the world's best supercomputer 10,000 years. The moment of quantum supremacy seemed to have finally come. Some even spoke of a "Sputnik moment" in information technology. Quantum computers will possibly provide people with almost immeasurable possibilities for processing information. What this technology could enable us to do is beyond our imagination. They might even be so powerful that there is barely a limit to their application.

Without the belief that initial breakthroughs in quantum computing have already been achieved, it is unlikely that so much money would be flowing into the industry already. These sums are likely to multiply again as further progress is made. One might feel transported back to the early 1970s, before commercial computers existed. Only this time, everything will probably happen even much faster.

Here, Europe should aim at catching up with the progress made by the USA and, increasingly, by China, as we saw in Chap. 5, both of which are investing significant money into developing quantum computing (China 10 billion US-dollars, the US intends to invest even more, if approved by parliament). Thus, EU has launched a very important research projects funding on the subject of quantum computing.[10] This is also particularly important, as quantum computing is seen as a new global "arms race" that can attack digital defence technologies. Another critical point is: Once quantum computers are available to us one day, it could very quickly lead to a strong form of AI with all the complex consequences of such an achievement. Could a quantum computer even help to achieve a super-intelligent AI which processes *everything* better than humans and think beyond what we can follow?

Nano Particles and Nanotechnologies

As we briefly saw in Chaps. 3 and 5, in 1959 Richard Feynman gave an often-cited lecture entitled "There is plenty of room at the bottom".[11] There he talked about the fact that there is great potential lying dormant in the world of molecules and atoms and could be the foundation for a new technological discipline. The ideas of his lecture were to become the basis of *nanotechnology*.

At present, the technological transition from the micro- (10^{-6} m) to the nano- (10^{-9} m) world is happening quickly. It moves from dealing with clusters of thousands of atoms for a microscopic device to one with only few tens or even less atoms in the nano-world. To provide some perceptions about nano metres: 1 nm is one millionth of a millimetre and about a factor of ten to twenty larger than a hydrogen atom, and four times larger than a water molecule. The DNA double helix has a diameter of 1.8 nm, and a soot particle, about 2000 times smaller than the dot at the end of this sentence,

---

[10] For details go to; https://digital-strategy.ec.europa.eu/en/policies/quantum-technologies-flagship.
[11] Richard Feynman, *There's Plenty of Room at the Bottom: An Invitation to Enter a New Field of Physics*, Lecture given by Feynman at the annual American Physical Society meeting at Caltech on December 29, 1959, 3rd published 2012 by CRC Press.

is 100 nm in size. The structures of the nano cosmos are significantly smaller than the wavelengths of visible light (380–780 nm).

The nano-size is the borderland between the world of atoms and molecules, where quantum physics applies, and the micro- and macrocosmic scale with the classical laws. In this intermediate area, scientists and engineers can specifically use quantum effects to prepare macroscopic materials with unique properties. Furthermore, at the nano-level particles consist almost entirely of surface, which makes them enormously reactive and gives them unexpected mechanical, electrical, optical and magnetic properties.

It was the development of the scanning tunnelling microscope by Gert Binning and Heinrich Rohrer in 1981 (1986 Nobel Prize in Physics that was ground-breaking for nanotechnology. This device enables the detection and indirectly the observation of *individual* atoms. Due to a special quantum effect (the tunnel effect) the current flow between the scanning tip and the electrically conducting sample responds to changes in their distance of a few hundredths of a nanometre, generating a 3D map of the surface on the atomic scale. And it is even possible to manipulate single atoms on the surface.[12] One result of it is that researchers succeed in building a computer memory cell for one bit out of only 12 atoms.

Today's nanotechnologies already find some very specific applications in our everyday life:

- In sun creams, nanotechnology offers protection against UV rays.
- Nanotechnologically treated surfaces enable self-cleaning windowpanes, scratch-resistant car paint and ketchup that flows completely out of its bottle.
- Nanoparticles can gather a large number of other particles around them, which is useful in applications such as: after a crack in the material, nanoparticles reassemble the paint like elastic rubber bands.
- Nano-treated textiles stop the smell of sweat. For example, antibacterial silver particles are used that prevent bacteria from decomposing the actually odourless sweat into unpleasant-smelling decomposition products.

A great future for nano machines is particularly predicted in medicine. Here are three examples that are already being implemented:

---

[12] There is a famous picture taken by researchers at IBM Zürich of the company's logo letters formed by xenon atoms on a nickel surface.

- The Israeli company "Nano Retina" has developed an artificial nano retina that could restore sight to the blind: A tiny, flat implant houses a high-resolution network of nanoelectrodes. The nano retina stimulates the optic nerve so that incoming light particles collected by the electrodes are transmitted to the brain as visual stimuli.
- "Lab-on-a-Chip": Nano-biosensors detect antibodies and specific enzymes in human body fluids. Only one thousandth of a millilitre (or even much less) of blood, urine or saliva is placed on a credit card-sized chip, and the nanoparticles integrated on it detect characteristic chemical, optical or mechanical changes. In this way, the chip determines diagnoses for numerous disease symptoms in just a few minutes.
- Nanoparticles transport drugs specifically to inflammation foci or mutated cells in for attacking them efficiently with active substances. For a long time, the question of how such nanostructures could be transported in the blood at all was unsolved, since blood proves to be as viscous as honey for such small particles. In the meantime, they can even be controlled, for example by magnetic fields. Their use is planned, among other things, in precise chemotherapy against cancer cells.

But nanotechnologies can go a lot deeper, as we already saw in Chap. 5. Potentially, they could even manufacture molecules that replicate themselves, i.e. our DNA.

In the 1990s, not more than a few experts believed that nanotechnology would develop into one of the key technologies of the twenty-first century. Today, there is no end to the large number of publications dealing with new findings and techniques in the nanometre range. In fact, researchers are already succeeding in producing basic building blocks for nano machines: rolling nano-wheels, nano-motors, even a nano-elevator, nano-gears that turn along a jagged edge of atoms, propellers, hinges, grippers, switches, tiny electronic sensors and much more. Furthermore: devices storing data on DNA, and nano-cars in which four individual motors have been mounted on a central support (with the tip of a scanning tunnelling microscope supplying the molecule with electricity and sets it in motion). These machines are all about one ten-thousandth of a millimetre (100 nm) in size, thus consisting of only a couple of hundred atoms. It was thus no surprise that the 2016 Nobel Prize in Chemistry was given "for the design and synthesis of molecular machines" to three pioneer scientists in nano-metre research: Jean-Pierre Sauvage, J. Fraser Stoddart, and Bernard Lucas Feringa. They had constructed nanoscale machines that can potentially lead to better drug delivery to our body, improved microelectronics, and more generally enhanced control of nanoscale matter.

And it goes even tinier: The world's smallest electric motor is just one nanometre in size and consists of a single bent together molecule sitting on a copper surface. In this molecule, two hydrocarbon chains of different lengths (one butyl and one methyl group) hang like little arms from a central sulphur atom. The whole molecule is freely rotatable and linked to the copper surface. It is driven by a scanning tunnelling microscope whose electrons excite the rotational degrees of freedom of the molecule through the (quantum mechanical) tunnel effect. The running speed of the motor can be controlled by the electron flow and the outside temperature.

Concerning our health: The check-up with the doctor every two years could in the future be replaced by a perpetual nano-check. Ultra-small nano-robots, nanobots, permanently roam our bodies and preventively detect pathogens, genetic mutations and dangerous deposits in the bloodstream. If necessary, they send the results directly to the family doctor, who then calls his patient and makes an appointment. Or they start therapy immediately by asking for and then delivering drugs specifically at the site of the disease, fight viruses, inhibit inflammation, remove cysts and cellular adhesions, and prevent strokes by opening blocked arteries and even performing surgical procedures.

Nanoparticles or nanobots could also be used in our diet: They could help us digest food so that the nutrients are optimally absorbed by our body. This would be helpful in the case of diseases that still require a strict diet today. Researchers are also working on providing healthy foods with nanoparticles on their surface that give our palate a great taste of crisps, chocolate or jelly bears and yet do no harm, and are even healthy.

The research in the range of nanometres paves the way for ever more complex medical tools and other applications. Nanotechnology thus creates enormous opportunities to improve our lives. Nevertheless, the prefix "nano" causes great discomfort in most people, just like the prefixes "gene" and "atom", which refer to something equally unimaginably small. All three, nanoparticles, genes and atoms, cannot be seen or perceived directly. Still, the technologies that rely on them have long influenced our everyday lives.

Of course, these nanobots could also be used as spying tools or even weapons. There are science-fiction movies that describes dystopic universe where such nanobots are chasing and killing individuals. Clearly, like in many other fields, we will need to think carefully about the development and use of such powerful tools. In fact, already today's nanoparticle, and in particular future ones bear risks. Nanomaterials can especially enter the body without our intention via a variety of pathways and overcome important protective barriers within the body. The result could potentially be damages to genetic material, inflammation, or organ harm. When consumers come into direct, uncontrolled contact with nanomaterials these can be absorbed by the body via different routes:

- Via the mouth, when they are eaten or accidentally swallowed (like in food or toothpaste). Via the gastrointestinal tract, the materials enter the bloodstream and from there penetrate into organs. Even the body's own barriers, such as the blood–brain barrier, offer no protection against these super-small substances.
- Through breathing, for example, when nano-cleaning or impregnating sprays are used. For workers, the lungs can quickly become the most critical intake pathway, as respective toxic dusts can be produced during production and disposal of their substances.

However, there is still too little data on the exposure of humans to nanoparticles. Some results came from animal experiments that show that certain nanomaterials do have a toxic effect: They caused damage to genetic material and organs directly as well as inflammations, possibly leading to tumours. Free nanoparticles, i.e. nanoparticles that are not firmly bound, such as fullerenes and carbon nanotubes, are of particular concern. However, many questions remain unanswered: For example, it is not known whether nanoparticles accumulate in our body and to what extent *long-term* damage is to be expected. In addition, an assessment of the risk valid for all nanoparticles is not possible. Each substance must be assessed on a case-by-case basis. Even different geometric shapes of the same nanosubstance can have very different effects.[13]

Where Are We Going?

It is remarkable: The fact that sciences and their technologies are the foundation of our modern life as well as its future risks are not widely acknowledged. Many people take the growth of our technological applications for granted but then still complain about what science says. The latter has become quite popular in recent years often even going into election campaigns (in the US).

---

[13] Panagiotis Isigonis et al., *Risk Governance of Emerging Technologies Demonstrated in Terms of its Applicability to Nanomaterials*, EU Project: RiskGone HORIZON (2020); https://doi.org/10.1002/smll.202003303.

Some actually ask with some vulgarity: What is the work on objects that are behaving according to quantum laws or the theory of relativity serving beyond "intellectual masturbation" of the scientist themselves? The obvious answer usually draws puzzled and surprised faces: Without the science of the last 100 years the boxes that most, if not all people look into (TV, computer, cell phone) and through which we all feel connected to the world would remain rather dark. The same applies to much of our medical diagnoses and treatments,[14] fast and reliable communication via writing or speaking words over large distances, quickly finding our paths in going from A to B, online-shopping or online-learning, having the ability to use CO_2-neutral technologies for environmental reasons, birth control and family planning, and many more things we do regularly and are depending upon. Furthermore, cars and trains would not drive, lights would not appear at night, and food not enough produced to feed us all even in the developed countries without the science in the last 200 years. Science and technological applications and our lives are inseparable today. It does not take much to state: What we discover in science today and apply by technologies tomorrow will determine the conditions of our lives in the following years - with all its challenges and opportunities.

A last point of this chapter concerns risks management: As we saw with the various new technologies, there are risks to each and all of them (the same applies to other technologies dealt with in the following chapters). In the future, we therefore need to develop a form of *active risk management* of technologies, i.e. analyse and deal with their risks actively and early on. It is only the awareness of the risks that will make scientific applications be actively accepted by society, and consciously minimize their risk of derailment. Unfortunately, today new technologies are mostly sold as advantageous only, which leads to a broad acceptance by the masses with no risk awareness. That is exactly what needs to be fought against as the risk is then often taken by those who know the least of it (many kids in Silicon Valley, for example, are not allowed to use any of the many applications available for cell phones, often not even a cell phone itself, as the parents know about the risks, e.g. of data collection and manipulation).

In order to be able to assess and limit future risks, we must make far more resources available than we currently do for.

a. understanding the risks better, qualitatively as well as quantitatively, and

---

[14] Some people claim that life was so much healthier in the past. However, the numbers show: Life expectancy has more than doubled since 1990, mostly due to medical research.

b. respective scientific-ethical monitoring of technological risks - preferably in an interdisciplinary fashion across natural sciences and humanities.

Simultaneously, risk management for technologies (not only for economic application) needs to become more scientific itself. It has passed new dimensions in the last twenty years, as there have been major mathematical progresses in understanding and modelling it - especially concerning extreme risks which are so important here. Still progress in risk management are due, in particular in terms of understanding the various dependencies among risk factors in extreme events.[15] Unseen or non-considered dependences between risks in times of crises would reduce diversification benefits obtained by pooling the risks. The most significant example: Undoubtedly, the way out of climate change will go through the development of science and corresponding technologies, as politicians have overall reached very little, so far, but it will have to be combined with a strong development of scientific risk management. This means considering simultaneously the risk and the reward of various projects and not to concentrate only on one of the aspects ignoring the other one.

What we need are what can be called "philosophical risk managers". They should not be technological doomsayers, but rationally calculating and competent evaluators of technological development potentials, comparable to risk managers at insurance companies who calculate extreme risks from natural disasters or accidents over decades and centuries. The challenges, however, are much higher, as many of the risks have never been present in the past.

---

[15] For a discussion on the importance of risk management in our times see: Michel Dacorogna and Marie Kratz, *Living in a Stochastic World and Managing Complex Risks,* (2015) available on the website of SSRN https://papers.ssrn.com/sol3/papers.cfm?abstract_id=2668468.

8

Biology from 1953 to 2023: Major Breakthroughs and Their Ethical Issues

How Biology Became the Centre of Science and Today Also Lies at the Centre of Ethical Concerns

In the second half of the twentieth and the first quarter of the twenty-first century biology developed from a science that was still searching for its foundation to one of the most dominant scientific disciplines and technology providers overall. At the same time biology has diverged into many different fields which cannot all be covered in one chapter. This chapter shall thus deal with the core developments in biology of the last 70 years, while the subsequent two chapters will then discuss some of the most important parts of biological research in more detail: brain research (including concepts of self-consciousness) and artificial intelligence.

The Second Foundation of Biology: Genetics

In 1943, the physicist Erwin Schrödinger gave his famous lecture "What is life?" in Dublin, in which he raised three central questions:

- What is the nature of heredity?
- What is the storage of information in biological systems?
- What is a possible genetic code?

With this lecture he gave biologists the vision of explaining life and its origin completely from scientific-naturalistic principles. Only ten years later, modern biology had its constituting moment, when the structure DNA as the storage device of genetics was discovered.

Before 1953 an important question remained: While it had already been understood that the genetic information was stored in some yet unknown structure of DNA, the question of how that storage works in detail, remained unanswered. Can a DNA molecule have enough information storage in the first place? This is where a young research duo who we briefly already met in Chap. 2 came on the scene: Englishman James Watson and American Francis Crick. Equally important but less recognized in the process of finding out the structure of gene storage in the DNA was the crystallographer Rosalind Franklin.[1] The story of discovering the gene code can be written as a thriller, which is to some extent what happened with Watson's autobiographical account of this time ("The Double Helix", however with a severe bias).[2]

James Watson began his scientific career in the field of virus research, more precisely in the field of bacteriophages. To do this, he had gone to Copenhagen, where, by chance, he met Maurice Wilkins from London, who introduced to him the idea of DNA for the first time. At that time, Wilkins was already working on methods to crystallise DNA by examining its structure using X-rays. Watson decided to devote himself to DNA and, being 23 years old, went to the Cavendish Laboratory in Cambridge for this purpose in 1951. There he met Francis Crick. A deep friendship and fruitful scientific collaboration began between the physicist Crick and the young biologist Watson. However, neither was assigned to work on DNA - that was Wilkins' territory. So, they did their studies in their free time. At least some things were already known about the DNA. The Austrian chemist Erwin Chargaff had discovered in 1952, that almost equal amounts of certain bases are present in DNA. This led him to the hypothesis that these bases always occur in pairs (adenine and thymine as well as cytosine and guanine). This so-called Chargaff rule was to be an important clue that the DNA would form a spiral structure.

At the same time X-ray images of the crystallised DNA molecule by Rosalind Franklin, Wilkin's colleague in London, additionally supported the assumption that this molecule has a helical structure. It was also already clear that DNA consists of a chain of individual nucleotides in which sugar molecules and phosphate components alternate. However, the question of how exactly the concatenation of the nucleotides occurs, so that the chain molecule is as well stable as capable of copying itself in the process of cell division and reproduction, was still open. How many helices were there, one,

---

[1] See Lars Jaeger, *Women of Genius in Science - Whose Frequently Overlooked Contributions Changed the World*, Springer (2023).
[2] James Watson, *The Double Helix - A Personal Account of the Discovery of the Structure of DNA*, (1968), Atheneum (1980).

as in proteins, or perhaps a double helix in which two strands are interwoven, or maybe even three interwoven helices? A double helix, in which two single strands are interwoven, fitted wonderfully into the picture in which DNA splits and copies itself. But how are the nitrogenous bases attached to the sugar molecule arranged in it?

Wilkins' and Franklin's lab was only a short train ride away from London, so Watson also attended lectures by Wilkins and Franklin in London. In one lecture, Rosalind Franklin presented her latest X-ray images of DNA. These revealed evidence of a double helix structure in which the phosphate groups and the sugar molecules were on the outside and the nitrogen bases were oriented towards the inside. Back in Cambridge, Watson and Crick began trying out various arrangements of the different molecular groups of the DNA in their construction kit, just as if they were building blocks for children, in order to test whether a stable form could be found. The components of the construction kit were the atoms of the different molecular components (phosphorus, carbon, oxygen, hydrogen), represented as small spheres. Narrow connecting pieces represented the chemical bonds between them. After a short time, they had found a stable model, and proudly presented it to Franklin and Wilkins. Franklin immediately recognised a mistake: they had arranged the nitrogen bases on the outside, not on the inside as her pictures showed. Watson had misremembered her lecture.

So now they had to assume that the nitrogen bases in the double helix structure were on the inside. In order for the double helix to remain stable and not fall apart, two opposite nitrogen bases had to be connected to each other. However, this connection was not allowed to be too strong, because the helix should be able to split during the cell division. Crick asked the mathematician John Griffith, a friend of his, whether he could mathematically determine the possibilities of linking the different bases together. Griffith found out that there are only two possibilities (like Chargaff did before him, which Crick did not know how to interpret correctly at the time): Adenine with thymine and cytosine with guanine. During a conversation with Chargaff still in 1952, he confirmed to Watson and Crick that the ratio of adenine to thymine and cytosine to guanine in the DNA of every cell was almost 1:1, regardless of the organism, whether bacterium, fish or human. This strongly implied: Adenine was opposite and connected to thymine, analogously cytosine to guanine. But Crick and Watson were not yet absolutely certain, not least because they were not given direct access to the measurement data of Rosalind Franklin, who jealously guarded her treasures and did not want to share them with anyone.

On Jan. 30th 1953, Watson tried to persuade Franklin to work with them despite their differences, as her X-ray data would likely enable them to find their solution. But Franklin refused. When Watson suggested that Franklin cannot interpret her own data correctly, the situation escalated. Watson angrily left Franklin's lab and bumped into Wilkins, who had often clashed with Franklin himself. Wilkins took sides with the Cambridge scientists and stole Franklin's research results from her office. What an unfair act and massive violation of good scientific practice, but it was to decide who discovered the structure of DNA in the eyes of the public to this day. From Franklin's data it was absolutely clear that the bases are on the inside.

Crick and Watson finally had their model ready. It consisted of two helical strands of nucleotide chains that are intertwined as in a spiral staircase. In it, the nitrogen bases are connected opposite each other on the inside of the double helix structure. Their bond consists of so-called "hydrogen bonds", a form of bonding that is much looser than electric ("covalent") bonds in a molecule. In the process of reproducing DNA during cell division, these hydrogen bonds break easily enough, and the entire chromosome opens up like a zipper. This frees the nitrogen bases to recombine with free nucleotides in the cell. Since the open-end pieces of the nitrogen bases in the now single-stranded chromosome half still determine the exact sequence of the connections, the free nucleotides form in the same arrangement as those in the original double helix structure.

It was the first days of March 1953. A milestone in the history of science, a key discovery of the twentieth century had been accomplished: the universal mechanism of heredity, DNA replication, had been found! And this would have major consequences in future biological technologies, as this was already very clear. Watson and Crick's scientific article of 25 April 1953 on this thus ended with the sentence:

> It has not escaped our attention that the specific pairing we have postulated here immediately suggests a possible copying mechanism for the genetic material.[3]

On the same day Rosalind Franklin's published her paper which essentially described the same structure, however with much more empirical evidences that Watson and Crick, besides their playground results, did not have. The credits she received were nonetheless marginal. Only many years later people

---

[3] James Watson, Francis Crick, *A Structure for Deoxyribose Nucleic Acid*; Nature volume 171, (25. April 1953) p. 737–738.

found out that actually she had found the structure before Watson and Crick![4]

The genetic model was wonderfully simple and at the same time plausible. On the one hand, the double helix explained how genetic information is stored in the cell: the special arrangement (order) of nucleotides with nitrogenous bases contains the genetic code. On the other hand, it showed how DNA replicates. It was to take a few more years before the genetic code could be cracked in its details. But, finally in 1962, the two, together with Wilkins, were awarded the Nobel Prize for Medicine for the discovery of the detailed structure of DNA. Rosalind Franklin, who had played such an important role in the development of the model, died of cancer in 1958, at the age of 37.

First Steps in "Genetic Engineering"

The process of understanding genetics did by no means end in 1953. Open were issues like: How does DNA decode its information within the cell? And how are proteins, the central form of connections in cells, then formed from it? As it soon turned out, another type of nucleic acid, RNA, plays a decisive role in these processes. In 1956, American researcher Mahlon Hoagland discovered a multitude of small RNA molecules in cells, each in specific triple combinations of the nitrogenous bases adenine, uracil (takes on the role of thymine in RNA), cytosine and guanine as their nucleotide building blocks. Hoagland found out that a certain triple combination of RNA nucleotides is always associated with exactly one corresponding amino acid. Why a triple? That was easy to answer. Let us do a little maths here There are 21 amino acids in the proteins of our body - the 20 long known ones plus selenocysteine. A combination of three nucleotides each with four possible bases results in 64 possible different combinations - four to the power of three. If one had only two combinations of nucleotides, there would be only 16 possibilities - four to the power of two, too few to distinguish 20 amino acids. Thus, at least 64 possible combinations of genetic bases are necessary to represent 21 amino acids. Over time the biologists found out which combination of the four bases corresponds to which amino acid. The represented amino acids now "dock" to "their" RNA bases in the correct order, which leads to the resulting protein with its specific amino acid sequence. The RNA in turn receives the sequence of nucleotides directly from the DNA. Through this process, the information from the arrangement of the nucleotides in the DNA is transferred via

---

[4] For more details, see: Lars Jaeger, *Women in Science – Hidden contributions and significant work*, Springer (2023), chapter 14.

the RNA to the protein: the specific combination of amino acids in specific proteins thus corresponds to a specific arrangement of the nucleotides in the DNA. In this transfer process, RNA serves as an intermediate transfer agent, which is why Hoagland gave it the name "transfer RNA" (tRNA).

However, the RNA was initially only found outside the cell nucleus. How does the information get from the double-helix chromosome strands of DNA in the cell nucleus to the RNA outside the nucleus? The details of the mechanism turned out to be even more complicated: In addition to the tRNA, there is another type of RNA that carries the information from the cell nucleus into the cell plasma. Biologists call this "messenger RNA" (mRNA). With it, protein synthesis proceeds as follows:

- Along the DNA in the cell nucleus, an RNA nucleotide strand forms according to the arrangement of the DNA nucleotides, thus creating the mRNA. For this purpose, the nucleic bases of the DNA are transcribed into the nucleic bases of the RNA - with uracil replacing thymine.
- In the now single-stranded RNA, three consecutive bases form a triple combination, a so-called "codon", which as we saw codes exactly one amino acid. There are many amino acids encoded in a row (that eventually create the desired proteins).
- As an information carrier the mRNA now moves out of the nucleus into the cell plasma (which double-stranded DNA cannot do).
- A specific amino acid must then dock to each of its codons, creating together the protein, which consists of the sequence of amino acids encoded in the order of the codons. This requires an intermediate RNA that binds the amino acid on one side and recognises the corresponding codon on the messenger RNA on the other (called the "tRNA" – transfer RNA).
- To be a little more detailed: This "translation" by the tRNA takes place on the ribosomes, which in turn consist of proteins and RNA. On one side of the tRNA, a corresponding triple combination of its bases, a so-called "anti-codon", matches exactly to one particular codon of the mRNA. Its other side is loaded with the matching amino acid.
- This way, two matching (following each other) amino acids are brought into spatial proximity to each other, allowing a bond (called "covalent peptide") to form between them. This also applies to all the other amino acids.
- Thus, the sequence of amino acids that is specific for the desired protein is created, or copied from the original DNA sequence. This amino acid

sequence (protein) eventually folds into a specific three-dimensional structure, which then defines the final protein and its biological activity.

Many details of the biochemical sequence of this entire highly complex protein synthesis process, including the process of protein folding, have only been studied and understood in more detail in recent years. Others are still unknown and subject of current research.

The role of RNA in the transmission of the hereditary through the building of proteins was theorized by Jacques Monod, a French biochemist. He proposed the idea of the RNA messenger and received the Nobel Prize in Medicine or Physiology in 1965, together with François Jacob and André Wolff, "for their discoveries concerning genetic control of enzyme and virus synthesis". Monod was not only a biologist but an esteemed musician and one of the few scientists who intervened in philosophy. His book "Chance and Necessity",[5] written for a non-specialist readership, was a very influential examination of the philosophical implications of modern biology.[6]

The Development of Life on Earth

According to the findings of astronomy and geophysics, our planet is about 4.54 billion years old. During the first 400 million years of Earth's history, the conditions (high temperatures, intense radioactive radiation, chemical composition of the atmosphere, etc.) made any development of life impossible. Then, 4.2 billion years ago, the Earth started cooling down. Earth volcanoes spew gases, including water vapour. As the planet then cooled, clouds formed and rain fell. It actually rained for many millions of years, eventually forming a world-spanning primordial ocean. Considering the developments in Earth history and the discovered oldest traits of forms of life, we can say that life on Earth began almost immediately after the circumstances on our planet had changed in such a way that life became even possible at all. The metabolism of the first life forms probably consisted essentially of producing organic substances from water and atmospheric carbon dioxide. Put simply, the basic synthesis of life consisted of bringing hydrogen and carbon dioxide into a more complex chemical compound. Early earth's atmosphere at that stage offered these starting materials in abundance in

---

[5] Jacques Monod, Chance and Necessity: *Essay on the Natural Philosophy of Modern Biology*, Harper Collins Distribution Services (1971).

[6] In the title of the book, "necessity" refers to the fact that the enzyme must act as it does, catalyzing a reaction with one substrate but not another, according to the constraints imposed by its structure.

in direct or indirect form (water). But they do not normally react with each other. Together with nitrogen compounds (from the ammonia present in the atmosphere), sulphur and phosphorus compounds (which reached the Earth's surface from the interior through volcanoes) and certain metal compounds (found in the oceans), a well-seasoned soup was created for the emergence of life, the "primordial soup". Various elemental organic hydrocarbon compounds - the first "molecules of life" - were indeed able to form under the prevailing conditions about 4 billion years ago. It is highly probable that the next step towards life was the aggregation (called "polymerisation") of the first biomolecules ("monomers") into chain-like macromolecules ("polymers"). This could only happen with the help of external energy sources, which were probably abundantly available in the ocean in the form of solar radiation, radioactivity, chemical processes and from volcanoes. For the latter, magma pushed from the earth's interior to the surface and gradually piled up to form islands and finally the first continent "Ur". Since the syntax of the genetic code is the same for every living being, it is more than plausible for biologists to assume that this early process must have been uniform for all life today, from the simplest bacteria to humans. In other words, there was a single process for the emergence of all life on Earth.

Some researchers believe that the first living cells developed in the volcanic springs (so-called "black smokers"), stone walls of chimney-like vents in the deep sea, places where the circumstances for the first molecules of life were almost perfect. Here, as many biologists assume, the "chemical gardens" developed in which the appropriate reactions took place that gave rise to even more complex "molecules of life", including the macromolecules that still play an essential role in the metabolic reactions of all life today. Among the descendants that grew out of these first microbes are two "kingdoms of life": the *bacteria* and the *archaea* (initially also regarded as bacteria, today sometimes called "extremophiles", as they can life in very extreme environments).

Besides the atmospheric circumstances the geological and astronomical conditions on Earth were also very different from today's environment. For example, the Earth rotated much faster around itself, so that a day was only four to five hours. The moon orbited much closer, creating higher and more frequent tides. Also, huge meteors hit the Earth, kicking up dust and water vapour that blocked sunlight for long periods - in addition to dust from volcanic eruptions. Under such circumstances, it is rather unlikely that life on land arose before about three billion years ago.

Early life forms had to solve two problems that all life is still based on today: First, the processes taking place in "their bodies", such as growth and reproduction, requires a constant supply of energy from outside. For physical reasons there can be no "metabolism" without energy: A living organism, if it wants to maintain its orderly state, must compensate the otherwise physically unavoidable increase in entropy inside its body by energy supplied from outside. Secondly, early life forms had to be able to reproduce themselves. While external energy taking on the increase in entropy was sufficiently available, reproduction proved to be much more difficult. Since in all living beings DNA and RNA are responsible both for storing vital information and passing it on to future generations, understanding the origin of life requires the understanding how carbohydrates and amino acids ultimately gave rise to RNA and DNA that do *contain information.* While they are abundant in the oceans, they are free of any information storage via the randomly distributed first polymers. So, where does the information structure come from?

This is still an open question in biology, if not the most important question overall. One opinion among today's biologists is that this process began at the aforementioned black smokers. Biologists assume that RNA, which is chemically more suitable for autocatalysis processes than the double structures DNA, first developed in a preliminary form of genetic information storage. Only then came DNA, equipped with the code syntax that is identical for all living beings. The process of biological evolution started. The raw materials for the first RNA polymers (long chains of individual RNA molecules) as well as the necessary energy for their production were probably sufficiently available in the primordial ocean: Phosphates were available in the hot springs; the sugar ribose and the other organic compounds could form in some "primordial soup" from the carbon dioxide-hydrogen compounds. But individual RNA molecules contain as little genetic information as a single letter from A to Z can yield a meaningful statement. Could purely random improvements in the structure, function and stability of such molecules really be transformed into information at all? This question resembles the "creatio ex nihilo" question about the cosmic origin - except that here it is not a matter of energy or matter, but of information. However, the probability that a complex information carrier such as RNA or DNA arose by chance out of individual RNA molecules is practically zero.

The Origin of Life

Here we have reached the "crux" of a physical explanation of the beginning of life on earth. In order for something like Darwinian evolution and thus the development of functional forms of life to get underway at all on the molecular level, a natural selection of molecules was required in which "better" molecules prevailed over "less good" molecules over time. The quality is measured by their ability to create, store and transport information, i.e. well defined structured of individual RNA molecules. Such a process presupposes a priori criteria for the "efficiency" and "fitness for life" of a respective molecule. In other words, it already needs "biological information", and for this in turn a genetic code. Only a corresponding genetic code, however primitive it might had been at first, made it possible for molecules to store the information necessary for further developments and selections and at the same time to be carriers of evolutionary changes. We do not know how this happened, just as in physics we have no fundamental law (see the list of the unanswered questions in science in the final part of Chap. 2, the third one listed there).

The first scientific experiment to create or simulate the origin of life on Earth was carried out in the same year the DNA was discovered by the young US researcher Stanley Miller and his supervisor, Professor Harold Clayton Urey. They knew that at the time of the emergence of life about 4 billion years ago, the atmosphere had a different composition than today. Instead of consisting essentially of nitrogen and oxygen, its constituents at that time were largely hydrogen-based molecular compounds, especially methane (CH_4), ammonia (NH_3) and molecular hydrogen (H_2). And the environmental conditions included violent lightning and thunderstorms, intense UV radiation from the sun, from which the earth was still largely unprotected without an atmospheric protective shield, and a combination of inorganic substances that rained into the oceans. These complicated organic molecular compounds were the origin from which the basic building blocks of today's life could have developed were supposed to have formed spontaneously: Amino acids, proteins and nucleic acids. Urey and Miller came up with the idea of recreating the conditions on our planet at that time in a simple experiment. To do this, they combined water with ammonia, methane and hydrogen in a vessel and exposed this solution to strong UV radiation and electrical discharges (lightning).

And indeed: After a while, they were able to identify simple amino acids in the solution. And after longer times of the experiment, even more complex amino acids were formed. Finally, the amino acids produced corresponded exactly to the 21 amino acids that we find in living organisms today! An approach to a possible scenario for the natural origin of life and its basic building blocks was revealed. A plausible answer had been found to the question of where amino acids could come from.[7] But even though this was a first significant step toward the scientific explanation of life, the road to its complete exploration was still very long. The details of the development from simple building blocks to cells and more complex life forms with metabolism, self-reproduction and evolution remained unexplained. And to date, no experiment of the Urey-Miller type has been able to demonstrate the emergence of more complex macromolecules that contain information. Although biochemistry offers some theories on the origin of life, it has not yet found an applicable unified and generally accepted explanatory model that includes the issue of the origin of genetic information.

How about the physicist's new laws of self-organisation in sufficiently complex systems (see Chap. 6)? The first impetus for a theory of molecular self-organisation of prebiotic open systems as the basis and possible beginning of life and molecular evolution, i.e. the creation of information, goes back to biochemist Manfred Eigen (who received the Nobel Prize in Chemistry in 1967) and chemist Ilya Prigogine in the 1970s. Eigen's so-called "hypercycle theory" describes in an initially abstract way how, far from thermodynamic equilibrium, something like a genetic code could have formed from simple macromolecules through self-organisation and subsequent selection in a prebiotic evolution. This hypercycle is a system of proteins and RNA molecules in which the proteins catalyse the formation of RNA molecules, which in turn catalyse the formation of proteins. In this cycle of reciprocal relationships, molecules are created that mutually catalyse their synthesis, which describes the process of positive feedback, a fundamental characteristic of self-organisation of systems. It could cause the described polymerisation from monomeric building blocks of biomolecules, which cannot be explained from experiments of the Urey-Miller type. The formation and selection of the molecules and combinations of molecules essential for life would thus not be something coming from outside, mystical or transcendent, but from the emergent inside. Matter can develop intrinsic properties that cause molecules to organise themselves into forms that become increasingly complex, and in which information is also stored at some point.

However, Eigen had to leave open exactly how this happened in detail. The hypercycle theory is based on highly simplifying assumptions, remaining at the level of the schematic. To this day, biologists are limited with such models and "plausible" simulations when researching the origin of life. Even

---

[7] The trail to the origin of life got even hotter when organic substances were discovered in extraterrestrial sources in later decades. In 1970, scientists discovered five different amino acids in a meteorite in Australia. They suspected that the amino acids found there had originated in a similar way to the experiment by Urey and Miller, but extraterrestrially.

together with experiments simulating the chemical composition of the Earth's atmosphere at that time, this is still not enough to formulate a theory that can explain how information storage and thus life really arose. The question of whether life arose through self-organisation or not can ultimately only be discussed on the basis of concrete physical and chemical processes. Secondary would be the question of whether this is possible in principle, i.e. whether a corresponding process could have taken place within the framework of known physical and chemical laws. Terms and concepts such as "self-organisation" and "emergence" must not be stopgaps for insufficiently realistic biological and chemical models, not substitutes for unknown mechanisms. Both terms are initially defined only in the context of mathematical models in which a corresponding dynamic of feedback and other non-linear processes is described as a prerequisite for self-organisation. The extent to which they will eventually be able to represent real processes is still to be determined. Plus, the components put into variables are in reality not isolated, substantial units, but can only be grasped contextually with respect to the entire system and the environmental conditions - another emergence effect that works against the modellers. We can even see parallels to the concept of reality the famous Buddhist philosopher Nāgārjuna (second century) provided: The fundamental level of reality in our world is not represented by individual independent substances such as atoms, molecules or complex molecular structures, but rather their components are in a permanent interaction with each other, from which they cannot be separated. Thus, the real historical development of very early life is dependent on many specific boundary conditions and parameters that can hardly be captured and represented in a model today.[8]

All living beings on earth are the result of an evolution that originated about 3.8 billion years ago. That should not be discussed any longer. But a controversy is quite appropriate when biologists claim that genetic information arose suddenly and completely by chance in the form of a first self-replicating RNA or a preform of it, from which terrestrial evolution then developed. As already explained, the probability of such a process is almost zero. And too much time was not available between the point at which complex macromolecules were first "survivable" (i.e. did not immediately disintegrate again due to the adverse conditions on Earth) and the earliest (bacteria-like) forms of life in which the genetic information processing was already taking place. The development of the first forms of information processing necessary for life must have been much faster than if it had been

---

[8] The term „Pratītyasamutpāda" (or "dependent arising") is originally referring to the dependence in the arising of dharmas.

based on simple statistical fluctuations. If the relevant information were to arise by chance, a simple calculation shows that even the smallest effective protein molecules, which already consist of about 100 amino acids, have a spontaneous probability of formation of more than one in 10^{100}. So even if thousands of tries could be made in a second, creating the right information would still take much, much longer than the age of the universe. It cannot therefore have been an a priori all too improbable process. Biologists therefore assume that a mechanism must have been at work very early on that "selected" the right molecules and allowed them to continue to "live" and develop. Representatives of religious movements invoke a divine act of creation as the only possible answer to the question of how this could have happened. One of the creationists' favourite arguments is the so-called "Boeing 747 argument" by Fred Hoyle:

> The chance that higher life forms might have emerged in this way is comparable with the chance that a tornado sweeping through a junk-yard might assemble a Boeing 747 from the materials therein.[9]

Naturally, scientists object such a statement focussing on "chance" only, thereby neglecting "necessity" as Jacques Monod would say. Necessity here refers to the iterative rules of evolution ("survival of the fittest"). Similar iterative rules exist for chemical reactions, e.g. for the forming of fullerenes (soccer ball like chemical structures). But until the exact sequence of the origin of life can be clarified and, if necessary, reconstructed in the laboratory, it will hardly be possible to completely refute religious explanations with the methods of science.

The reductionism suggested by biologists like Craig Venter, according to which life and its origin can be explained solely from a description and combination of the individual components and their corresponding historical evolution, encounters severe methodological limitations. They themselves are actually at the centre of a potent as well as complex biological environment. But a unified model for the very early chemical evolution is lacking. The consideration of the existing models illustrates in fact how little we know about the very beginning of life on Earth and thus how difficult it is to describe the actual emergence and development of life. Even Richard Dawkins, probably the most prominent anti-creationist and atheist, despite all his eloquent (and sometimes all too polemical) explanations, cannot eliminate this problem and find a definitive answer to the question of "design (God) or coincidence"? Is this perhaps where the scientific method itself

---

[9] Fred Hoyle, *Hoyle on evolution*, Nature, Vol. 294, No. 5837 (November 12, 1981), p. 105.

reaches its limits? Is a combination of chance (arbitrariness) and necessity (laws and circumstances) like in evolution at all principally capable to explain the very origin of information creation itself for such an information-rich biological complexity we have today? (This is the very assumption defended by Jacques Monod in his book "Chance and Necessity", see below). As in physics, also biology, besides much effort in the last 70 years, has not answered its key question: What is the origin of life? And an equally important question is: Can we humans even understand the origin of life? Or is it possibly a phenomenon the origin of which lies in transcendence? Let us remember the statements of the nineteenth century German scientist Emil Du Bois-Reymond about questions science will never be able to answer:

1. How did the first life come into being?
2. Where does the purpose ("the deliberate purposeful arrangement") of nature come from?

These are actually the key questions here and have so far kept the truth of Du Bois-Reymond's statements. However, scientists like Monod say:

> The first scientific postulate is the objectivity of nature: nature does not have any intention or goal.[10]

In order to avoid natural local increase of entropy in an open system (like the Earth is existing in) it is natural to increase the order in the sub-systems at the expense of the entropy outside of them. So, one could say that the "purpose" of nature on Earth is to dissipate best the energy received from the Sun.

However, all attempts so far to find a clear and unambiguous explanation for the origin of biological information and thus life have not succeeded. This failure tempts a few biologists to look in the opposite direction of where they have looked so far. The phrase "In the beginning was the Word" in the prologue of John's Gospel then becomes "In the beginning was information". The future discussion between natural sciences and faith should start at this point.

---

[10] Jacques Monod, *Chance and Necessity: Essay on the Natural Philosophy of Modern Biology*, Vintage, New York (1971).

Genetics Since the 1970s

Today we distinguish the era of finding the details of the structure of DNA, which is referred to as the "DNA era" or "genetics", from the area of "genomics", i.e. the field focusing on functions of the DNA as well as eventually the editing (changing, engineering, manipulating, often referred to as "genetic engineering") of genomes (often in practise also called genetics as well). It was in the early 1970s when the first era transformed into the second constituting today's barely controllable spectrum of applications and product generation opportunities in the area of genetics/genomics.

Genomics rests on a systematic analysis of the complete genome of all active genes, e.g. those functioning in a cell, a tissue, an organ or an entire organism. The leap from the study of individual genes (genetics) to the analysis of whole genomes (genomics) is inextricably linked to genetic engineering with its ever-new methods. This required dealing with more and more data on genes and their functions. The development of bio*informatics* was thus a prerequisite for coping with the flood of data that emerged as of the late 1980s. The goals of genomics generally included a better understanding of.

- The evolution on a molecular level,
- Pathogens and hereditary diseases and their possible treatment,
- The development of diseases or the reaction of the organism with respect to the environment,
- How to use DNA as fingerprints (standard today, e.g. in forensic science).

In 2003 the genetic researchers celebrated the successful completion of the Human Genome Project with 99% of our standard genome being sequenced to a 99.99% accuracy. However, besides the "better understanding" of genome structures and its application to specific human diseases based on them, the most important development since the early 1970s was the application of genetics to create corresponding products based on changing specific genes in plants and animals for various purposes:

- As early as 1980, it was possible for the first time to integrate a new DNA section into the genome of an animal, a mouse (since then it became possible to integrate an entirely new genome into a mouse, see below). Two years later, researchers announced that their transgenic rodents grew faster and larger thanks to additional genes for growth hormones. Experimental animals that have been genetically modified in this way to test new drugs or therapies are now frequently used in medical research as "animal models". Another application is to increase the meat ratio of animals to be eaten. However, this still stands at a preliminary point, as governments regulate that area heavily (for good reasons!).

- "Green" genetic engineering. Such applications serve to produce plants with higher nutrient content. The two best-known examples are the "golden rice", a genetically engineered rice variety with additional beta-carotene, and phytophthora-resistant potatoes. Genetic engineers have recently been given a new, powerful tool for plant breeding (CRISPR, see below).
- "White" genetic engineering: In many industrial sectors, conventional chemical processes are increasingly being replaced by genetic engineering methods, because biological processes consume less raw materials, energy, and water, and generate fewer waste products than chemical processes. The production of new groups of substances is also possible through genetic engineering; for example, biopolymers similar to spider silk are used in medicine as well as in the cosmetics and textile industries.

The application to treating *human* diseases through gene treatments and modifications is today called "Red" genetic engineering: With the help of it, active substances can be produced for drugs and other substances needed in medicine. In 1978, the first genetically engineered drug, human insulin, came onto the market. The much sought-after substance for treating patients lacking insulin in their bodies no longer had to be extracted from millions of pig pancreases but was now produced by bacteria. Today, every second new drug is produced by genetic engineering. During the Covid-19 crisis of 2020/21 it was possible to develop a very effective vaccine against the virus within a very short time (effectively a few weeks; the rest of the time was for testing it).

In many Western countries, experiments on human embryos are prohibited, however in others (like in the UK), the so-called 14-day rule applies, according to which embryos left over from artificial insemination are allowed to be grown in the laboratory for two weeks. In the USA genetic engineering on human embryos is not explicitly prohibited, either, but equally forbidden after a few weeks. But even this is not the case in other countries. In spring 2018, a Chinese researcher published a study in which he had introduced a modified gene into a fertilised human egg cell that made the embryo immune to AIDS. The baby was eventually born. However, as it turned out he came with a variety of other genetic issues. Gene scientists thus realize that even the most modern gene methods still work far away from perfectly, with unexpected errors that occur due to unknown genetic structures and interactions. Gene technology is therefore not yet mature enough for responsible researchers to dare to be attempted to grow a human being from a genetically modified egg cell (besides the severe ethical issues involved, see below).

That the discussion on how to use the capacity of genomics is anything but finished, is shown by the situation in Germany. For several years the German Ethics Council had been working on a declaration about the possible modification of the human genome. Finally, in a statement from May 2019 it assessed such interventions in the germ line still as too risky. However, and that is nothing less than a change of paradigm, the council does not want to exclude them ethically *in principle* for the future.

At the same time, modern genomics develops rapidly and enables scientists more and more to apply genetic changes, which increasingly includes potentially modifying the human genes. And a specific technique has dramatically accelerated the possibility to manipulate genes at determined points: CRISPR.

Revolution of Genetics in 2012 – As Amazing as Scary New Technologies

The most important bio-medical and genetic engineering breakthrough of this century (so far) is CRISPR, a powerful new gene technology discovered in 2012, largely unnoticed by the public. CRISPR stands for "Clustered Regularly Interspaced Short Palindromic Repeats" and describes sections of repeated DNA or RNA fragments in the genetic material of bacteria. When infected with phage (viruses), the bacteria are able to integrate parts of the viral foreign DNA into the CRISPR regions of their own DNA. The integrated DNA part then functions like a mug shot: As soon as viruses with this DNA attack the bacterium again, the cell recognizes the exogenous DNA and can immediately build up the desired protection. In this way, the bacterial cell becomes immune to the viruses.

When joined by the enzyme Cas9 (a so-called "endonuclease"; Cas stands for "CRISPR-associated") bio scientists Emmanuelle Charpentier and Jennifer Doudna recognized in 2012 that DNA targets in a genome of any being (including human ones) can be aimed at and then manipulated. For this they shared the Nobel Prize for Chemistry in 2020. The complex works like a Lego brick finder and scissors at the same time. It allows genetic engineers to directly access and manipulate individual genes. For that, they populate the CRISPR/Cas9 enzyme complex with a sequence that is exactly complementary to the desired DNA target sequence. The complex then finds the desired target sequence in the DNA and disconnects the target sequence exactly there. Any new gene sequence can then be inserted or the old one be removed without replacement. This method is therefore also referred to as "word processing in the genome". While previous gene editing techniques were more like a shotgun that simply aimed shoots at the DNA hoping to hit the desired gen, CRISPR/Cas9 is more like a precision rifle or scalpel that can be used to remove or insert specifically targeted genetic building blocks. And despite its incomparable potency, the CRISPR technique is so easy to handle that it could be made available to any gene laboratory, and soon perhaps even to high school classes.

The technology has already been applied in practice on a large scale, especially in the modification of the genetic make-up of industrial plants. By 2015, for example, biologists were already able to use CRISPR to modify the genes of a cultivated mushroom in such a way that pressure points do not turn brown quickly. In spring 2016 the long-lived mushroom became the first CRISPR product to be approved by the U.S. regulatory authorities. With CRISPR, however, genetic engineering on animals and humans is also entering a new phase. Interventions in the human genome will no longer be a *technical* problem. For some medical applications, the technology has already reached the stage of clinical trials. This will affect the treatment of numerous hereditary diseases caused by genetic defects, which have so far been considered incurable, but also of human plagues such as HIV, malaria or even diabetes, cancer and other age-related diseases.

But what happens when this technology is applied to embryonic cells, egg or sperm cells? Then not only the individual is manipulated, but subsequently all of his or her descendants. Scientists today no longer ask on which genes the new method *can* be used, rather they ask: On which genes *should* it be used? Should CRISPR maybe not only be applied to treat diseases, but also to genetically influence human characteristics such as eye colour, height or even intelligence? CRISPR could ensure that designer babies do not remain a utopia (or dystopia?) any longer. With its help, parents will be able to put together, i.e. design the characteristics of their children as they desire. In addition, genetically optimized people could soon be cognitively and physically superior to "normal people". We would enter the era of genetic human breeding.

In May 2015, the scientific journal Nature asked: "Where in the world could the first CRISPR manipulated baby be born?" With regard to the respective legal situation, the answer was: Japan, China, India, or Argentina. As we saw above, in November 2018, the race was decided: China. It had become clear: With CRISPR we have finally arrived at the age of human experimentation and designer babies.

Like so many other modern technologies CRISPR reveals both the curse and blessing of technological progress. It is a great blessing for the individual, when CRISPR can be used to repair the genome of his or her "broken" cells. CRISPR offers enormous opportunities for the treatment and prevention of diseases that have been incurable up to now. But what happens if this technology is used to change the healthy human genome, in worst case with consequences for all future generations? Its incomparable potency has led philosophers, theologians, ethicists and, last but not least, politicians to discuss alongside bioengineers about the consequences of the CRISPR technology in human hands. The weighing of these conflicting goals must go through the filter of reflection and, as we outlined in the Chap. 4, philosophical reflections of what applications can and should be done, and what should be forbidden and abandoned without hampering the acquisition of new knowledge. Here too, as for physics applications, comes the task of risk management dealing with the ethical consequences of new applications and assessing the risk of undesired consequences.

Synthetic Life

Everyone is talking about the CRISPR gene scissors. However, next to her a new biological field that is equally as terrifying as it is awe-inspiring has established itself alongside it: the production of artificial life, specifically the synthesis of artificial gene sequences, i.e. *synthetic biology*. Like CRISPR, this holds enormous biological potential - to unlock life's remaining secrets, to open up whole new technological horizons - as well as it raises a host of disturbing questions, next to scientific and technological ones, important philosophical and ethical ones.

When the controversial genetic engineering pioneer Craig Venter announced in 2010 (see Chap. 5) that he had for the first time created an organism from the bacterium *Mycoplasma capricolum* with a completely artificial genome (there were still important functions in the cells from natural life), the media response was still rather contained. But it was a milestone in modern genetic engineering. In fact, it sounded quite simple: The complete genome was successfully synthesized from a computer record and transplanted into an existing cell that had had its DNA removed. Since then, the goal of genetic researchers has been "life from scratch". Barely four years later, Jef Boeke succeeded in reconstructing one (out of 16) complete chromosome of yeast with a few artificial modifications. Yeasts are so-called eukaryotes. Their genetic material is much more extensive and complicated than that of the bacteria and viruses in Venter's studies. Humans are also eukaryotes; basically, the jump from the bacterial genome to the baker's yeast genome is greater than that from the baker's yeast genome to the human genome. So, the first steps towards artificially producing more complex life forms have been taken. Boeke and his team have by now achieved the critical milestone of the completion of synthesis of the first functional eukaryotic chromosome in yeast.

How exactly does the creation of artificial genomes work? From a database of many millions of genes, genetic engineers use computers to simulate the properties of a large number of possible gene combinations. From this, sets of genomes with certain desired properties are identified, then chemically synthesised before they are finally introduced into the cell nucleus of the target organism, which has been freed from its original genome. The chemical synthesis corresponds to the assembly of Lego bricks, just like in the discovery of the DNA structure: putting together the calculated sequence of the four nucleic bases adenine (A), cytosine (C), guanine (G) and thymine (T).

Thus, after his success, Venter spoke of a new "digital era" in biology, in which DNA, as the "software of life", can be programmed at will to create e.g. microorganisms as needed. These could produce precisely desired amino acid sequences, i.e. proteins, which would, for example, enable the production of new drugs, which until now have been very difficult and expensive to produce. But the genetic engineers' ambition goes further: they want nothing less than to learn to use the programming language of life to produce genomes better than nature. This could lead to entirely new organisms, and these with very real benefits: In addition to applications in medicine, one hopes for functions in energy production, breaking down oil spills in the world's oceans, decompose plastic in the oceans, develop CO_2 "eating" microbes from the labs, increase agriculture performance and thus food production, and others.

The next step in artificial life out of synthetic biology will be genetically modified real (complex) animals. As mentioned, the step from yeasts to larger animals is not as large as intuitively assumed. However, that does not mean that everything is simple from here. In August 2022 scientists around Magdalena Zernicka-Goetz in Cambridge (UK), after more than ten years of heavy research, announced that they have created synthetic mouse embryos, not from eggs or sperm, but from stem cells. The work could decisively shape the idea of how a human being is created. Even if the method is still far from perfect (only a very small proportion of the synthetic embryos grew as desired, and even the best results still differ in important details from the natural model) the work enabled the researcher to observe the development of organs in a mammal in unprecedented detail – in order to eventually change it.

Specifically, the group around Zernicka-Goetz imitated natural processes in the lab by directing certain types of stem cells found in early mammalian development in the right proportions and environments to promote their growth, to the point where they begin to cooperate themselves. The researchers were indeed able to get the stem cells to communicate with each other properly. Thus, the stem cells self-organised themselves into structures that went through the various stages of development until they had a beating heart, the foundations of the brain, as well as the part from which the embryo develops and receives nutrients in the first few weeks (yolk sac). Unlike other synthetic embryos, the processes developed at Cambridge reached the point where the entire brain began to develop. Experiments and equally detailed observations of the very early stage with *human* stem cells are already planned.

This technique could one day replace experiments on living animals in developmental biology research or be even used to grow organs and tissue for transplantation into humans. Latest then, it should be clear that important ethical questions arise around the embryo, and it being grown into particular organs. However, it is still unclear whether *synthetic* human embryos growing into particular organs are even to be considered embryos in the legal sense, or under what conditions that must be the case. In other words: When is an embryo an embryo? That has remained an open question … so far.

Life Prolongation – When Humans Want to Play God II

It lives in the Mediterranean, bears the name *Turritopsis nutricula*, and is not much more than a floating watery jelly disk. But it has an amazing property that we already saw in Chap. 5: It is immortal. This particular jellyfish has a cell program which reverses the usual translation of young cells into differentiated cells. It thus constantly rejuvenates its cells.

Even some forms of unicellular life like the *paramecium* have the chance to live billions of years, as they continue to divide into new cells. Many creatures are thus potentially immortal. Man, however, ages until he finally dies, very latest at an age of little more than 120 years. Is there a chance that in the future we become immortal like the jellyfish or the paramecium?

Well, the life expectancy of humans can never exceed 64 years and 9 months, as US-American demographer and statistician of the insurance company Metropolitan Life, Louis Dublin, calculated in the late 1920s. But, contrary to Dublin's prediction, the average human life span in the most highly developed countries has by today increased to more than 80 years. And it continues to grow with an average of 2.5 years per decade. Its doubling over the past 150 years is, as we already saw in Chap. 5, almost entirely due to advances in medical and natural sciences, including better hygiene standards, more suitable nutrition, and much broader medical emergency care. But how far can the human life be prolonged? And why are we even getting older and irrevocably die at some point in the first place? Surprisingly, science cannot yet provide a satisfying answer to this question. In simple terms: Our cells and organs simply lose their capacity to function.

Some of today's geneticists as well as doctors and biologists assume that this process can be stopped or even reversed. They believe that the possibilities in genetic manipulation might make a primordial dream come true: the fountain of eternal life. According to them there is no insurmountable biological limit for the human age, for aging is ultimately nothing more than the succession of defects in cell division and repair - caused by increasingly frequent genetic copying errors. But if the damaged genes could be "healed" through genetic editing techniques such as CRISPR/Cas9, we should witness the decisive breakthrough in man's fight against aging or even death. The internet giant Google has already invested more than a third of its investment budget in biotechnology, into various companies dedicated to the extension of the human life span. This is led by Ray Kurzweil, one of the biggest believers in the moment when humans - with the aid of technologies - can potentially live forever.

How would that work in detail? The most popular theory of aging is that it has to do with the ends of each DNA strand. Geneticists call these gene regions "telomeres". Telomeres can be compared to the plastic sleeves at the end of shoelaces, which protect those from unravelling. Biologists have observed that the telomeres shorten each time the cell divides. This continues until chromosomes have become too short and thus cell division is no longer possible. As a result, the cell dies. However, if the cell has a specific enzyme, the "telomerase" it maintains their telomeres' length by adding genes with the repeats "TTAGGG" to the ends of chromosomes. For gerontologists (scientists who deal with the process of biological aging), CRISPR/Cas9 and the possibility to edit genes like texts in a Word document constitutes an amazing opportunity. These could enable the cells to produce this particular enzyme and thus continue to divide further indefinitely.

Research in the field of gerontogenes (genes, which control the aging processes) also has the goal of extending our life span. The geneticists have already identified some genes that control the aging process in lower organisms, such as the "age-1", the "2daf-2", the "bcat-1", and the "clk-1" gene. The "FoxO3" gene broadly referred to as the "methuselah gene" is too a member of this group. By deliberately inserting, changing or blocking these genes, genetic researchers have already massively increased the lifespan of animals in the laboratory.

Parallel to research on the level of cells, biologists and doctors also work to breed entire substitute organs. As soon as an existing organ loses its function in our body, respective replacement organs could be implanted. Thus, the cultivation of organs in animals has long been on the researchers' agenda. Already over a hundred years ago, zoologist Ross Harrison was able to get nerve cells he had cultivated outside the body to divide. In 1972, Richard Knazek and his team were able to grow liver cells from mice on hollow fibres. And just ten years later, burn victims were transplanted skin which had previously been bred from body-borne cells. A last example: In 1999, biologists were for the first time able to breed nerve cells from embryonic stem cells of mice. When these were injected into other mice which were infected with a kind of multiple sclerosis (brain cancer), the animals recovered.

We can furthermore simply print organs. This has already been done with a small tissue sample and a 3D image of the corresponding organ. The organ is then built up layer-by-layer with body-specific "ink cells", which are produced from stem cell cultures (in the terminology of 3D printing this technique is referred to as the "rapid prototyping method"). Already today, hip bone and foot bone transplants are printed in 3D printers with an accuracy unimaginable just a few years ago.

With this type of "tissue engineering" another powerful method can be made available to doctors. In the past, one way of therapy has been to take differentiated cells from a donor's organism and multiply those in the laboratory with the goal of replacing a diseased tissue in another patient. The persistent problem with this method, has so far been the rejection by the receiver's body. Here the stem cells come into play. Tissues bred with them are not classified as foreign bodies by the patient's immune system and are thus not rejected. Adult stem cells are multipotent. For example, an adult stem cell from the skin can generate all cell types, a simple liver cell or blood cell is not able to do that.

The combination of genetic engineering, stem cell research and nanotechnology (3D-printing) could increase our physical and mental well-being and, last but not least, our life expectancy into yet unimaginable dimensions. If we specifically edit and re-program those genes that control the aging of our cells, breed substitute organs in the laboratory (or in animals), or use stem cells to heal our ill cells and organs, the dream of a further prolongation of our lives or even reaching human immortality no longer seems that utopian. Even though human immortality is unlikely to be realized very soon, this project will certainly remain on the radar screen of scientific efforts.

But do we really want to live in a society in which humans are immortal or live, let us say 200 years? Among other considerations, we would surely have to avoid an overly high number of humans living on Earth. But perhaps by then humans have found ways to live on other planets (though possibly as unlikely to happen any time soon)...

Ethics for Today's Biology

As we have seen in the last chapter, ethics has become the primary interaction between physics and philosophy. While biology is much younger than physics and had in fact not been confronted with ethical issues until the 1970s, when genetics did reach first application opportunities, its ethical issues grew as much and as quickly as they did for physics with the explosion of nuclear weapons onto Hiroshima and Nagasaki (also physicists had little ethical discussions before WWII).

Bioethics is about appropriate behaviours in dealing with living beings and with nature. It is primarily concerned with the question of which interventions and experiments we regard as "reverence for life", and which of our potential actions and usages should be prohibited. Special attention is increasingly given to genetic engineering. There is an agreement that studies on human beings require the consent of the persons concerned, and that experiments which threaten the life and/or the dignity of test subjects are to be abstained from. Empathy for the suffering of animals and respect for the lives of laboratory ones, especially primates, have also increased recently. It has become a consensus in biomedical research to spare animals avoidable sufferings and to limit the number of animal experiments to "indispensable" levels.

However, these views are not shared to the same extent internationally. Difficult issues remain controversial across countries, such as abortion, the use of early stages of human development kept in vitro in medical research, human cloning or the legalisation of active euthanasia (the practice of ending the life of a patient to end his or her non-curable suffering). In the USA, for example, no law prohibits privately financed human cloning – which does potentially disturb the dignity of the involved person. Genetic manipulation of a human embryo is also allowed in the US. In Europe there is a strong prohibition on both.

However, we should recognize that bioethical considerations and their implementation in practical regulations do require a sense of proportion and the ability to compromise. An overemphasis on bioethical arguments could bring important branches of biological, medical and pharmaceutical research and development to a standstill - with unforeseeable consequences for human life and health. Should we thus treat bioethics as a special - less strict - morality? No, a too soft implementation of core bioethical standards comes again with severe ethical problems. We see, bioethics is one of the most difficult ethical areas in science today. However, it should be clear: There is a need to develop a new ethics that include the latest scientific findings. These are by no means absolute, but always relative truths which need to be evaluated through an *objective analysis* of different potential situations. Constant progress on ethical issues are essential to cope with bioethics.

Already in 1975, it was 140 scientists involved from 17 countries who, in their meeting in Asilomar, California, drew attention to the risks of their molecular biology experiments. As a result of the now legendary Asilomar conference, it was the researchers themselves who had drawn up guidelines for working with genetic engineering methods. These guidelines then formed the basis for national genetic engineering laws in many European countries as well as the US. Although since then numerous breakthroughs have been achieved in genetic engineering, these guidelines are still not outdated today because they are periodically adapted to the latest findings and developments. It is in the nature of things that these adjustments can only be primarily worked out by scientists who are involved in this research. However, they can surely not do this alone. The public needs to have a strong word to say. Politicians should then integrate themselves in that discussion and create the respective laws - along with experts, of them not too many economists, who have way to little knowledge about the subject, as unfortunately it is often the case today. The development since 1975 shows that self-regulation together with legal control has worked reasonably well. The fact that the necessary discourse among scientists is possible still today is also shown by two groups of leading biologists in spring 2015. They called for a stop (or at least a pause) to the biotechnological and genetic engineering possibilities newly offered by CRISPR and check the respective ethics involved in detail.[11]

Another issue is prenatal genetic analysis: Molecular diagnostics allow the detection of many hereditary defects in the prenatal state with a high

---

[11] Jennifer Doudna, Samuel Sternberg. *A Crack in Creation: Gene Editing and the Unthinkable Power to Control Evolution*, Houghton Mifflin Harcourt, Boston (2017).

reliability. Genetic counselling could be developed based on these aims of informing parents about expected hereditary diseases. In most countries (right now changing in the US, though) it is exclusively the responsibility of the parents whether a serious genetic defect is used as a medical reason for termination of a pregnancy.

Dealing with genetic diagnoses in living humans also requires a high degree of ethical sensitivity: a respective diagnosis (even without the breakout of the disease) can be misused to put the person affected at a disadvantage in professional life, even in daily issues like insurance matters. Yet another issue in biological ethics is species protection. Currently, we already see a major disappearance of many species caused by human activity. Species protection requires some significant investments in order to change the food or energy production so that the diversity of species can be kept. As we have seen in Chap. 3, capitalistic structures as well as corrupt governments are the greatest threat to environment and nature.

Humans as a Bull in a China Shop

But will humans be able to create an ideal world in balance with animals, with ecological reconciliation of nature and living, with appropriate environmental energies and with correct applications of genetic engineering knowledge, or is this all an illusion? In reality, throughout history, man must be seen as a disaster with respect to dealing with nature (as well as violently with himself). Ever since modern homo sapiens has existed on this planet, he has brutally intervened in the "order of nature", demolishing the diversity and beauty of nature to make room for himself and his culture (and killed himself in numerous wars) - which goes back to antique times, perhaps even to prehistoric eras. Here is a short list of examples (there are many more) of disappeared animals due to human activities:

- Shark Eagle. Extinct since fourteenth or fifteenth century
- Schomburgk deer. Extinct since 1938
- Tasmanian wolf. Extinct since 1936
- Aurochs. Extinct since 1637
- Chinese river dolphin. Extinct since probably 2004
- Barbary lion. Extinct since 1920s
- The Dodo. Extinct since 1690

Even further back, but also - at least partly - exterminated by humans (possibly also partly by climate changes; were all extinct around 10,000 years ago):

- Mammoth
- Wool rhino
- Cave lion
- Megafauna

The current extinction of many animals due to human activities is likely to constitute the sixth animal wildlife extinction period on Earth (the last one was 65 million years ago). The most drastic wave of extinction is to be found among insects, with possibly catastrophic consequences for the balance of nature. Scientists say that without the human destruction of nature, such rates of animal extinction that we have seen *in recent years* would have taken many thousands of years, if it had even happened. Is this going to continue in the future? There has been a long historical tradition since Machiavelli to assume man must be bad to survive. However, modern psychology would actually claim the opposite. Man is, for example, much "nicer" than monkeys (who kill each other significantly more often when they do not belong to the same tribe). That said, most humans lack a long-term view, one beyond their individual lives, required for dealing with e. g. climate questions. But humans can learn from past mistakes. For example, after bringing the Central European forest down to about 20% in a period of around 1000 years, humans stopped reducing the forest at late Medieval times, as they noticed the importance for their lives. On the other hand, the Romans went on killing trees in an equally very tree rich country until all trees were basically gone, as we can still see in today's Italy.[12]

The possibly most severe problem for the climate today, after all, is the number of humans living on our planet. In 2000, there were about 6 billion people; 90 years ago, there were 2 billion; today, only 23 years later, there are 8 billion humans living on Earth, with every day about 220,000 people being added. Over 93% of this growth takes place in developing countries, while in many developed countries the number of people is actually decreasing. And soon some developing countries will have advanced enough to have the number of their people go down as well: As of 2023, China's population will decrease, just as India – still fast growing – will overtake China with the number of people living in their country. Under the conditions today, what

---

[12] For an elaboration of this, see: Rutger Bregman, *Humankind: A Hopeful History*, Little, Brown and Company (2020).

is the capacity of the Earth for our species? What is the number of people in order to achieve a stable state (at least in the medium term)? An estimate is naturally difficult because it depends on framework conditions. Many experts in the 1990s thought that 6 billion already far exceed the medium-term carrying capacity of our planet. We are now at 8 billion people and will likely hit 10 billion by roughly 2060. However, there is a chance that growth patterns will change. The key important continent for this - responsible for almost the entire growth - is Africa, a continent with great potential for an increase in the wealth and rights of women. Both, a richer life and women having more to say (among other aspects) about becoming pregnant and have babies, are the two main drivers of lowering birth rates. Both economists and biologists agree upon these conditions.

9

Brain Research Since the 1990s

Significant progress in understanding human (self-) consciousness or a scientific attack on something outside of science?

We now dive into the brain research of the last 30 years, the period with the most significant progress in this scientific area compared to any period before. We will also automatically get into philosophical, in particular ethical discussions, which we will deal in more detail with in the second half of this chapter.

History of Brain Research Until 1990s – A Rather Short Story Compared with What Happened Thereafter

Very few naturalists or philosophers of antiquity attributed a significant impact in understanding to the functions of the brain. For Aristotle e.g., the brain was simply no more than a cooling unit for the blood heated by circulation. He assumed that consciousness and the place of thinking were in the heart. Among the ancient researchers who at least partially recognised the importance of the brain for our thought processes were Herophilos and Erasistratos, the founders of medicine in Hellenism. 250 years BCE, Herophilos already described the anatomy of the nerves and developed an idea of the functioning of the central nervous system.

However, it was not before the sixteenth century that the brain was investigated further, when the Dutch anatomist Andreas Vesalius recorded more precise anatomical structures in the brain. In the seventeenth century, René Descartes formulated the "mind–body problem", which is still hotly debated among philosophers as well as scientists today. This is the debate about how mental states of the mind, consciousness, the soul, or the psyche relate to physical states in the material body and brain. Descartes expressed a clear dualism between both, i.e. a separation of body and mind. From the seventeenth century onwards, researchers examined the brain more and more closely using scientific methods. Anatomists suspected that the brain was divided into a multitude of different regions, each of which performing different tasks. In 1664, English physician Thomas Willis provided the first detailed description of the anatomy of the brain and nervous system in his work *Cerebri anatome* (Anatomy of the Brain). In it, he also coined the term "neurology" (teaching of the mind). Willis discovered two types of substance in the brain, a grey one which comprises the outer area of the brain, and a white one which consists of nerve fibres. He also recognised that the seat of mental functions is in the cortex and not, as Descartes had claimed, in the ventricles (small cavities in the brain). He nevertheless maintained Descartes' strict mind–body-dualism.

During the eighteenth century, brain researchers came across further deficiencies in Descartes' dualism. Growing evidence came up that the brain is the centre and starting point of *all* mental functions. Still, many of the brain researchers' assumptions could only be examined experimentally and (some of them) verified during the following nineteenth century. In 1861, French physician Paul Broca, after examining the brain of a patient with severe speech disorders, located the speech centre in the left frontal lobe of the brain. Shortly afterwards, German physician Carl Wernicke, also after studying patients with speech disorders, determined the sensory speech centre in the left temporal lobe. But what was the connection between brain, muscles and other parts of the body? Physiologist Emil du Bois-Reymond, who we already know from previous chapters in the context of what questions science will never answer, the physicist Hermann von Helmholtz, and his assistant Wilhelm Wundt showed that the brain transports information to corresponding parts of the body by electrical signals. Helmholtz was also able to determine the speed of propagation of these signals (from about 1 to 100 m/s).

Carriers of electrical impulses are the nerves already mentioned by Herophilos and Erasistratos. But how are the impulses being transported in the brain? It was a Spanish and an Italian researcher who succeeded in establishing the neuron theory in detail. With the help of silver nitrate salts, in 1873 Golgi had invented a way to bring nerve cells into clear contrast with other cells and to observe them in good resolution. This enabled him to recognise that the brain is a large network of interconnected cells. Using his method, Golgi was also able to detect small cavities between individual nerve cells but did not attribute any great significance to them. Santiago Ramón y Cajal came to a slightly different conclusion than Golgi: Instead of recognising a network of directly interconnected neurons, which, like the threads of a spider's web, could only have a functional meaning as a whole network, Cajal noticed that the neurons have actually no direct contact with each other. He suspected that the fine open contact points between them observed by Golgi were of great importance. They were to be termed "synapse" by the English neurophysiologist Charles Sherrington in 1897. Cajal formulated the hypothesis that the individual neurons communicate with each other through this contact point. In a large network, the neurons can thus process and store or object information depending on the individual synapses letting through electrical signals or not. At the end of the nineteenth century, the first light had fallen on the structure of our brain. But researchers still did not know any details of the processes involved in the neurons' signal processing. In particular, the transmission of electric excitation from one neuron to another remained unexplained.

In the first half of the twentieth century, neuroscientists succeeded then in gaining an increasingly precise picture of nerve cells and the information processes that take place within them. The result of the processing in their different parts and then in interaction with other cells (a single cell can be connected to 200,000 other ones) determines whether a neuron itself becomes active and sends out a signal further or not. But what kind of signals are transmitted during neuronal information processing? The electronic conductivity of biological cells and their biochemical components is too low for direct electron transport as in metals. According to a model established by the physiologists Alan Hodgkin and Andrew Huxley in 1952, electrical signal transmission in nerve cells occurs through the transport of certain ions, especially sodium, potassium and calcium ions. This flow of electrically charged particles from one cell to another is controlled in the cell walls ("membrane") of the nerve cell by the so called "neurotransmitters" at the synapses: When an electrical signal is transmitted between two neurons, the receiving one determines whether the corresponding ions continue to flow or are inhibited from flowing. Hodgkin and Huxley received the Nobel Prize for Physiology or Medicine in 1963 for their information processing model for neurons.

In 1929, the German neuroanatomist Hans Berger succeeded in recording the electrical signals of neurons in a living human brain. This technique, now called "electroencephalography", EEG, became a key technique in brain research to gain a deeper understanding of the complex interaction of the many neurons in our brain. Increasingly, the scientists' growing knowledge would now also influence thinking about mental disorders and serve the purpose of treating respective diseases.

With the neuron theory introduced by Cajal, neuroscientists had come to a far-reaching understanding of the properties of individual neurons. But how do entire neuron clusters interact with each other? How do the individual parts of the brain communicate with each other? How does the brain function as a whole? And how does (subjective) consciousness arise? Only in recent years have these questions become the focus of brain research again.

In the first 65 years of the twentieth century, the question of consciousness was almost banished from the psychological sciences, in contrast to the pioneering ages of psychology, brain research in the nineteenth century, and in the last 25 years. Its focus was lying on the *un*consciousness on the one hand and the behaviourism that predominated in the Anglo-Saxon world, which only wanted to accept observable behaviour, on the other hand. It was not until the 1970s and increasingly since the early 1990s that brain researchers began to ask again whether consciousness could be linked to certain states in the brain. Can it perhaps be reduced entirely to the material properties of our central nervous system? After the discoveries that the earth is not at the centre of the cosmos (also see Chap. 12) and that humans are the result of a natural evolutionary process, another core question of human self-understanding is at stake here. In contrast to the other two, this one remains open until now.

Early Brain Research as of 1990 – First Insights and Many Problems Left

As of about 1990 the increase in more detailed knowledge about our brain exploded. Neuroscientists thus also refer to the 90 s of the twentieth century as the "decade of the brain". In these 10 years, brain research discovered far more about the brain and its functioning than in the entire history of neuroscience before. It was the time, when scientists successfully started penetrating more deeply into our brain and its structures and have thus come to further understand the physiological basis of mental processes.

Today's brain researchers study the brain on many different levels, from the function of individual molecules in neurons to entire sections of the brain. Better and better EEG recordings of the electrical activity of neurons started giving the scientists an insight into the processes that take place in the brain, for example for the control of simple muscles. More sophisticated imaging techniques were then added to the brain research as of the 1980s. A game changer was "functional magnetic resonance imaging", fMRI, which enabled neuroscientists to visualise working regions in the brain in real time. Now, you can watch the brain at work, so to speak. Another technology that was developed in 1990s and was first used in medicine in the early 2000s is "positron emission tomography", PET. (This technology has a history going back to the 1930s, however, its medical use as well as for deeper science did not happen before the early 2000s). With these new techniques brain research achieved to obtain many more new information.[1] The coloured images of the corresponding brain areas obtained with this method are now also used in numerous popular representations.

There is a rising agreement among brain researchers that the functions of the nervous system are based on communication and information processing between countless individual neurons. In other words, our brain is a huge information-processing machine in which extensive data exchange takes place between cells and cell assemblies by chemical and electrical means. With the growing understanding of how our central nervous system works, brain researchers have begun to ask whether the processes of information processing in the brain cannot also be recreated artificially. In other words, can we simulate the functions of the individual neurons as well as the information processing of the entire system in an artificial network? A new research direction emerged: the simulation of the nervous system by means of calculation across artificial "neural networks" (via a computer), with first efforts in this direction going back into the 1950s. Research on neural networks is closely linked to research in mathematics and computer science.

However, biological neurons are much slower in their information processing compared to electronic circuits: While the processes in the latter today have a switching time of a few nanoseconds (billionths of a second), neurons only work in the range of milliseconds (thousandths of a second). And yet the human brain can solve problems that are - at least for the time being - completely unreachable for artificial electronic machines, and many problems that both are able to solve the brain solves them much faster. How is this possible? The answer to this question lies in the way the many neurons in the brain interact. Their massive interconnectedness is the fundamental prerequisite for thinking, complex behaviour and consciousness. Unlike computers, information processing in the brain does not proceed according to

---

[1] Influential papers by Francis Crick and Christof Koch in the early 1990s are often credited for instigating this turn of events. In particular, they are credited for having defined an empirical approach to consciousness. By focusing on visual awareness, progress was made on consciousness by looking closely at the brain's visual system.

the principle of "one after the other" (the sequential "Von Neumann architecture"), but highly parallel in that countless neurons communicate with each other at the same time. The special abilities of our brain are therefore not reducible to a rigid, rule-based process, but are based on a flexible and adaptive input–output system the components of which are highly interrelated - and which can also change their interaction patterns, i.e. the functions of the respective synapses in letting ions through or not, at any time and many times.

The last point above is another important difference to conventional computers: A network of neurons does not run a finished and fixed programme entered from the outside. Instead, the "programme", the sequence of information processing, must constantly be learned and re-learned by the neural network itself. A particular process of learning discovered in 1949 by the Canadian researcher Donald Hebb plays a central role here: the connections *between individual neurons* change over time. When they are "used" frequently, i.e. the signals go through two neurons via their synapses more often, the connection strengthens; when they are used less frequently, they weaken. Concretely, the nervous system implements this principle by changing the permeability of the neurons' synapses to signals depending on how often the two cells the synapse connect fire at the same time, or how often the firing of one follows the firing of the other. The more frequently two neurons are active with each other, the more closely they are connected, i.e. ions pass through the synapse. To put it succinctly: "What fires together, wires together". Hebb's rule manifests the fundamental importance of synapses recognised by Cajal.

Since the late 1980s, brain researchers have developed an increasingly detailed idea of how this mechanism of information storage and processing, which is so important for living beings and human learning, takes place at the biochemical level, i.e. by the formation and action of neurotransmitters, small signalling molecules transmitting nerve impulses from one nerve cell to the next (initiated by ions).

Research About Our Consciousness - How the Brain Generates Our Mind

Many millions of years of evolution have given rise to very sophisticated systems of perception and information processing in animals. Animal brains are able to analyse complex information from their environment in a fraction of a second enabling them to react accordingly. In terms of size, complexity and performance, however, the human brain is once again far above the most powerful animal brains, and this despite the fact that, according to modern neuroscience, there is no "uniqueness" to be found in our brains. We have nothing that is not already present in preliminary stages of non-human brains. The human brain, with its enormous capacity and the possibility of linguistic communication, is characterised simply by "more quantity" (more neurons in relation to body size) and more "complexity" (more neuronal connections) rather than by exclusive qualities of the neurons.

An essential characteristic of our species compared to (other) animals is that we are self-aware - although this may also apply in weaker forms to highly developed animals as scientists believe today. When we feel something, think about something or do something, we are aware that it is "me" who have these feelings, thoughts or actions. We are also aware of our boundaries. We know what is "me" and what is the World outside "myself". We knowingly ("subjectively") have mental states such as perceptions, emotions, memories and thoughts and can in turn recognise these states as such by looking at them a level higher, i.e. as an observer. Some see this as the root of all kinds of arts. This ability is limited to us humans only. Conscious states occur especially when we compare complex information about the environment and our own body with memories of previous experiences and use them for flexible behaviour planning, e.g. what I want to do tomorrow or next year. This leads to the formation of a mental world in the brain in which we perceive, feel, think and plan with a "sense of self". Such an *inner world* can even be entirely independent of the *outer world*, as we can imagine a world with features that do not exist in reality. For our ability to create an inner world in which I can imagine all kinds of things, however, neuroscientists cannot find a real equivalent, a centre, an "actor" of any kind in our brain. In the opinion of many neuro-researchers, this "I" is therefore nothing more than a fiction - albeit one that is difficult or impossible to recognise as such. Furthermore, our brain generates all kinds of constructions, but processes and presents information on them to our consciousness only very selectively.

Even if today's brain research cannot yet describe or explain the functioning of conscious mental states in detail, it is able to make more precise statements about their functions. There are processes in our brain for which consciousness is required and which therefore always go hand in hand with conscious experience. We can only react to an ever-changing, complex event in a sufficiently complex and variable way if we become aware of it, just as we can only grasp the meaning of a complex sentence by consciously (in this case, "attentively") reading or listening to it. In this case, consciousness is associated with a linguistic expression of our thoughts, feelings, inner states, etc.

Although (or because) they have no universally accepted definition of consciousness, philosophers and scientists distinguish different types of consciousness, often in distinction to more theological terms such as "soul". The most important distinction is between the following three types of consciousness:

1. *Phenomenal consciousness* refers to the pure experience of sensations and emotions. Someone who is in pain, happy or cold experiences this consciously. In a scientific context, this elementary subjective experiential content of mental states is called *qualia*. In colloquial language, qualia describe that perceptions feel a certain way.
2. *Mental consciousness* refers to thought processes. Someone who possesses mental consciousness is able to imagine things, and to distinguish between these exclusively imagined things and real, "objective" ones, thus between illusion and reality, and in turn to reflectively encounter the representations of these within himself. This person can think, remember, evaluate, go through different event scenarios, plan into the future and have expectations of something. In this context, one also speaks of *intentionality*, the capacity of a mental state to refer to a state of affairs (a concept that dates back to the German philosopher and psychologist Franz Brentano in the nineteenth century). This reference can be directed towards an action goal, but can also be evaluative or assessable, i.e. true or false.
3. In addition to phenomenal and mental consciousness - qualia and intentionality -, living beings can also be aware that they have such forms of consciousness. There is a "self" in them that experiences and thinks. This "*self-consciousness*" enables an awareness of oneself as an individual and of his/her physical limits.

Living beings with intentionality and self-consciousness have the capacity to think about and reflect on themselves, to face themselves in thought as in a mirror. As far as we know, this is only possible for humans - while the basics of sensual experience and thus of a purely phenomenal consciousness, for example fear of pain, are very likely also present in more highly developed animals. Only a human can detach him from himself in his thinking, and "abstractly" reflect on him independently of his existence. Once we had this ability (probably around 80,000–100,000 years ago in our evolution), this ability made us go through a cultural evolution of our own, which was recorded in one form or another in "stories" in the various human societies,[2] in pieces of art, in writing systems etc. Lastly, it also leads us to what we broadly call "philosophical questions". Collectively, philosophers also refer to this quality as the "human mind". In the fundamental question of how consciousness acts and how it relates to our body, especially the brain, we recognise Descartes' mind–body problem.

About Our ("Ego-")consciousness – Fundamental Open Problems

Leaving aside medical research for the purpose of treating brain diseases, it can be said that research into the neurobiological basis of consciousness is one of the last remaining major tasks of neuroscience. Brain researchers are optimistic about this. With the materialistic view characteristic of natural scientists, they are convinced that everything in nature, including our consciousness, can eventually be reduced to material connections (this idea is not necessarily new: it already appeared in antiquity with Democritus and Epicurus). However, this does not mean that our consciousness can necessarily and solely be explained by the firing of a few neurons. Rather, we should assume that consciousness is a special physical - albeit very, very complex - state in our brain. It obviously only occurs under certain material, energetic and functional conditions, such as those that prevail in sufficiently complex brains. Thus, consciousness could be the result of a mechanism that, although it runs according to known physical and biochemical laws, causes phenomena and processes that have their own lawful behaviour. As we saw in Chap. 6, scientists speak of "emergence" - referring to a principle from the physics of complex systems or biology - an entity that is observed to have properties its parts do not have on their own. Accordingly, neuroscientists call their position "emergent materialism" (a very highly emergent materialism).

However, numerous sceptics contradict the optimism of neuroscientists. They consider it impossible to grasp the essence of consciousness scientifically. Their essential argument is as follows: While natural science can deal with all phenomena that occur objectively (i.e. independently of the observer), consciousness harbours a *subjective world*. Such a world cannot be grasped by the methods of science. No one can describe the consciousness of another person, because they are all subjective, i.e. an inner state with no objectivity.

---

[2] For more details see David Graeber, David Wengrow, *The Dawn of Everything - A New History of Humanity*, Macmillan USA (2021).

Nor can we imagine the perceptual world of an animal - to which the philosopher Thomas Nagel referred in a seminal article he wrote in 1974 under the title "What is it like to be a bat?".[3]

In reference to the first two forms of consciousness above, sceptics name two fundamental problems: the *qualia* problem and the *intentionality* problem. The former is that there is no simple discernible connection between neural states and subjective mental contents of experience (qualia). But without this connection, the obvious question arises as to how it is that we experience anything at all (subjectively) when certain neuronal processes take place in the brain. When pain arises in the body, certain stimuli are sent to the brain, which are processed there and finally trigger corresponding reactions in the body. However, it is by no means necessary and compelling that also a conscious experience of pain arises in the process. In other words, the conscious subjective experience of pain cannot be explained by the properties of the corresponding neuronal state of the pain. Intentional consciousness cannot solely be described by corresponding neuronal patterns: Intentionality refers not only to the "directedness" of consciousness but also to the "truth-value" of mental states. Thoughts can be true or false. Neuronal states, on the other hand, cannot be described with truth criteria. They simply exist and disappear. Accordingly, it will be difficult to find a neuronal correlate to intentional consciousness. Neither qualia nor the intentionality of consciousness should thus ever be fully explained in reductionist terms by natural science, say the sceptics. They see in these problems the expression of a central obstacle to any purely materialistic worldview of natural sciences.

Their counterparts from the camp of neuroscientists argue against the sceptics that their position has many historical antecedents, each of which failed sooner or later. All too many problems in the history of natural research have been deemed "inaccessible to any systematic science", just to be proven later, as soon as science had developed sufficiently. Examples are the atoms, the history of the origin of man and the structure and development of the universe. And already today there are no doubts as to whether it is possible to explain phenomena and capacities such as seeing or memory neurologically. Although neither the qualia nor the intentionality problem has yet been solved scientifically, brain researchers claim that one day they will be able to explain consciousness in all its facets by scientific means.

But can the points of the sceptics be dismissed that easily, i.e. by only looking at parallel developments in the history of science? Neuroscientists know that every form of information processing in the brain is related

---

[3] Thomas Nagel, *What is it like to be a bat*, The Philosophical Review, 83, 4 (1974) p. 435–450.

to specific patterns of neuronal activity, "neuronal correlates". A central motive of neuro-research is the search for the neuronal correlates of the different forms of consciousness. Neuroscientists devote themselves with particular priority to researching conscious sensory perception, i.e. phenomenal consciousness. The first task is thus to clarify the qualia problem. One can divide it into two sub-problems: The first is to grasp the neural correlates of sensory perceptions, i.e. how the feelings connected to seeing, hearing, smelling, tasting and touching are mapped in the brain. As a rather simple problem, it does not yet involve the core of the qualia problem. The more difficult part of the problem involves the direct neural correlate of the *consciousness* of such sensory perceptions. Seeing, hearing, smelling, tasting and touching could theoretically (and practically does in many animals) come without a *human* form of consciousness. The researchers are still quite far away from solving this qualia problem.

More Methods and Results of Research on Our Consciousness

In the study of conscious sensory perception, neuroscientists have two different research methods at their disposal. The first comes from brain pathology. People with brain damage can suffer from numerous impairments, some of which touch on the brain's ability to generate conscious experiences. The affected patients then no longer consciously experience certain perceptions and are thus massively restricted in their phenomenal experiences. The most famous person to have survived severe damage to the brain but continued to live is probably Phineas Gage, who on the 13th September 1848 had by an accidental explosion a tamping iron gone through his head. He survived, and a few months after the accident, he went back to work. Before the accident he had been their most capable and efficient foreman, one with a well-balanced mind, and seen as a shrewd smart businessman. Afterwards he was fitful, irreverent, and grossly profane, showing little deference for his fellows. He was also impatient and obstinate, yet capricious and vacillating, unable to settle on any of the plans he devised for future action. He is the first patient from whom we learned something about the relation between personality and the function of the front parts of the brain. Today, such functional limitations of the brain can be reproduced externally by strong magnetic fields. This makes it possible to directly examine the significance of the affected areas and to localise the regions in the brain specific to certain consciousness processes - as well as possible connections of conscious experience to other, for example emotional states.

As of the second method, researchers explicitly measure neuronal brain activity. In doing so, they benefit from the ever-improving imaging techniques that have been developed over the past 20 years, giving researchers increasingly detailed insights into the activity patterns of neurons in the living brain. In particular, the above mentioned fMRI techniques are instrumental in describing the local neural processes involved in experiencing sensory experiences and emotions. Still, the temporal and spatial resolution of these procedures is not yet sufficient to explore the highly complex states and complicated interactions between different, often non-adjacent, brain regions in detail. Here, the researchers are counting on future technological advances.

The neuroscientists benefit from the fact that the brain is largely functionally structured: Its individual parts are each responsible for very specific tasks (something Broca and Wernicke had already discovered in the nineteenth century) - even if ours has amazingly great flexibility about the functions of individual brain areas. Most experiments investigating the neural correlates of sensory perception involve researchers setting an awake person certain mental tasks, such as recognising colours, and using imaging techniques to measure the patterns of neuronal activity that then occur. The excitation patterns of individual or entire groups of neurons can then be directly related to certain sensory perceptions, such as the recognition of colours and shapes or the hearing of a sound. If the test person registers a certain external pattern, e.g. a certain angle in a geometric shape, or a certain colour, a certain group of neurons becomes more excited. Damage to the corresponding areas in the brain leads to the inability to perceive the respective shape or colour. Other mental phenomena such as learning, memory, emotions and problem solving can also be clearly identified: There is a direct connection between them and a certain activity of neurons.

However, the vast majority of processes in the brain take place without an accompanying conscious mental state. A major challenge remains to distinguish these from neuronal processing directly linked to conscious states. It is thus precisely the distinction between conscious and unconscious states that proves to be particularly difficult in the experiments. So far, neuroscientists have only been able to find out how a certain sensation or a certain cognitive function, e.g. emotions, a particular language or a personal decision, is reflected in the neuronal patterns. They were able to record how the brain distinguishes stimuli, processes information, determines reactions to it, produces verbal reports of experiences, and so on. For instance, they could show that musicians activate other parts of the brain when hearing music than non-musician.[4] But they have not yet been able to answer the question of when, how and why all this is associated with conscious experience. Thus, the qualia problem remains largely unsolved, not to even mention the intentionality problem. Whether neuroscientists will be able to provide insights into this crucial question in the next few years is open - and decidedly exciting.

First steps in researching the neural correlates of conscious sensory perception are already emerging. One example is the phenomenon of "binocular rivalry". It occurs when the two eyes of a subject are presented with two different images that cannot be integrated into a unified image, such as a horizontal red bar to the left eye and a vertical green bar to the right eye. The result is that the person alternately perceives consciously a red horizontal bar or a green vertical bar in total, but never both together. Our brain seems to be unable to consciously control changes in the perception of the red and green bars. Measurements of the neuronal events during the binocular rivalry provide insights into how the (well known) neuronal correlates of the respective perceptions "behave". It is particularly interesting to see how the change between unconscious and conscious perception of the red or green bar is represented at the neuronal level. A large part of the involved neuronal activity remains unchanged. This shows that the neurons map both images at the same time, even if only one is consciously perceived. The neuroscientists succeeded in identifying the neuron groups that become active precisely when the spontaneous switch takes place. They displayed the patterns that must be specifically responsible for the respective conscious perception of one bar. It will be up to future studies to identify the more complex and large-scale neuronal patterns in the brain that map the numerous other processes of consciousness.

Some physicists suggest (rather speculatively) a connection of consciousness and quantum physics. Based on the subjectivist interpretations of quantum mechanics, the possibility that consciousness could have an influence on the state of matter via the measurement process was considered early on. The English mathematician and theoretical physicist Roger Penrose has turned the tables. Since the 1990s, he has been developing a theory together with biologists according to which consciousness only arises thanks to quantum effects. This still is an open discussion with ongoing, newer experimental studies (e.g. a 2022 experiment that might indicate quantum entanglement in certain areas of the brain[5]) - but no clear evidence, yet. Most

---

[4] Lutz Jäncke, *Macht Musik schlau? Neue Erkenntnisse aus den Neurowissenschaften und der kognitiven Psychologie.* Verlag Hans Huber, Bern (2008).
[5] https://iopscience.iop.org/article/10.1088/2399-6528/ac94be. Quote from the abstract: "Those brain functions must then operate non-classically, which would mean that consciousness is non-classical."

brain researchers, however, are sceptical about such explanations. They see no need to speculate about the significance of quantum effects in the brain whose even smallest functional components (cells) are way beyond the sizes on which quantum effects occur. Consciousness should thus be detectable without quantum theory.

The problem of consciousness has yet another dimension that shall be mentioned here and elaborated further in this chapter: How, where and why did consciousness or even *self*-(ego-)consciousness arise in the first place? From previous research into sensory perception, we know that human and animal brains are able to map perceptions and experiences from the external world in neuronal structures. What role do subjective feelings such as fear, aversion, disgust, or happiness, which often involuntarily accompany such perceptions, play here? What are their evolutionary advantages? Furthermore, as we have already emphasised, the human brain is capable of abstractly recognising its carrier, i.e. the individual him- or herself, outside of his own body, and thereby of recognising the body as being endowed with its own consciousness. Both, the feelings associated with sensory perception and our ability to abstract ourselves (our *self*-consciousness) must have been a significant advantage in the evolution of man (see below).

Last but not least, consciousness is also of great importance in the sexual drive and the selection of a reproductive partner. Both are mostly conscious in humans and involve deep feelings. What the poets like to call love between two people would hardly be possible without self-consciousness and feelings. Thus, it appears quite advantageous for both survival and reproduction to have a self-consciousness.

And Yes, It Does Change - The Plasticity of Our Brain

Until the late twentieth century, brain researchers attributed plasticity and the possibility of changes in the brain on a larger scale only to the immature brains of children. The adult human brain with its many neurons and their connections they regarded as statically organised. Once formed, the neurons and synaptic connections should be fixed and hardly subject to renewal. Already Cajal had advocated this thesis. In the last 25 years, this view has been strongly challenged. In 1999, the neuroscientist Elizabeth Gould succeeded in demonstrating that new neurons are generated in adult primates.[6] The

---

[6] Elizabeth Gould et al., *Hippocampal neurogenesis in adult Old World primates*, PNAS, 96, 9 (1999) P. 5263–5267.

debate about neurogenesis (the formation of neurons) in the adult brain has not yet found a conclusion. But evidence that neurons are continuously produced even in human adults has grown in recent years. At the same time, studies with patients with severe brain injuries show that the human brain and nervous system have an astonishing regeneration potential. Damage to specific functions in the brain can be compensated over time with other parts taking over. These insights led to numerous follow-up studies and finally to a shift in our understanding of our thinking organ, away from a static to a dynamically organised nervous system. Today, neuroscientists know: The adult brain is much more flexible than what researchers had assumed for a long time. In possible functional and structural reorganisations of the brain, they recognised a basic principle of its operation. This provides our brains the distinctive ability to adapt to new challenges and accordingly a potential with the name "neuroplasticity". Like the body in sport, the brain depends on how active it is and how much it is used. It can be trained: "Use it or lose it" is how neuropsychologists sum it up today. This principle can also be observed directly with the ever-improving brain imaging techniques that have been available for some years now: Reuse of mental functions leads to an enlargement of the responsible brain areas and thus to improved functioning.

A comparison with a computer illustrates the significance of neuroplasticity: A modern computer with a clock frequency of several gigahertz (one billion operations per second) is vastly superior in speed to the human brain with only one kilohertz clock frequency (1000 operations per second). Nevertheless, our brain is unbeatable in terms of performance, as unlike the computer, it is not based on rigid circuits and connections like the "electronic brain". In the human control centre, the connections between nerve cells are rather being constantly changed, newly created or broken down. The strengths of their connections at the synapses adapt to the strength of the stimuli. The brain's plasticity is thus a crucial prerequisite for an independent learning - without external programming like for computers - and a flexible action-oriented memory (and not a static one). However, the neuroplasticity of our brain comes with an unpleasant consequence. Just as new ways of thinking and behaving can be trained, destructive and disturbing thoughts and behaviours such as addictive behaviour or certain emotional patterns become entrenched through constant and repeated occurrence. Trauma, as can occur in childhood after e.g. severe war experiences, chronic pain, or domestic violence can be deeply imprinted in the neuronal linkage structure of our brain.

Key Technologies – "Improving" Our Minds with Neuro-Enhancements

The more we learn about the genetic, chemical and neurological backgrounds of feelings such as trust, compassion, forbearance, generosity, love and faith, the more we will be able to use this knowledge to manipulate ourselves and others using respective technologies (see point VII in Chap. 5 which we shall elaborate here in more details). The more deeply neuroscientists understand our brain's functioning and the processes of our decision making, the more precisely someone can potentially influence how we feel, experience and think. Let us take a closer look at some of the implications of this:

- Today, mental illnesses are already being alleviated using neurotransmitters' manipulating techniques. However, this does not necessarily end at people with illnesses. The mental performance of heathy people can also easily be manipulated - keyword: brain doping (the public media has already picked up the issue of electrical manipulation of the brain with this resounding phrase). Brain doping with "smart drugs" aims at enhancing concentration, alertness, attention, memory, learning, mood and other mental functions in us. In general, it aims at improving our cognitive skills or psychological well-being in a targeted manner with the goal of increasing these beyond a "normal" measure. Furthermore, the idea is to minimize certain difficult personality traits, like shyness, low self-esteem, etc., or, most prominently, to compensate stress. Brain doping is often used, for instance, among students and people working in demanding professions. It is also used "for improving athletes'" performance in their respective discipline - where psychological effects play a major role. Unfortunately, all these attempts at doping today still have significant side effects (that are disclosed in small writings at the packaging of the medicine) such as confusion or paranoia, drowsiness and balance headache, insomnia, abdominal pain, diarrhoea, vomiting, itching, and many more, sometimes even strong ones like psychosis, heatstroke, dangerous arrhythmias or even death. We have seen young football players dying suddenly of heart strokes. What they had taken (or if their cases were just statistical outliers) often remains in the dark. In the future, more and more sophisticated as well as manipulative smart drugs will be available, and more and more people can be expected to take it.
- Direct stimulations of desired areas in our brain by neuro-electrical impulses are equally already available (as have been tried for two centuries). Besides being used in clinical practice during surgical evaluation of patients

(e.g. epilepsy patients), it can, also change or controls moods, attention, memory, learning, self-control, willpower, comprehension, sexual desire and much more. It can also be expected that electrical stimulation will advance significantly in the next few years and also be available for healthy people. This creates the same potential influences as brain doping.

- Scientists are working on micro- (or nano-) chips that can be implanted in our brain, so that they can permanently improve our state of mind, raise our sense of well-being, increase intelligence, memory and concentration, or even provide lasting happiness. It is in fact not improbable that one day people will wear technical components in their bodies and brains in order to improve their performance. Conversely, machines could contain biological components (called "wetware"). Thus, they will go far beyond the soma drug from Huxley's *Brave New World*.
- Brain researchers are already making brains and machines interact, as we have seen already in Chap. 4. Using, for example, brain-computer interfaces (BCIs), they are transferring content from a person's brain (given by electrical impulses) to a machine so that the machine in turn assists the person with various tasks. It is possible to create messages purely from the thought of a person: An EEG electrode cap records the brain activity of the sender while she browses through the relevant letters on a screen. Her brain signals are then converted into the desired message. Such "brain-typewriters" are already commercially available. This technology will be very useful for entirely paralysed people for them to communicate with the outside world. But it will possibly first be broadly used in video games to make those even more experiential with artificial sensory impressions projected directly into the brain. Even given their still experimental state, we can expect that in the future these technologies will be sophisticated enough to influence and manipulate our brains very deeply.
- It is even conceivable to bring different human brains into direct contact with each other via neuro-interfaces with electrical signals and thus connect several brains into a collective consciousness. It is hard to imagine what type of influence on our life, such direct thought and behaviour interaction with others will be like, if our brain does not act independently anymore. Who can deal with not having any thought or emotions that today he or she would rather keep to him/herself?[7]

---

[7] Andreas Eschbach describes such future scenario in a novel: In his trilogy "Black Out", "Hide Out" and "Time Out" (2010–2012) Eschbach describes a world full of people whose minds interact directly via brain implanted chips. The individuals are forced to exist within a central hyper-consciousness, the so-called "coherence", which simultaneously controls all their thoughts).

These interventions in our brain by respective tools have a significant impact on our perceptions and at last on our entire lives. Throughout human history how we look at and feel about ourselves and our environment has been solely dependent on stimuli from the external world. Light waves reach our eye, sound waves our ear, smells our nose, touches our skin, food our taste. New technologies of consciousness no longer need a "real" reality. By playing virtual realities (VR) into our brain, our perception can be reshaped at wills of others. We might even live in entirely new worlds (there has already been "land/ground" sold in such worlds). Direct access to the neuronal structures in our brain will in the future enable advanced targeted manipulation of emotions, thoughts and intellectual functioning.

However, virtual reality is already a powerful augmentation of the internet, particularly in online games. Our manipulability through VR grows dramatically once it ensures that our brain releases the appropriate combination of neurotransmitters that evoke in us what we crave most: Rewards, pleasures, and feelings of happiness. This goes straight into our innermost being: our Ego. By intervening directly in the functional layers of our self-model, evoking entirely new states of consciousness, we can not only lose ourselves in it, but we can also lose our selves. External algorithms then no longer control only our actions, but likewise our thinking, feeling and experiencing. And finally, even our will. This and the permanent connection of our brain to machines could eventually take away our independence. The ideas of Aldous Huxley for the year 2540(!) will already be possible to be implemented only about a hundred years after he published the *Brave New World* (1932).

More Philosophical Questions

As we already saw in Chap. 4, the neuroscientists dealing with consciousness cannot avoid some serious philosophical questions, epistemological as well as ethical ones. However, the turn of the brain researchers towards consciousness in the late 1980s has conversely also brought some important movement into philosophy. A new philosophy of mind directly related to empirical scientific tests and results can be seen as the first one of its kind in this field. But how can we ever put subjectivity, the phenomenological level, i.e. our inner experience as such with all its multi-layered facets and variations, into a scientific and therefore objective model? Both, scientists and philosophers, mostly agree that the problem of consciousness represents a completely new type of scientific – as well as philosophical - question.

This begins with the dualism between subject and object that we are so familiar with: Us here, the objective world out there, separated from us. The philosophical problem is to understand how we are able to experience subjective events "objectively". What is it that we experience when we describe ourselves as dynamic, subjective "universes" that have something like a centre, namely our own awareness, and see ourselves facing an objective universe? Why do we experience anything subjectively at all? In other words, what is it exactly that gives itself to be shown in us as "I"? Who or what is this entity that has these subjective experiences? And how can subjective sensations be described objectively at all? How can I feel to be hurt when I bump my head?

We know that subjective consciousness has an objective history of origin: its historical evolutionary development. In the long process from single-celled organisms to homo sapiens, mind and consciousness emerged at some point. The brain growing more and more complex was most likely a gradual process. The emergence of consciousness was then either a sort of phase transition or equally the result of a gradual process. There are important differences with particular human genes to the monkey's genes that have otherwise between 96 to 99% (depending on the species) congruence with our genes. On one hand, this speaks for an evolution of our consciousness. On the other hand, how small kids get to human consciousness happens in steps (new-borns clearly do not have it). Thus, consciousness in new born comes as "phase transitions" (see the work of Jean Piaget and his theory of cognitive development[8]). But how this happened some 70,000–100,000 years ago and eventually led to our self-consciousness today is still entirely unclear.

The ability of evolution to explain the human consciousness is particularly the interest of philosopher Daniel Dennett. His view is materialist and scientific. He even presents an argument against using the word qualia arguing that its concept is so confused that it cannot be put to any use or understood in any non-contradictory way, and should thus not be used in scientific research.[9] But does that really approach the structure of our mind?

With the help of today's imaging techniques, neuroscientists have meanwhile reasonably well been able to grasp the internal processes in the brain that take place when we feel pain. But they do not have the slightest idea *why* something hurts, why I experience the subjective pain when I bump my head. Why do all these processes not just happen without consciousness? In simple words: Conscious experience has subjective components that need to

---

[8] Bärbel Inhelder, Jean Piaget, *The Growth of Logical Thinking from Childhood to Adolescence*, Basic Books, New York, (1958).
[9] Daniel Dennett, *Sweet Dreams: Philosophical Obstacles to a Science of Consciousness*, MIT Press (2005).

be explained objectively. That this experience along with a "Me-feeling" is not self-evident, is illustrated by numerous examples of neurological diseases and exceptional mental states such as perceptions under the influence of drugs and hallucinogens. The qualia must be explained by a scientifically objective theory. However, no physical or neuronal event can make comprehensible how and why this experience happens. All previous attempts to link physical and mental states have an "explanatory gap" - just like in the evolutionary beginning with an information in genes, as we saw in the previous chapter. The question of the connection between subjective and objective states is what the Australian philosopher David Chalmers calls the "hard problem of consciousness";

> ...even when we have explained the performance of all the cognitive and behavioral functions in the vicinity of experience—perceptual discrimination, categorization, internal access, verbal report—there may still remain a further unanswered question: Why is the performance of these functions accompanied by experience? A simple explanation of the functions leaves this question open.[10]

He compares this (similar to strong and weak artificial intelligence) to the "weak problem of consciousness":

> The easy problems of consciousness are those that seem directly susceptible to the standard methods of cognitive science, whereby a phenomenon is explained in terms of computational or neural mechanisms.

One example of the weak problem is the above-mentioned internal processes in the brain when we feel pain, while the *subjectivity in our experience* of pain, i.e. the I that subjectively feels is a hard problem. Hard problems relate to the intentionality problem discussed before, weak problems more to the qualia problem (although that classification is not always 100% applicable in that scheme). But, other than the weak form in AI, the qualia problem is anything but solved or even close to it, as we saw above.

Some philosophers and scientists argue that consciousness is an emergent phenomenon of the human brain and therefore cannot be reduced to the fundamental laws of its neuronal structure. They thus advocate a new "non-reductionist physicalism", a "property dualism", or an "emergent materialism". According to them, consciousness is based on a mechanism that runs

---

[10] David J. Chalmers, *Facing Up to the Problem of Consciousness*, Journal of Consciousness Studies 2(3) (1995) p. 200–219; also available at https://www.consc.net/papers/facing.html.

according to physical and biochemical laws, but in its dynamics has states and properties of its own. In other words, the activity of consciousness emerges from the common patterns of activity of a very large number of neurons, whose complex dynamics enable a new form of collective self-organisation. Instead of acting locally in a particular part of the brain, it emerges through the entire complex interplay of some ten to 15 billion neurons with their ten to a hundred thousand times more numerous connections. This would ultimately also explain the "unity of experience", i.e. the fact that we perceive the most varied and diverse simultaneous sensory impressions as a *single coherent unit*. Capturing this precisely with comparatively simple models is extremely difficult.

Our Inner Model as Virtual Reality

An important scientific as well as philosophical insight comes from our experiences with VR: From this we already know that the human mental self-image (the philosopher Thomas Metzinger calls this the "phenomenal self-model" (PSM)[11]) is anything but stable. It is relatively easily shaken up, as shown by mental diseases, brain injuries, phantom limb pains, VR games, or fairly simple deceptions of our self-model as experiments like the rubber hand illusion (see below). Incidentally, experiences with hallucinogenic drugs also lead to this insight.

The insights of recent consciousness research go even further[12]: Our entire inner model of ourselves can be seen as of a virtual nature, and thus also the reality that our sensory impressions create for us. This idea is not brand-new. Various philosophers have emphasized the virtual character of our experiences, as the following examples show:

- The oldest known philosopher who dealt with this problem is Parmenides (sixth/fifth century BCE). He expressed the view that changes or destructions of that which exists is unthinkable (as well as that from nothing existing precisely nothing can emerge, which is often summarised by the formula *Ex nihilo nihil* - "Nothing comes from nothing"). Any change we perceive for what exists is thus just a subjective (i.e. virtual) experience.

---

[11] Thomas Metzinger, *The Ego Tunnel: The Science of the Mind and the Myth of the Self*, Basic Books (2010).
[12] Olaf Blanke, Thomas Metzinger, *Full-body illusions and minimal phenomenal selfhood*, Trends in Cognitive Science, 13 (2009) p. 7–13.

- In the seventh century the Buddhist epistemologist Dharmakīrti associated ultimate reality only with perceptions of our senses.
- In the eighteenth century the English philosopher George Berkeley said that reality is constructed entirely of immaterial, conscious minds and their idea.
- The most famous philosopher of recent centuries, Immanuel Kant, in his transcendental idealism equally defined the world as we perceive it, as we can only have knowledge of things we can experience. The phenomenal world of objects bearing spatial and causal relations is merely ideal and not necessarily absolutely real.

Neurobiological research today gives us clues to a similar conclusion: The "Ego" seems to be nothing more than a mental construction - albeit one that is difficult or impossible for us to recognise - that is created by our brain in order to selectively and effectively represent and process information. What we perceive as the external world is a projection of what is going on inside us. There are no colours,[13] sounds, etc. in the "real world", our brain rather transforms electromagnetic waves in the frequency range of 400 to a 700 nm into the colour spectrum, which is then seen intersubjective (cannot be entirely subjective). Our brains produce what modern philosophers call a "model of the world" (which is inter-subjective). This should not be considered as a modern version of Plato's cavern, as a dynamic representation of an assumed external reality - as every model is, while reality per se is not necessarily known here (or at least not fully, we would like to leave one of the most profound question in philosophy open in this book, as it is not the main subject here[14]). We do know however, that we constantly modify our perception through our interactions with some reality outside us.[15] Moreover, this model of the worlds has some universal features we all feel. Coming

---

[13] As Ludwig Wittgenstein famously noted in his *Tractatus Logico-Philosophicus*, published in 1922: (Statement 2.0232) „Roughly speaking: objects are colourless."

[14] The question is much discussed in philosophy, from Plato onwards. Kant's philosophy was a turning point. Popper turned again to direct realism. However, the question remains open. The idea that our senses provide us with direct exposure to objects as they really are, i.e. we directly experience reality per se is often called "naïve realism". Naïve realism is often put as a contrast to "indirect realism", the idea that we do not perceive the world externally to our perception, as it really is, but know only our ideas and interpretations of the way the world is. Naïve realism is, however, the basis of the scientific method as we construct the theories based on our direct perception of some reality and we correct our theories based on experimenting within that reality.

[15] George Berkeley in the eighteenth century, thinks that nothing material exists outside what we directly perceive, as it never exists independent of our perception, i.e. nothing exists without being perceived by us. A pure subjective idealism like this one sounds quite absurd, it is, on the other hand, impossible to state that we grasp "reality" per se. All stays on the modelling side of "some" reality.

back to the example of colours, when we all wonder in front of a beautiful painting by van Gogh, it becomes clear that he was able to choose in his brain colours that speak also to ours. Similarly, when we find beautiful a statue of a pharaoh sculpted 3000 years ago, the artist had a similar sense of beauty to ours.

At the same time, any model of the world creates a conscious self, the Ego, as the content of a self-model that helps us to process information as efficiently as possible, to make appropriate predictions and to interact socially. In doing so, our brain allows us to be "naïve realists", i.e. we believe that the world is in reality as our brain represents it. All of this is happening inside us, but interacting with something not inside us, which we call "the world". There is no such thing as an intrinsically existing, irreducible Ego. What is, is the experienced sense of "I-ness" and the constantly changing contents of consciousness of ourselves, generated by a material entity, our brain. Still, we find a general "Ego" that is present in every (mentally healthy) human being who can thus communicate about it. We need to further understand the properties of this general Ego, in order to deepen the understanding of our self-consciousness, because the model of the world is not simply Plato's cavern, but is likely directly connected to some reality (which we cannot experience in all aspects, see above) and has universal properties transmissible among human beings (i.e. being inter-subjective).

Our Mind and Self-Consciousness – More Empirical Studies, Dramatic Applied Technologies and - yet Again - Ethical Issues

Nowadays, theories about our mind can at least partly be investigated and founded empirically. This is most often done by connecting the conscious self-model in our brain via brain-computer interfaces (BCIs) with external systems, for example with virtual bodies. Such connections are increasingly realised experimentally. In fact, the human phenomenal self-model can by now attach itself in many ways to artificial senses and organs. We then actually feel the artificial object as a part of our body. Neuro-researchers speak of a "virtual embodiment".

The most well-known experiment is the third-hand illusion. Here the participant observes an external rubber hand being touched in synchrony with touches applied to his real hand. The latter is hidden out of view. This creates an illusory experience that the applied touches are felt on the rubber hand and that the rubber hand is thus one's own hand.[16] This experiment clearly shows how our perception can be manipulated. This does not only apply to hands, but the entire body: The "body transfer illusion" defined by the sense of owning either a part of a body or the entire body other than one's own has been experimentally shown in various ways.[17] In Such experiments the visual perspective of the subject was manipulated. This is what results in an illusion of transfer of body ownership: People can in their experience be put into the body of a child or that of a six-armed creature, their visual perspective can be separated from their body, or our own heartbeat can be made visible. This demonstrates that visual and sensory signals correlate to the perceptions of the subject's body. Thereby not only becomes the identification with situations or people in the virtual world stronger over time, such experiences also change the perception of oneself in the real world. This appears to be a revolution in our views on ourselves as well as the world. As a consequence, we might mix virtual reality and the world with each other.

We thus recognize a new powerful development from neuro-research: A "consciousness-technology" (as we may call virtual realities technologies). In there, virtual reality and neuro-technologies can cause new states of consciousness by acting directly upon the functional layers of our self-model. Virtual embodiments outside our biological bodies may soon enable fascinating applications in which our self-model couples to corresponding artificial body images, the avatars we know already from Chap. 4. Avatars and the recording of brain activity will then enable the thought-based control of all sorts of robots and other machines. We might then regard avatars as forms of out of body experiences, which people can also experience spontaneously (and for which we see evidence in much of the mystical and religious literature).

The fact that with a suitable setup we can very easily identify ourselves with an avatar puts a new light on us as human beings. The transmission of specific information directly from our brain to a machine or robot/avatar is no longer pure vision or science fiction but might soon become an everyday reality. The next human generation will do tasks, learn new skills, and pursue certain pleasures (including sexual ones, possibly even among the first ones, as that was the case with the internet). Companies could use virtual reality to make products palatable to their customers, customers would be allowed to

---

[16] Matthew Botvinick, Jonanthan Cohen, *Rubber hands 'feel' touch that eyes see*. Nature, 391 (1998) p. 756 online accessible https://www.nature.com/articles/35784; Arvid Guterstam, Valeria Petkova, Henrik Ehrsson, *The Illusion of Owning a Third Arm*, PLoS ONE 6(2) (2011); online accessible: https://doi.org/10.1371/journal.pone.0017208.

[17] See e.g. Olaf Blanke, Thomas Metzinger, *Full-body illusions and minimal phenomenal selfhood*. Trends Cogn. Sci. 13 (2009) p. 7–13. https://doi.org/10.1016/j.tics.2008.10.003.

try them out. In particular, so-called augmented reality (AR) could play an important role. It consists of a real environment replicating and improving VR with virtual elements, in the future likely across all five sensory modalities: visual, auditory, haptic (sense of taste), somatosensory (sense of touch) and olfactory (sense of smell). For example, in AR, information about objects or operating instructions for devices can be superimposed on the field of vision. A future virtual reality will rival the real world - whether we like it or not.

By intervening directly in the functional layers of our self-model, VR and AR could soon give rise to entirely new states of consciousness. What we see, hear, smell and feel is then no longer bound to external influences ("realities"). Virtual embodiments outside our biological body and "real" perceptions make fascinating applications conceivable in which our self-model is coupled to artificial sensory and action organs. First approaches to controlling all kinds of robots and other machines directly with our minds using an avatar and recording our brain activity already exist.

As fascinating as these technologies are, as intimidating are some of the conceivable scenarios they lead to.

> The possibilities of soon being able to move in virtual environments almost like in the real world, comes with yet unimaginable consequences for our psyche and self-perception.

says Thomas Metzinger in his "Ego-Tunnel" and lists possible associated risks, such as psychological manipulation, hallucinations, personality changes, or influencing the subconscious, all of which our society has paid far too little attention so far. As our society responds hostile to the use of mind-altering substances (drugs), we will need to test corresponding reaction patterns also regarding virtual realities. "Which brain states shall be legal in the future?", Metzinger asks specifically.

There are in fact some severe ethical questions that come up: What happens, for instance, when with the help of avatars and purely through power of your thought robots do something, you consciously would not do or want to happen, e.g. become violent because the robots realize your innermost fantasies of violence? Are you then responsible for this? Or may we soon control drones or war robot with the help of avatars? This would be a new kind of warfare. With the US drone attacks in Pakistan and other countries, the first directions of such have already begun.

However, the most dramatic development would be, when several people can get mentally connected via BCIs and virtual embodiments. This is not as far away as many would think. With monkeys this has already been achieved. The Brazilian neurobiologist Miguel Nicolelis had monkeys networked via

BCIs in separate cabins and control a virtual arm solely by their thoughts (this the monkeys learnt quite quickly) to work on a joint task. Each of the three monkeys was able to control only two of the three dimensions, which were displayed to him on a screen. In order for the arm to move in the right direction, the three monkeys had to learn to work together in their minds. In the monkeys' effort to reach a certain goal the neurons in their brains fired soon *in resonance* as the scientists were able to measure. This means nothing less than that the animals had synchronized their brain activities. The neurons in the three monkeys' brains behaved as if they belonged to one brain.[18]

There are concluding statements based on these technological developments: If we believe that all projections of our inner world are exact copies of the outer world, that all this is also happening outside and independent of us, we find ourselves in what we should call - following Thomas Metzinger - the "ego trap". But can we even do otherwise? Or is the perceived reality not after all really a part of the reality per se? Our brain and presumably also self-consciousness are, as we saw already, the product of an evolutionary development. During the developmental past, the external circumstances for our ancestors and thus also their sensory impressions and sensations were anything but constant. The particular world that our present-day perceptions impose on us is thus the one that allowed us to survive (and reproduce) best in our evolutionary past. Our self-models enable an evolutionary view on their own origins, by recognizing them due to their survival (including social) power as specifically advantageous in the "cognitive arms race" across species. In other words, phenomenal and intentional self-models gave our ancestors great advantages in the battle for survival and reproduction, as well as perceptions, emotions, and processing within us that corresponded to the conditions for a quite particularly perceived reality which arose as optimal during our evolution. They made our perception of the world highly selective such that we can live best, so that today our sense organs and our brain only allow very specific experiences from all possible external influences to become impressions. For example, visually we only recognise a very limited range of the electromagnetic wave spectrum (the visible light; the seen frequencies are those of maximum of intensity sent by the sun). Other examples are the frequencies from the auditory spectrum.

Humans also perceive (or generate?) a particular concept of time: We can experience seconds to years and decades, but nothing significantly smaller

---

[18] Arjun Ramakrishnan, Miguel Nicolelis et al., *Computing arm movements with a monkey brained*, Science Report, 5, 10767 (2015); www.nature.com/articles/srep10767/h. Miguel Nicolelis et al., *Building an organic computing device with multiple interconnected brains*, Scientific Reports, Volume 5, Article number: 11869 (2015).

than a second or larger than a few decades. We can also imagine besides the "now moment", past and future. With this capacity for abstraction, humans can have an idea of themselves at other times, which enables them to plan future actions and reflect on past experiences (which animals cannot, or only to a very limited extent). Furthermore, the self-aware organism is able to direct a greatly increased attention to external warning signals. Our emotions and feelings play a major role in this. The Portuguese neuroscientist António Damásio distinguishes between emotions and feelings[19]: Emotions, he says, are bodily reactions that follow a stimulus automatically and are without us becoming aware of or being able to steer them, physically rooted survival mechanisms, so to speak. Feelings, on the other hand, arise when the brain analyses and consciously perceives the reactions of the body. While animals can certainly show emotions, can they have actual feelings[20]?

The various possibilities of Ego-consciousness that biologists have meanwhile recognised as evolutionarily advantageous are numerous, including those of various forms of social cooperation. With all these subjective perceptions and potential candidates for illusions, it nevertheless must have been some kind of objective reality that somehow exists and evolutionarily created our ways of subjective perception. Or is there another explanation possible? We see that quite a few research and science problems remain unanswered. Will science really answer them in the future?

Summary: Scientific Knowledge, Philosophical and Ethical Questions, and Remaining Openness

Let us synthesize the currently prevailing view of most neuroscientists and at the same time the beginning of a neuro-philosophical explanation of consciousness as follows: The (self-)consciousness of ourselves, our "Ego", first manifests itself in subjective perceptions and feelings. It is an experience-constituting component of our world, i.e. we cannot discard it, which is why its objective recording, a scientific understanding, is difficult. Conscious experience ultimately consists of a representation in our brain that creates a model of reality for us. It is a kind of "biological data format" for humans (and can be quite different for other beings/animals in the world), a certain way of representing information about the world, focusing our attention,

---

[19] António Damásio, *The Strange Order of Things: Life, Feeling, and the Making of Cultures*, Pantheon (2018).
[20] The movie: *Cody - The dog days are over*, by Martin Skalsky is a good reflection on this problem (2019).

creating an inner representation of the world and finally simulating different temporal scenarios. This inner holistic representation results from the assembling of many individual pieces of information that reach the brain from the sense organs. The resulting unity in our perception, i.e. the fact that we recognise our experience with all the many sensory impressions as a union (neuroscientists speak of the "coherence" of consciousness), is one of the most amazing achievements of our brain/mind. Another is the "temporal unity of the Ego", i.e. that we believe being the same person as yesterday, last year or decades ago, and this despite the fact that over time all the cells in our body, including those of our brain, are replaced or altered. Exactly how this coherence happens is scientifically still largely unclear. Neuroscientists also speak of the "binding problem". Particularly popular among them is the hypothesis that this coherence of experiences goes hand in hand with synchronised (concurrent) actions of neurons or neuron associations that result from rhythmic, oscillatory (periodically oscillating) or other types of discharges of neuronal electrical potentials. Our "I" would thus in the final analysis presumably be a complex, emergent physical event, while at the end still an activation pattern in our brain. The fact that we do not realise that the Ego is a certain kind of representation (some would say an illusion) is part of the scheme. The neuro-philosopher Thomas Metzinger speaks in this context of a "transparent representation": This creates the phenomenology of naïve realism, the robust and irrevocable sense that we are directly and immediately perceiving something which must in our view be objectively real.

With all the current findings about our brain and experiments about our Ego, scientists speak of the theory of "psycho-neuronal congruence", i.e. the correspondence of brain and mind as two different aspects of the same entity, an "ontological equivalence" of conscious processes and corresponding physiological brain processes. But as we have seen, there is no deeper knowledge of what exactly the mind is. That said, there are numerous applications beyond the therapeutic field for the still young research branch of applied neuroscience. In some researcher's opinion, so-called "neurochips" have a particularly great development potential even for healthy people, which are supposed to improve the state of mind, increase the sense of well-being, bring about lasting happiness or even increase intelligence. Such chips in the brain or drugs that make us more intellectually potent, more attentive or happier at the push of a button are still pure utopia. Further steps towards mechanization and digitalization of our bodies and thus minds are not only realistic but have already been initiated. They would thoroughly change our view of the world and of humanity.

With all this, the most important question still remains open: How is the emergence of consciousness and subjective experience possible in an objective, i.e. subject-independent physical universe? How can one explain that something like subjective, conscious experience could arise on the basis of objective physical processes? Are subjective feelings and the emergence of an inner perspective even conceivable in an objective natural order of things - or are we confronted at this point with a final mystery, with a fundamental white spot on the map of the scientific view? We will certainly experience exciting discussions on these questions in the coming years. They will also lead us to the core of an age-old philosophical question, which is also the core of the Enlightenment project: What is the human being?

New Questions on Social Relationships

Most recently, consciousness research and the development of respective technological applications have also opened up a broad catalogue of *social* questions. For we not only experience ourselves as conscious of ourselves, but also recognise that other people are conscious of themselves. We can put ourselves in their shoes, (possibly) understand their thoughts, emotions and actions, just as we can understand our own. In this context, neuroscientists and humanists speak of us carrying a "theory of mind" within us. In the first approaches, this social cognition has already become tangible for brain research. Neuroscientists have recognised how similar brains mutually exchange information, to the point that they can even oscillate in unison (see also explicit experiments with monkeys above). One recent discovery has made particular headlines: Brain researchers have discovered neurons that respond directly to other people's actions and expressions. When we look at a person with a certain reaction (e.g. fear), these neurons in us fire exactly as if we experienced this emotion ourselves. Brain researchers call them "mirror neurons".[21] The self is thus not only a window into the inner workings of one's ego, but also a gateway into the social world. Hence, neuro-research ultimately also touches on ethical questions of group behaviour.

There is one field in which the question about possible applications is particularly sensitive, especially in relation to consciousness technologies and brain-computer interfaces: the military. With the use of American drones in Pakistan or Somalia, controlled from tens of thousands of kilometres away in Utah, we experienced the first beginnings of warfare through avatars. There

---

[21] Giacomo Rizzolatti, Laila Craighero, *The mirror-neuron system*, Annual Review of Neuroscience. 27, 1 (2004) p. 169–192; also: https://doi.org/10.1146/annurev.neuro.27.070203.144230.

is no way around an international moratorium here. Only in this way can we limit the corresponding global arms race, which has long since begun. The experience with the restriction of nuclear weapons can serve as a model here.

On the other hand, virtual/augmented reality (VR/AR) technologies could also help us achieving a growth of our empathic abilities in addition to an increase in our biological, neurological and cognitive potential. We could, for example, by means of a virtual embodiment, also more profoundly understand and sympathise with these others. With VR/AR technologies we face the mammoth task of deciding which applications will benefit humanity and which will harm us and/or provide advantages only for a very small group. In order to make these decisions, we have to ask and answer philosophical, especially ethical, and even spiritual questions. If you are a young person today and are thinking about which professions will be significant in the future and will not be taken over by intelligent machines, you will find what you are looking for in philosophy. Government agencies as well as companies like Google and Myriad Genetics (an American genetic testing and medicine company) are already, in addition to well-trained programmers and technicians, desperate for philosophically well-trained people.

What is a Human Being and What Should a Human Being Be?

As we saw, developments towards virtual embodiment also raise massive ethical questions. What is the responsibility of our actions in virtual spaces? Is there any independence in our autonomy when we connect with avatars? How can we control that these avatars do not act on the basis of our unconscious impulses? And last but not least: The question *What is Man?* now comes with an entirely new question: *What is Man supposed to be?*

So, from what we saw in this chapter (as well as already in Chap. 4), it is evident: Not only do existing or imminent technological developments need to be politically recorded. There is also a need for awareness of the risks that various technological developments will bring in the future – and here as well especially of "extreme risks" - even (and especially) for scenarios with a low probability of occurrence, but significant consequences. (Of course, high probability of occurrence, and significant consequences, are even more key to consider. Best example here is the climate.) For example, the creation of an artificial intelligence with self-awareness: This remains a very unrealistic scenario, i.e. comes with a rather small probability. However, it is still a possible future development - and could have severe implications. We have seen that such consciousness "emerged" during our evolutionary history from the ever-growing complexity of our brain operations. Another point: what if quantum computers come into play that can potentially deal with significantly higher complexities? Because of the dramatic nature of its effects - it would threaten the very existence of humanity - this scenario must be included in the discussion on the consequences of brain and AI technologies. With such developments life could potentially become impossible to be defined as human and thus worth of corresponding dignity.

10

Artificial Intelligence from Its Origins Via Today to the Future

Significant Progress in Understanding, Replicating, and Changing Us Humans or Solely Technological Advances Contained to Optimising Certain Processes?

A discipline that is, as we have already seen in the last chapter, strongly related to the study of consciousness is artificial intelligence (AI). It has developed dramatically, particularly in recent years, in fact so dramatically that we can get scared when going intellectually a little more deeply into it. Not less than 30 years ago, in 1993, the mathematician Vernor Vinge published the prediction that "within 30 years we will have the technological means to create superhuman intelligence".[1] Superhuman intelligence is an intelligence that is much smarter than the best human brain in *every* field, including general wisdom, scientific creativity, and social skills. It must thus be seen as a strong AI. Interesting - or scary? - is what Vinge wrote thereafter: "A little later, the era of humans will be over." For a possible point in time when machines could be more intelligent than humans in everything, futurologists use the term "technological singularity". Such a high level of artificial intelligence superior to humans, could in turn drive science forward even faster, accelerate technical progress even more massively, and in turn create even more intelligent artificial systems. With this feedback loop and such rapid technological progress we would eventually no longer be able to follow the superhuman intelligence of the machine (the "superintelligent" computer) intellectually.

However, how can computers evolve in that direction? Most likely by developing similar structures as our brains. As early as the 1960s, the American computer science pioneer Joseph Licklider described the vision of a

---

[1] Vernor Vinge, *What is the Singularity?*, Omni magazine (January 1993).

massive network of human brain and computer - as well as that of a global computer network - which is surprisingly close to today's internet.[2] Nowadays, the term "cognitive computing" is used to describe a possible future generation of computers that have human capabilities such as adaptive learning based on feedback loops, reasoning experiences, developing flexible reactions to external sensory inputs, and employing social interaction with other systems including humans (all with the help of deeply functioning neural networks). You could argue that these abilities are all still quite far away from today's computers, but they are moving fast towards such goals. However, in order to get these features, their architecture possibly needs to mimic even much more closely that of the human brain.

30 years after the statement by Vinge there is still a long way to go before fully making computers super-intelligent. Nevertheless, the development of AI has been amazing in recent years in a manifold of applications. Just a few examples: the handling of a vast amount of data with complex interrelationships, whether in medicine and our body's metabolism, or in the analysis of the huge amount of data at the Large Hadron Collider (LHC) at CERN, creating virtual realities, estimating traffic in urban planning, image, music, poem or story creation, or understanding the dynamics of global capital markets. All this requires increasingly powerful computing skills and handling larger and larger sets of data, goals that modern computers have reached at unimaginable levels. Two examples: Translation was one of the first goals of Artificial Intelligence, which failed in the 70's giving rise to the stopping of financing by the DARPA (American Army). Today Google Translate or DeepL are routinely used and produce better and better results. On the level of chess games and Go games algorithms beat the World Champions, which was deemed impossible thirty years ago (for Go even still before 2016, when it happened). By now the broad public users have realized what nice pictures or music AI can create. AI is thus finally reaching the general mass of people (most without much reflection).

History of Artificial Intelligence

The (theoretical) idea of making a computer act like with (simplified) neurons as binary classifiers based on external stimulation goes back to 1943, when McCulloch and Pitts developed what today is called a "perceptron". In 1950,

---

[2] Joseph Licklider, *Man-computer symbiosis*, IRE Transactions on Human Factors in Electronics. HFE-1 (1960), p. 4–11; https://sci-hub.se/10.1109/THFE2.1960.4503259.

Alan Turing explains in his article "Computing Machinery and Intelligence"[3] how intelligent machines can be built, and their intelligence be tested. As early as in 1951, Marvin Minsky built the first wired learning machine called SNARC.[4] Five years later, the first AI programme was presented at the Dartmouth Summer Research Project on Artificial Intelligence (DSPRAI), an eight-week summer workshop. This was the foundation of AI research for the next decades. Also at this 1956 conference, John McCarthy first recorded the phrase *artificial intelligence*. In 1958, Marvin Minsky and John McCarthy initiated the MIT AI Group that evolved into what is now known as the famous MIT Computer Science and Artificial Intelligence Laboratory. Starting in 1958, John McCarthy developed Lisp, the second-oldest high-level programming language supposed to be fitted for AI. He and his partners said – intriguing from today's view: "Artificial intelligence is not, by definition, simulation of human intelligence".[5] Later on, McCarthy moved to Stanford and founded the not less famous Stanford Artificial Intelligence Laboratory (SAIL) in 1963.

Generally, in the 1950s researchers were driven by an almost limitless expectation of the capabilities of computers for AI. They believed that within as little as ten years a computer could do similar things as humans. Their optimism received then, however, a strong damper when they recognized about two decades later that the structure of sequential information processing in conventional computers (the so-called "von Neumann computer architecture", see below) is completely unsuitable for the development of AI. This gave rise to what is today called the "AI Winter" of the 70's.

In the 80s and 90s, significant advances in AI occurred again by computer programmers understanding the dynamics of massive parallel processing of information in our brain on the one hand and the possibilities of modelling the structures of our thinking organ by so called "neural networks" on the other hand. This triggered a new wave of enthusiasm. As early as 1976 the first significant mathematical theorem was proven with the help of a computer; the four-colour theorem. And in 1997, AI received particular public attention, when the computer "Deep Blue" by IBM managed to beat the world chess champion Garry Kasparov. Today even the best human chess players have little chances against commercially available chess programmes.

---

[3] Alan Turing, *Computing Machinery and Intelligence*, Mind, Volume LIX, Issue 236, (1950) p. 433–460.
[4] A few years later, in 1969 he wrote with Seymour Paper the book *Perceptronx*, which became the foundational work in the analysis of artificial neural networks; Marvin Minsky, Seymour Papert, *Perceptronx - An Introduction to Computational Geometry*, The MIT Press, Cambridge MA (1969).
[5] John McCarthy; Marvin Minsky, Nathan Rochester, Claude Shannon, "*A Proposal for the Dartmouth Summer Research Project on Artificial Intelligence*" (1955).

However, despite these successes it had already become clear that the structure and functionality of our brain has an unmanageable complexity. And the dimensionality of the state space of programmed neural networks can quickly, i.e. with only a few neurons, assume astronomical values. So, it seemed impossible to simply "re-program" our brain on a computer. A second downfall of excitement, i.e. "second AI winter" of hibernation, was the consequence, starting in the mid80s.

However, equally already in the 1980s, John Hopfield introduced "deep learning" techniques that are supposed to enable computers to learn more complex structures from data of experience, called "feature" or "representation learning". The idea of deep learning is to use neural network architectures which employ a high number of hidden layers with a particular structure to get them closer to the human brain (that is what "deep" refers to). Traditional neural networks until the 1990s only contain 2–3 hidden layers, while deep networks can have as many as 150 of them. Only in the 2000s, when computers became powerful enough to run deep learning networks (which they surely were not in the 1980s), was this technique applied. This then started in the following years to many of the AI landmark goals being achieved.[6] Thus, AI thrived again (for long despite little government funds and public attention) which has not stopped until today. AI finally received some public attention again in 2016 when AlphaGo, an AI-driven system playing the Chinese game "Go", beat Lee Sedol, the then human Go world champion (see below for more details). Today's particular deep learning technique is called *Long Short Term Memory (LSTM)*, a neural network model described by Jürgen Schmidhuber and Sepp Hochreiter already in 1997,[7] which greatly improved the efficiency and practicality of recurrent neural networks. With deep learning technology neural networks moved closer to the way the highly-multi-layered human brain works. This makes it unlikely that the new excitement wave will soon be over.

History of Computers and Computer Science

The history of artificial intelligence is strongly linked to the history of computers. Today's capability of AI systems rests on the ability of today's computer in processing information in size and velocity that only 25 years

---

[6] Igor Aizenberg, Naum Aizenberg, Joos Vandewalle, *Multi-Valued and Universal Binary Neurons: Theory, Learning and Applications*, Springer Science & Business Media (2000).
[7] Sepp Hochreiter, Jürgen Schmidhuber, *Long Short-term Memory*, Neural Computation, 9(8), (1997) p. 1735–80.

ago looked like beyond imagination of what is possible. Moore's law from 1965predicted a doubling of computer cells' speeding power every two years, turned out to be quite correct, also because the computer cell producers oriented themselves around it. The calculation capacity of computers in 2023 is around 3,000 times higher than in 2000.

The origin of the concept of programmable machines dates back to Charles Babbage joined by his student and then partner Ada Lovelace in the first half of the nineteenth century. The woman Lovelace, daughter of Lord Byron, designed during that time the first actual computer code.[8] The first forerunner of later computers in the twentieth century was the electrically driven mechanical calculating machine "Z1" constructed by German researcher Konrad Zuse in 1938. Z1 worked with binary numbers and was able to execute programmes encoded on perforated cinema strips. It already described all the elements of the later Von Neumann architecture. While the Z1 wasnever fully functional due to problems with manufacturing precision, a few years later, in 1941, Konrad Zuse built the "Z3", the world's first functioning computer. In March 1945, "Z4" was completed, which was the first commercial computer in the world and still the only functioning computer in Central Europe in 1950.

However, the design, features and structures of modern computers was almost entirely developed in the US. The *Von Neumann architecture*, published in 1945 by John von Neumann in his report "First Draft of a Report on the EDVAC"[9] – the acronym stands for Electronic Discrete Variable Automatic Computer – was the decisive breakthrough in the development of programmable calculating machines. In computers built according to this principle, data and programme are stored in binary code in the same memory. This means that the installed programme can be changed during a running calculation process (something we are completely used to today) – which was impossible before leading to severe limitations in the calculation.

The von Neumann architecture became the basis of all modern computers. Specifically, it consists of the following building blocks:

- Processing unit consisting of a logic unit that processes instructions such as AND, OR, IF/THEN, and at the same place a processor register that serves as a temporary buffer for results of partial steps,

---

[8] See Lars Jaeger, *Women of Genius in Science - Whose Frequently Overlooked Contributions Changed the World*, Springer 2023 – Chapter 7.
[9] John v Neumann, *First Draft of a Report on the EDVAC*, Moore School of Electrical Engineering, University of Pennsylvania (1945).

- a control unit consisting of an instruction register (programme) and a programme counter which ensures the sequence of the programme steps,
- a memory unit that stores data and instructions,
- an external memory, and
- input and output mechanisms for the programme and the data to be processed.

Even though today computer solutions of complex differential equations are indispensable for weather forecasting, climate modelling and many other things, we should not forget: The original purpose of von Neumann's computer and its algorithms was to perform calculations about the explosions of hydrogen bombs.

The "Electronic Numerical Integrator and Computer" (ENIAC) developed by John Eckert and John Mauchly at the University of Pennsylvania on behalf of the US Army was presented in February 1946. It was the first electronic Turing-powered universal computer and thus a direct competitor to the Z4. However, the ENIAC never made it into commercial series production. It was not until 1951 that a US public computer came onto the market, the "Universal Automatic Calculator" (UNIVAC), which was also designed by Eckert and Mauchly. The UNIVAC was the first to use magnetic tape as external memory. It consisted of 5,200 electron tubes, 18,000 crystal diodes, weighted 13 tons, required an electrical power of up to 125 kilowatts and could perform 1905 computing operations per second.

But then, on 23 December 1947, the Americans John Bardeen, William Shockley and Walter Brattain achieved something central for computers to this day: They succeeded for the first time in controlling current flows in a system of semiconductors. These components made of semiconductors, the today so-called transistor, proved to be far superior to the vacuum tubes used until then.

- They can be controlled in a very finely graduated manner by external parameters such as the applied voltage, temperature or even the introduction of different atoms into their crystal structure (doping).
- They consume much less energy, mainly because they produce less heat.
- The first transistors were only 1/50 as large as the vacuum tubes. With that (and better thermal management options), the miniaturisation of computers began.

When the engineers were looking for a suitable name for their new invention, they considered terms such as semiconductor triode, crystal triode, solid-state triode, iotatron and numerous other names. Finally, a combination of the words *transconductance* and *resistor* won the race: Transistor.

The discovery of the transistor effect paved the way for the realisation of von Neumann's dream of handy and super-fast calculating machines. From now on, the transistors became smaller and smaller, and from the 1970s onwards they were packed together on microprocessors and memory chips to form integrated circuits. Today, individual elements of chips comprise only a few dozen atomic layers (less than 10 nm).

In 1955, Bell Labs constructed for the US Air Force the TRansistorized Airborne DIgital Computer (TRADIC), the world's first computer entirely equipped with transistors instead of vacuum tubes. It was the miniaturisation of components on semiconductors that led to the formulation of Moore's Law, a prediction that still sets the pace of digitalisation today. Moore's Law is a rule of thumb based on an empirical observation by Gordon Moore, one of the co-founders of INTEL. The number of functions realised on a single computer chip doubles approximately every two years. As a consequence of his observation, Moore predicted a drastic increase in the performance of data processing with a simultaneous decrease in relative costs – both on an exponential scale. And he was right: It was the miniaturisation of technical components that made ever more powerful computers possible, i.e. doubling their efforts in less than two years (it was actually 18 months on average, but getting slower as of about 2010). In 1971 Intel produced the first mass-produced microprocessor. It consisted of 2250 transistors. Fully in line with Moore's law there are billions of transistors on today's processors about one square-centimetres large microprocessors that run with a speed of up to eight Giga-Hertz (one GHz is one billion processing units per second).

In parallel to the computer development was the progress of *computer programming*. In 1951 the Swiss mathematician Heinz Rutishauser first found a way to the today so common "automatic programming" of computers. Important milestones were then the development of Fortran (from "FORmula TRANslation") as the first higher programming language in 1957, ALGOL (from "ALGOrithmic Language") from 1958 to 1968, Lisp (from "LISt Processing") and COBOL (from: "COmmon Business Orientated Language") in 1959, BASIC (1960), C (1970), Pascal (1971), Prolog (1972) and SQL (1976), Ada (after the woman above, Ada Lovelace, 1977), C++ (1985), Python (1991), Java (1995).

Where AI Stands Today

Let us be honest: This paragraph cannot age well. The developments in AI are too rapid. And this is just why we live in a fascinating era: Anyone who follows the discussions among AI researchers receives quickly the confirmation that their discipline has entered a third phase of excitement. The basis of this new elation is the ability of more and more powerful computers to extract relevant information from large data sets, and the possibility to handle increasingly complex information efficiently on the basis of deep learning.

The aim of contemporary AI research is to develop machines with intelligent behaviour that exhibit similar functional and behavioural characteristics to humans or animals. Researchers as well as philosophers in discussing artificial intelligence distinguish between two types of AI, we have already seen in the previous chapter:

- Strong AI is a form of intelligence that can think creatively and solve problems like humans, and which is characterised by some form of consciousness. Despite years of research, the goal of creating strong AI remains unachieved to this day.
- In contrast, weak AI is concerned with solving specific problems. Here, the goal is not to imitate higher intelligence, but to simulate intelligent behaviour using mathematics and computer science.

This includes problems that are generally understood to require some form of rational intelligence to solve, such as recognising patterns, mastering languages, playing chess, searching web pages on the internet, proving mathematical theorems (by pure trial and error), learning and automating simple actions, or as of latest search specific information from large, disordered data sets. Currently, its centre is increasingly the systematic handling of information, especially its automatic processing with computers. Modern industrial production such as car manufacturing would be inconceivable without weak AI (which in the near future is likely to be able to fully detect the road and then self-drive cars).

The modelling of artificial neural networks illustrates how a in principle simple basic structure can develop complex solution procedures and perform intelligent processes for pattern recognition such as the recognition of texts, images or faces. With the aim of systematically developing such processes, weak AI based neuro-informatics has by now emerged from a new scientific to a technological discipline. In this process, it became increasingly clear that the development of solutions for such complex problems is only achievable by the possibility of endogenous (coming from within) changes in the information-processing system, in other words, with the help of independent learning.

In all of this, the AI's performance matches or even exceeds in many areas that of human experts. For example, AI pattern recognition algorithms have become just as good in the diagnosis of skin cancer as skin doctors. AI-controlled machines are quite hands-on: The surgical robot "Smart Tissue Autonomous Robot (STAR)" already outclasses human surgeons when it comes to precision of surgical procedures on pigs.

As already mentioned, in 2016, AI received particular popular attention again – by the victory of the Google computer AlphaGo over the World Champion of Go, Lee Sedol. Go is a much more complex game than chess with roughly estimated $3^{19 \times 19}$ possibilities per move, which means that for mastering it creativity and intuition is required rather than pure calculations. Suddenly, many realized how far the learning algorithms of artificial intelligence systems have developed. AlphaGo is not just a fast computer, but rather encompasses an unimaginably large number of possible learning steps and thus the ability to perfect neural connections which mimics the learning and thought processes of the human brain. The computer was fed countless historic Go games and from those learned entirely new strategies of the game that even experts had never seen before. When looking for the reason of the last move of AlphaGo that disconcerted Lee Sedol, the engineers were surprised that AlphaGo looked for a winning move that had never been played by humans before. AlphaGo seemed to present totally original moves it creates itself. This was very unsettling for his human opponent. And it made it clear that the underlying learning and optimization means of today's AI enabled a massive and broad increase in machine intelligence. It is obvious: Future computers will no longer deal solely with the particular purpose they were designed for, e.g. play chess or browse through databases, but will be able operate on much broader scales.

Somewhat less in the public eye was a new version of AlphaGo just 18 months later (AlphaGoZero): In 100 games it played against the world's best chess computer up to that time. The later had been fed with data from millions of historical chess games and thus the centuries-old experience of chess-playing humans; its computing power was 70 million positions per second. AlphaGoZero won 28 times and drew 72 games - in other words, it did not lose a single time. The amazing thing was: AlphaGoZero had only been fed the rules of the board game four hours (!) earlier, and it could also "only" evaluate 80,000 positions per second. Four hours were enough to turn AlphaGoZero from a beginner into the unbeatable, best chess machine in the world! In this short time, it optimised its neuronal connections and learnt openings and game strategies. The new AlphaGoZero also learned to play Go in almost no time at all: This more powerful, self-learning version also beat its predecessor AlphaGo 100:0 in a hundred games, after equally only four hours of learning by playing against itself! One of the creators of AlphaGo and AlphaGoZero, Demis Hassabis, said that AlphaGoZero was so powerful because it was "no longer constrained by the limits of human knowledge".[10] It is thus becoming increasingly clear: Artificial intelligence research aims to reach broad based outperformance of machine intelligence versus human intelligence. Due to AI the game Go today is very differently approached in games by humans than during the entire 2,500 years of tradition.

What most people barely have on their radar screen yet: AI is already part of our everyday life. Computers with AI software optimize energy consumption in industrial plants, create cancer diagnoses, write journalistic texts at the touch of a button, advice bank customers in their optimal investment strategy perform facial recognition in our phones, and support lawyers dealing with complex legal cases. They read customer letters and e-mails and recognize the degree of annoyance of the sender, computers in call centres chat on the phone like human employees, and, last but not least, computers perform cognitive cooking creating new culinary trends. AI even writes poems, entire plots for films, composes symphonies and operas, and paints in the style of true masters. AI is starting to get seen by us directly, e.g. in creating nice pictures for Christmas according to our individual requests. And in Japan, a country with a very high belief in the functionality of AI, a candidate for mayor of the city of Tokyo, Tetsuzo Matsumoto, presented himself with an artificial intelligence claiming that he would govern based on the decisions of the algorithms.[11] He attracted thousands of voters.

Introduced to the public in late 2022, the chatbot ChatGPT (GPT stands for "generative pre-trained transformer") has caused quite a stir. Developed by Open AI, a company that originally started as a non-profit research lab, ChatGPT for some is merely a funny toy, for others a useful tool. And some see severe risks for the world we live in. While students might gladly use it to do their homework (or see it being done) within seconds, teachers are afraid of just that. ChatGPT still has obvious limitations. The concluding lines of an essay by Noam Chomsky, Ian Roberts and Jeffrey Watamul (March 2023) read as follows:

---

[10] https://www.telegraph.co.uk/science/2017/10/18/alphago-zero-google-deepmind-supercomputer-learns-3000-years/.

[11] https://www.newsweek.com/ai-candidate-promising-fair-and-balanced-reign-attracts-thousands-votes-tokyo-892274.

Given the amorality, faux science and linguistic incompetence of these systems, we can only laugh or cry at their popularity.[12]

But with GPT-4, released only days later, the bot has received significant improvements in the fields of reliability and creativity. However, the technical paper by OpenAI for the release of the new version[13] discloses nothing about the underlying architecture - much to the disappointment of AI researchers. The laws of the market have taken over. It remains to be seen how many books, songs and screenplays will be written with or by ChatGTP (or similar engines), and if we will even be aware of the author's nature.

The emergence of ChatGPT-4 is undeniable evidence of the remarkable advancements achieved by generative AI in recent years. Although the inner workings of the algorithm are not fully understood, it is based on stochastic language models (LLMM). GPT stands for Generative Pre-trained Transformer. Essentially, the model doesn't comprehend the content it generates, but rather makes statistical predictions for word sequences using a vast text database and a combination of supervised and reinforcement learning algorithms. While this model type is not new, it has faced criticism for introducing cognitive biases that reflect human biases like gender stereotypes, religion, and race.

Fortunately, it appears that ChatGPT-4 mitigates the most obvious biases by incorporating a human classification system to guide the algorithm's responses through reinforcement learning. Furthermore, it is plausible that in the near future, these algorithms will be further improved through meta-algorithms that analyze responses for meaning and assign scores to the generative algorithm. The developers of ChatGPT are actively pursuing this approach by granting free access to the system and allowing users to rank the responses, transforming them into optimizers of the model. The remarkable growth in user numbers serves as evidence of the tool's success. Since its launch on November 30, 2022, it reached one million users by December 5, 2022, 100 million in January 2023, and a staggering 1.6 billion by March 2023, making it the fastest-growing application to date.

Simultaneously, this success has brought the issue of generative AI and its potential excesses to the forefront. We stand at the precipice of a transformative period that is challenging to fully comprehend. As users ourselves, we have realized that ChatGPT-4 is unmatched in certain applications but severely limited in others. Nonetheless, it is undeniably becoming an

---

[12] Noam Chomsky (et al.), *The False Promise of ChatGPT*, guest essay for The New York Times (March 8, 2023).
[13] https://cdn.openai.com/papers/gpt-4.pdf.

indispensable tool, akin to web browsers or Wikipedia. Recognizing its significance, Microsoft, a significant investor in OpenAI, is currently in the process of integrating ChatGPT-4 into its browser, a move that has the potential to revolutionize internet searches.

Consequently, the regulation of generative AI has suddenly become an urgent matter. The European Parliament is already working on legislation parallel to the General Data Protection Regulation (GDPR), with the intention of enforcing it in 2025.[14]

The field develops very fast. Every few weeks a new generative AI system is announced. Recently, there was turmoil within OpenAI about the future of their new software Q-Star. Disagreement about it was apparently the reason of the dismiss of Sam Altman by the board of directors. He was reinstated in his role and the board was reorganized under the pressure of the shareholders (particularly Microsoft). It shows that the movement forward of this technology will be almost impossible to stop. We can expect new progress in generative AI every few months.

The Current Interaction of AI and Our Brain – Does that Eventually Lead to Superhuman Intelligence?

The rapid technical progress of AI contrasts to the comparably constant performance of the human brain in the last 50,000 years or even longer. Nevertheless, the average human IQ has actually risen in the last 200, more rapidly in the last 100 years, by about three points per decade, which is considered the "Flynn Effect".[15] One major implications of this trend is that an average person of today would have an IQ of 130 by the standards of 1910, placing them higher than around 98% of the population at that time. However, more recently (since the late 1990s), a decline of average IQ has been observed.[16] The discussion about that trend is ongoing, as such an apparent reversal may be due to cultural changes (e.g. internet) which render certain parts of intelligence tests obsolete. However, compared with the multiple orders higher rate of change in the capability of computers, the

---

[14] By the way, to show the reader the power of ChatGPT-4., these paragraphs (starting at "The emergence of") were first written by the authors and then given to the tool for rewriting them. It is easy to see how good the software can be when properly guided.
[15] Richard Herrnstein, Charles Murray, *The Bell Curve - Intelligence and Class Structure in American Life*, Free Press (1994).
[16] Bernt Bratsberg, Ole Rogeberg, *Flynn effect and its reversal are both environmentally caused*. Proceedings of the National Academy of Sciences, 115 (26) (2018), p. 6674–78.

rate of any change in our intelligence is of minor significance. Will we soon reach superintelligence by the computer, a state in which computers can solve *everything* better than us?

It seems to be only a matter of time before the computing power of machines surpasses that of the human brain. One particular trend: The brain's computing power is estimated to be around 20 petaflops (1 petaflop equals 10^{15} - that's a 1 with 15 zeros - operations per second). In June 2022 Hewlett Packard Enterprise Frontier, or OLCF-5, became the world's first exascale (10^{18} operations per second) supercomputer. This is around 50 times as fast as our brain works, and it is to be expected that the 100 times barrier will soon be reached (possibly already before publication of this book).

However, hardware alone is not enough for true superintelligence. Every computer needs software. And for a technological singularity in the form of superintelligence, real strong AI would have to be created, which can think creatively about *all* kinds of problems and solve them independently if necessary. This would have to be programmed accordingly, most likely in the framework of deep learning. Due to the fact that research in the field of strong AI has in fact been stagnating for years, the singularity idea is quite controversial, whether it will ever be reached. In contrast, the development of weak AI is less about creating superintelligence or even forms of consciousness, but about mastering concrete application problems. AI systems have by now reached a point where they can create their own software. For example, the firm DeepMind (that also created the AlphaGo player) created an AI capable of writing codes to solve arbitrary problems posed to it. It participated in a coding challenge and by 2022 reached a placing in the middle range. Will AI in the future do our (and their) programming jobs? Is this possibly a path to reaching strong AI?

A major threshold for achieving a strong form of artificial intelligence is the still broadly insufficient knowledge of brain researchers about the functioning of the human brain. This has also put research policy makers on notice. About ten years ago, in 2013, US President Barack Obama launched the "BRAIN" (Brain Research Through Advanced Innovative Neurotechnology) initiative to map the human brain and thus "aimed at revolutionizing our understanding of the human brain". An alternative path to this, financially supported by the European Union, was the complete computational simulation of the brain in the "Human Brain Project".[17] Henry Markram, director of the Blue Brain Project at EPFL in Lausanne, had announced in 2013 with great fanfare in the press that within ten years his research group will

---

[17] See https://www.humanbrainproject.eu/en/ for more details.

be able to entirely simulate an entire rodent (mouse) brain. He was given one of the one billion Euro programmes of the EU. Quite quickly, though, it became clear that these were some very strong exaggerations and left a trail of angry neuroscientists across Europe. By 2016, Markram was removed from the project leadership, and the announced project support was consequently broadened across many neuroscientists. And now, ten years later, there is no full simulation of the mouse brain yet.

Looking at the various approaches towards optimizing AI further and further and thus asking for the achieved intelligence to be larger than the natural human one, it is unclear whether this is achieved at any level any time soon. That being said, one thing is very likely to occur/being developed, and this in possibly only a few years: the *improvement of human intelligence by an interaction with AI*. Such a technical implementation of a direct intelligence enhancement of humans, based on suitable computer-brain interfaces, is much easier than understanding the entire brain and solving the strong AI problem. The former could (realistically) increase the performance of the human mind to an extent that people would not be able to keep up without this upgrade. The advantage of this technology is that it does not require strong artificial intelligence as it builds on human intelligence. Nevertheless, two things are required at least for the technological linking of humans and computers:

- First, a reliable and low-risk interface between the organic and technical worlds - a so-called neuro-interface, for example in the form of special microchips or electrode systems - implanted in the brain, together with a connection to the outside world for writing information in and out.
- Secondly, a mechanism of transferring what is to be learned into the "language of neurons".

Solutions to both problems are already emerging. The first interfaces that deliver electrical voltages to the brain via electrodes have already been used in deaf patients as well as depressed ones. An even finer electrode system could be constructed with nanotubes, small carbon tubes with a diameter of 100 nm. The transfer of content from the brain to a machine has already been partially successful. For example, signals from dozens of neurons in the motor cortex can now be "read", enabling paraplegics to control prosthetic hands by "thought".[18]

---

[18] Already for the 2014 Soccer World Cup in Brazil, neuroscientists arranged a special PR coup: by means of implanted chips in his head and a metal suit on his body controlled by signals transmitted from his brain by a computer, a paraplegic kicked off the opening game.

Some neuroscientists are already talking about an artificial hippocampus (an area of the brain important for memory) that generates and processes the same signals as its biological counterpart. To do this, researchers would have to crack the code that the brain uses to represent complex information such as words and sentences. There is no significant hindering factor why scientists should not achieve doing this in the future. Finally, for most people it is unimaginable to see a development in which we could bring different human brains into direct contact with each other via neuro-interfaces and thus make a single brain a part of a collective consciousness of many brains. But, as we saw in the last chapters, scientists have already succeeded doing the brain connection for monkeys. What should prevent them doing this with human beings?

The more complex the neuronal representations of the information mapped and processed in the brain, the more distant the goal of artificially influencing them from the outside. Barring ethical and legal considerations, it is far from unlikely that humans will one day carry technical components in their brains that improve their mental performance, i.e. make them more intelligent. Conversely, machines could carry the biological components alongside their "hardware" (the wetware we already briefly saw in the previous chapter). In the foreseeable future, which is only maybe three to five years, it is, however, mainly patients such as the deaf, the blind, the paralysed, paraplegics and people with memory problems who will benefit from the latest developments of interaction between neurobiology and computers.

Have the first (very small) technological steps towards a strong AI (still in interaction with the human or animal brain) already been made due to the interaction of neurons and computers and the connection of different individuals' neurons (of monkeys)? Based on the current possibilities of computer-, bio- and neurological technologies and their developmental dynamics, it is by no means absurd to assume that the creation of a *from humans independent* artificial super-intelligence will occur in the course of this century, or even the first half of it. We should rather count on it. Such interaction tool could potentially be unimaginably powerful and would come with

1. powerful improvement of our intelligence (maybe just for a few humans who are the rulers then?); and
2. enormous monitoring and control problems for us humans.

Compared to this, our current economic, environmental, and social problems would be almost negligible. What is thus needed is a reasonable and rational risk analysis, where the possible areas of danger are clearly identified - ideally in the form of specific scenarios or use cases that need to be prevented. And they must also here include low probability but very severe risks. Based on that we must come up with a clear defence strategy of respective risks.

How AI Shapes Our Society

It is evident: Artificial intelligence is a prime example of the double-edged nature of technological progress. Next to its exciting new opportunities for our perception and new types of social (virtual) interactions, AI also brings to us entirely new forms of personal surveillance and massive invasions of our privacy, not to mention the increasing dependence of our entire infrastructure on it, which makes us vulnerable to cyberterrorists.

The learning and optimization processes today's AI is based upon (deep learnings), enable a massive increase in machine intelligence on a broad scale. Individual AI systems are now in a position to no longer only handle a specific purpose for which they were designed, such as playing chess or browsing databases, but to operate in far more diverse areas, this by *learning actively*, just like us humans. Sooner or later, with the continuing growth of computers' calculation power, artificial intelligence will be superior to human intelligence in most, perhaps even in all cognitive areas. Incidentally, this also applies to areas that most people today still consider to be the incontrovertible domains of human ability: Intuition, creativity and the ability to grasp the emotions of other people (the first two we saw in Go). The latter in particular, is likely to become a standard capability of AI systems within the next few years. AI programmers speak of "affective computing".

These developments are potentially very positive, for example, when it comes to caring for people, or in therapeutic treatments when it is important to react appropriately to the emotions of the patient. However, there are also threatening scenarios, like when machines recognize and manipulate our emotions, and this more effectively and more regularly than humans do. Former Google employee and Silicon Valley expert Tristan Harris reports that YouTube already has digital simulations of almost two billion people and their online behaviour based on their viewing habits. The company uses those to optimize its algorithms to determine which videos can be used to keep individual people on the platform the longest. "Silicon Valley is hacking into our brains," says Harris and other AI experts.

Even by analysing the most common form of our communication, speech, AI will soon be able to interpret us better than we can ourselves. The German company Precire Technologies, for example, has developed a software that can create a very accurate personality profile of a person based on their voice alone. For this purpose, the AI experts trained neural networks on voice data together with the psychological profiles of about 5000 test persons. And very successfully so: Companies are already using this software in HR application procedures. You do not need much imagination to envision what a powerful tool this software is in the hands of totalitarian regimes.

But for all the futuristic dreams (or nightmares), the developments in the field of AI are already imposing dramatic social and political shifts and distortions on us with the potential for massive social crises. A large number of people are threatened to lose their jobs, and thus social status, through AI. And this not only in the lower income segment: Academics such as doctors, lawyers and teachers will also be affected. In just 25 years, up to 50 percent of today's professions are likely to have become superfluous. David Graber said in 2018 that many jobs are already superfluous.[19] High-ranking economists are warning of great social inequality between a few winners and many losers in an economy of the digital world in which "the winner takes it all". In the future, social inequality will thus no longer be caused by the exploitation of people, but simply by social insignificance of many groups.

Besides the social problems: Is this what we really want? Implicit in the notion of "learning" is the idea that one can discern what is "right" from what is "wrong". In other words, the algorithm is able to do that as we give it a sense of value or an "ethics". Currently, programmers are aiming at producing some results, without thinking very much on the philosophical a-priories of values and ethics behind this aim. This ends up being a very utilitarian (non a-priories) ethics that is programmed in today's AI algorithms (greatest possible happiness for the greatest possible number). Perhaps, we should start thinking about the question: Which one is the ethics we should teach our computers?

Last but not least, the country with the most developed AI will most likely rise to become the dominant economic and military power on this planet – even possibly in combination with quantum computers. Currently, it looks like two countries are fighting for global AI supremacy: the US and China. China has caught up strongly in the last few years and is even about to leapfrog to the first rank. The Europeans have long since been left behind in this race and degraded to standbyers. Where Russia stands in this respect is less clear. The lead of the Americans and Chinese is, however, not based on smarter researchers, better AI algorithms or better computer programmers,

---

[19] David Graeber, *Bullshit Jobs: A Theory*, Simon & Schuster (2018).

but simply on the availability of data, the oil of the twenty-first century. That the collectors of large databases have great power is already shown by the enormous commercial success of firms like Facebook and Google, or Tencent and Baidu. These companies earn tens of billions of dollars with personalized digital advertising and have long since become dominant forces in election campaigns and other political disputes. (Again, the role of Russian involvements is still debated.) Such companies collect data about us without our knowledge, sometimes at places and from devices that seem completely unsuspicious like TVs or Xbox consoles. In 2016, for instance, a (female) consumer sued the manufacturer of a networked vibrator who had collected and stored highly intimate data about theuse of that device.

With the appropriate software for face and image recognition and a dense network of cameras, creating motion profiles of individual people in real time - which was one of the key challenges in the late 1990s - has long since become an everyday achievement. Soon, we will hardly be able to do anything without anyone knowing about it. Eric Schmidt, the former CEO and chairman of the board of Google, put it already quite early this way:

> If you have something that you don't want anyone to know, maybe you shouldn't be doing it in the first place.[20]

AI pioneer Stuart Russel paints a more drastic picture of us humans driving in a car towards a cliff, hoping the gas tank will be empty before we plunge. Russel claims that AI can be as dangerous to humans as nuclear weapons. Some experts are therefore begging for governments providing frameworks and regulations. They are driven by serious concerns that policy makers oversee dramatic technological developments, but do not take them seriously enough - or simply do not understand them. The scientific and technological progress has developed with such rapid and complex dynamics that it simply exceeds the imagination of the vast majority of political and social decision-makers.

The leading AI researchers also appeal to the public to start dealing with their research. It is, they claim, imperative that society begins to think about how it draws the maximum benefit from AI research while avoiding the dangers that emerge from it. Machine learning has made dramatic progress in the past two decades, which is why AI is one of those technologies that will very soon change our lives drastically. Eric Horvitz of Microsoft Research provides an example of a computer program which through the analysis of publicly available data can come to the conclusion that a particular person

---

[20] Interview aired on December 3rd 2009, on the CNBC documentary "Inside the Mind of Google".

might suffer from certain diseases in the future. He points out that intelligent algorithms by analysing personal Twitter and Facebook messages may identify candidates for an incipient depression - even before the persons consider this themselves. And that is just the beginning, according to Horvitz. Large data processing "can extend our knowledge and provide new tools for enhancing health and wellbeing. However, they raise questions about how to best address potential threats to privacy while reaping benefits for individuals and for society as a whole."[21] It makes it increasingly difficult for individuals to know what is known about him or her. Horvitz thus calls for appropriate legislation as "an important part of the future legal landscape - and will help to advance privacy, equality and the public good."

We should thus listen carefully when AI experts themselves express warnings and demand regulation. As early as in 2015, leading AI researchers have appealed to the public, especially politicians, to start dealing with their results.[22] They claimed that "the progress in AI research makes it timely to focus research not only on making AI more capable, but also on maximizing the societal benefit of AI". This is still a very mild articulation. In fact, the scientific and technological progress has meanwhile progressed with such rapid and complex dynamics that it not only exceeds the mental, but increasingly also the ethical radar of most people.

Who Should Deal with the Decline of Our Privacy?

As late as 1983, there was fierce opposition to the planned census in the Federal Republic of (West) Germany. It involved 36 questions about the housing situation, the people living in the household and their income. "Privacy is sacred", was the motto of the activists against those questions to everybody. The Federal Constitutional Court ruled in favour of the plaintiffs and stopped the project. Only one and a half generations later, we carelessly hand over the supermarket chain's bonus card with every purchase, surf in the internet without caring about what Google might know about us, and shop, email and chat, use every app without hesitation and thus allow many

---

[21] Eric Horvitz, Deirdre Mulligan, *Data, privacy, and the greater good*, Science, Vol 349, 6245 (July 2015) p. 253–255.
[22] Stuart Russell, Tom Dietterich, Eric Horvitz, Bart Selman, Francesca Rossi, Demis Hassabis, Shane Legg, Mustafa Suleyman, Dileep George, Letter to the Editor: *Research Priorities for Robust and Beneficial Artificial Intelligence: An Open Letter,* Ai Magazine 36(4) (2015), p. 3; https://www.researchgate.net/publication/318497640_Letter_to_the_Editor_Research_Priorities_for_Robust_and_Beneficial_Artificial_Intelligence_An_Open_Letter.

times more insight into our private lives than the government wanted to know 40 years ago. If we also wear digital wearables in the future, and a more powerful AI is able to detect ever more subtle patterns in our behaviour, we will no longer have a lot of secrets against Google and Co. Imagine a world where many of your daily activities were constantly monitored and evaluated by AI:

- What you buy at shops and online,
- where you are at any point in time,
- who your friends are and how you interact with them,
- how many hours you spend watching movies and what sorts of movies, or how much you are playing video games and which ones,
- what bills and taxes you pay,
- who you date,
- and many more private points.

It is not hard to picture these, because most of them already happen, thanks to all those data-collecting behemoths like Google, Facebook and Instagram or health-tracking apps such as Fitbit - and most of us happily being data donors.

In Europe, this development has brought legislators on the scene, albeit very late and still too timidly. With the General Data Protection Regulation (GDPR) passing the European parliament in 2016 and becoming European law in May 2018, personal data protection became valid legally. In the USA, by contrast, AI and data protection are not regulated by any federal law; even an independent data protection supervisory authority is lacking. But the topic is being discussed. Among other things, this has been warranted by (even though already since a few years ago, without anything happening since):

- The data protection scandals surrounding Facebook and Cambridge Analytica, as well as recurring data thefts, such as at Yahoo at the end of 2017.
- Political interventions such as the GDPR has made US politicians think. The state of California e.g. introduced its own data protection law in mid-2018, which is partly based on the European GDPR.

All this takes place within democratic structures. Let us imagine a system in which there is no democratic control of the government and where all people's behaviours are rated as either positive or negative and distilled into a single number, according to rules set by the government. That would create a

"Citizen Score" and it would tell everyone whether or not one is trustworthy. Plus, the rating would be publicly ranked against that of the entire population and used to determine one's eligibility for a mortgage or a job, where the children can go to school - or even just one's chances of getting a date. This could very much be a Star Trek like scenario, a totalitarian society, the opposite of what the "Federation" (the free world) stands for.

But: This is exactly what is happening in China. Particularly, if one behaves against the government's doctrines he or she is being punished by a low index. Everybody's rating is then publicly ranked against that of the entire population and used to determine a person's eligibility for almost anything in his life. The Chinese even use the index against each other. With the help of AI and huge amount of data, Chinese government is thus massively expanding its control over its citizens' behaviour through that Citizen's Score. It was voluntary for a while, but by 2020 it became mandatory, so far for the inhabitants of Beijing only, but possibly soon for all. And by 14 November 2022, the government released a draft law "Establishment of the Social Credit System" for fine-tuning the system. The arguments of the Chinese government for it: honesty in government affairs, commercial integrity, societal integrity, judicial credibility.

In the West, we are faced with the task of developing *global* standards for how to correctly deal with AI technologies. At the political level, the effort to develop global standards seems to be at an impasse. But experts are putting on pressure, as we saw with the statements above. And human rights groups have realised that a new and extremely important field of activity has opened up. Another example of a discussion: In December 2018 (revised April 2019) the EU Commission set up an expert group that addressed the public (without the latter noticing much of it) with a paper entitled "Ethical Guidelines for trustworthy AI",[23] which explicitly asked for public input (which the first author turned in in details, but never received any feedback). The experts unequivocally argue that we need an anthropocentric approach in order to shape the development of AI in such a way that:

- the individual human being, his dignity and freedom are protected,
- democracy, law and order, and civil rights are upheld,
- equality, protection of minorities and solidarity are promoted.

---

[23] See https://ec.europa.eu/futurium/en/ai-alliance-consultation.1.html.

To achieve what the authors call "trustworthy AI", the paper identifies a wide range of measures at both technical and non-technical levels. Basic guidelines for any AI technology must:

1. do good to people,
2. do not cause human suffering,
3. promote human autonomy,
4. apply the principle of fairness; and
5. achieve transparency and traceability in all effects.

All these rather vague expressions have to be made significantly more concrete. Currently, the EU parliament is working on the EU AI Act, which is poised to be the World first regulation on artificial intelligence. This Act will categorize AI systems based on varying risk levels, enabling tailored regulations that correspond to each level of risk. The discussion about human rights in connection with AI has different components. As it can threaten our independence, it often also has the potential to enable us to receive proper treatment in difficult situations, which is surely supported by human rights, e.g.

- quick and reliable answers to health care questions,
- save costs and time dealing with complex issues and data,
- invest cost efficiently and more successfully our money without the banks ripping us off,
- determine what data we can give out to whom online (what we actually need is much more anonymization when dealing with answers to questions, our shopping and other things we send out to the internet),
- make decisions in many areas faster and smarter with less errors.

How autocratic countries like China will develop increasingly powerful AI is hard to foresee.

The Development of Big Data

The concept of big data itself is relatively new, as only since recently computers are powerful enough to process these amounts of data. However, the conceptual origins of large data sets go back to the 1970s when a world of data was just getting started with first data centres and databases. Data storage was still very expensive in the 1980s and early parts of the 1990s, while today the costs of packing up huge data sets has declined so heavily that they are seen as next to nothing. Around 2005, IT companies began to realize just how much data users generate through Facebook, YouTube, and other online services. Besides falling costs, the following developments led to more and more data storage:

- Advent of the Internet of Things (IoT), in which more and more devices are connected to and stored in the internet,
- Ability to gather data on customer usage patterns,
- Possibility of product (selling) performance measurements,
- Emergence of machine learning,
- Prevent fraud and improve compliance,
- Maximise operational efficiency.

Processing large amounts of data has also become an important part of science as well. It enables highly efficient ways to plan, perform, publish and peg research activities, as large data storage devices have provided new ways to produce, put in storage, and probe scientific data. Across any discipline scientists have built the ability to link and cross-reference data from diverse sources and thus improve the accuracy and predictive power of scientific findings. In fact, an important part of science has become modelling the scientifically examined processes on the computer. The processing of big data and then employment of it for modelling the reality has become - next to experiment and theoretical formulas - a *third feature of science,* which creates new discussion what science really consists of.

What we are witnessing today is a continuing "datafication" of our social life. Our activities and interactions are being monitored and recorded with ever increasing effectiveness. This means that all of us generate an enormous digital "footprint". The resulting huge loads of data are a treasure trove for research, but also for sales and return generation, with ever more sophisticated tools being developed to extract knowledge from such data. The search in big data sets is more efficient when combined with artificial intelligence. What sounds appealing and efficient and promises to make our life more comfortable has like so many other developments also a negative side: In a world in which AI with Big Data performs more and more functions and extractions of us, not just our privacy, but also our independent thoughts, doings and appearances are increasingly disappearing. This is as dramatic as it sounds.

In dealing with big data, one has to process high volumes of low-density (i.e. not very highly concentrated information) and unstructured data. For some organizations, this might be tens of terabytes (10^{12} bytes) of data. For others, it may be hundreds of petabytes (10^{15} bytes) or even exabytes (10^{18} bytes). Global data volumes reach new unprecedented highs every year. It is estimated that over 2.5 quintillion ($2.5*10^{18}$) bytes of data are created every day. And as a result, there are over 100 zettabytes (10^{21} bytes) of data created in 2022, after 79 zettabytes in 2021 and 41 zettabytes in 2020 – which highlights an increase of almost 500 times from 2005. In 2025 more than 200 zettabytes are estimated. These larger and larger datasets are (most often, but not always) produced in a digital form and can be analysed through computational tools.

But how is that amount of big data created and *concretely* dealt with? Well, since the 1990s we are all living in the *Information Age*, an era, where we can analyse, review and create data in all areas of our lives. Today, from the moment we wake up to the second we go to bed we constantly create data that can be assessed. Just an obvious example: Every time we pick up our phone and use a search engine, data is being created: With an estimated 3.7 billion humans using the internet, there are on average a staggering 5.6 billion searches made every day with Google alone. Every time a doctor puts in some research results into the internet the data will be stored forever, and every time we buy something from the internet, the data of what we buy stays for all time accessible for many producers, vendors, or whoever.

Thus, by now, collecting and processing big data has, as we saw, become a core business of large companies like GAFAM or the Chinese companies Tencent and Baidu. The development of open-source frameworks like the one offered by Amazon or Dropbox was essential for the growth of big data, easily accessible and ready to be used. One particular example from research, however dealing with a key issues of today, is the following: At the department of geography of the University of Hong Kong, Professor Peng Gong leads a multi-million dollars project to integrate all the pictures taken by satellites since 1985 all around the World into a database. With the help of AI, his team and other academic researchers are thus able to detect alterations of the Earth due to climate changes and the population expansion.[24]

Today, the following three "v-points" most popularly describe big data:

- greater and greater variety,
- very high volumes,
- more and more velocity.

---

[24] Liu Han, Gong Peng, Wang Jie, Wang Xi, Ning Grant, Xu Bing, *Production of global daily seamless data cubes and quantification of global land cover change from 1985 to 2020*, iMap World 1.0., Remote Sensing of Environment (2021).

Variety refers to the many types of data that are created, processed and then made available. Traditional data types were stored and processed in a relational database, i.e. in collections of data in predefined relationships (e.g. in Excel). In other words, data was collected from one place and delivered in exactly one format. That has to be different for big data. Besides that traditional data processing software simply cannot manage its massive volume, big data also comes in new, diverse unstructured types such as texts in.csv or access, audios or videos, clickstreams on a web page or a mobile app, Twitter data feeds, or sensor-enabled equipment. It thus does needs pre-processing to be stored and exploited.

The *volume* of big data is what defines it. We have already talked about USA's Frontier, which has exceeded the exaflop limit with 1.102 exaflops (10^{18}); even desk top computers process about 100 terabytes (10^{14}) per second (as of 2023). So, faster and faster computers can process more and more data in less and less time.

Velocity essentially measures how fast the data is coming in and being processed and (often) acted upon. However, the incoming data is by no means regular. Some data comes in in batches. Others, like many internet-enabled products, operate in real-time and will require real-time evaluation and action.

There are two more characteristics of big data that have emerged over the past few years and need to be added to the v-list:

- aiming for high levels of veracity,
- the given value.

How truthful is the dat - and how much can one rely on it? *Veracity* describes the extent to which the quality and reliability of big data can be guaranteed. Especially data with high levels of volume, velocity and variety are coming with significant risk of containing errors, inaccuracies and biases. In the absence of appropriate validation and quality checks, this could result in a misleading base for claims of knowledge by the users.

Data has intrinsic *value*. But it is of little use until that value is discovered. For this we need to assess the multifaceted forms of significance attributed to big data by different sections of society, as it can be quite different across the users. It depends as much on the intended use of the data as on historical, social and geographical circumstances. Alongside scientific (and thus the effort of objective) value, users may assign reputational (often politicians), financial (investors), ethical (philosophers), and even affective value (searchers of pleasure) to data. The companies storing and processing the data also have ways of valuing data (most often financial ones), which may not always overlap with the priorities of the data users.

Just as for AI, data regulations have to be put in place providing policies and laws that processed data is governed and shared appropriately. In the USA, as we have seen, this goal is expressed much weaker than in Europe. It is clear that data usage will have to be regulated much more, otherwise we risk ending up in an environment close to the one described in Orwell's novel 1984 and managed by huge monopolies who will rule our private lives.[25] This is reinforced also by the rapid development of cloud computing that implies the transfer of data to big computing centres, often abroad. Cloud computing is equally heavily concentrated in the GAFAMs (plus IBM). Amazon in particular has enormously developed its offer in cloud computing through Amazon Web Services (AWS). This service now represents one of the group's main sources of profit. In Switzerland, a great controversy emerged in 2022 after the government's decision to contract a foreign consortium to manage federal administration data in a cloud.

Artificial Intelligence's Possible Consciousness of (Strong) AI

A final step of AI is to come with an artificial consciousness: a non-human being that has an inner life with subjective sensations. AI would then "feel" - just like us - how it is to be this being. Researchers refer to this as "synthetic phenomenology". How subjectivity can emerge from matter is exactly the "hard problem of consciousness" as David Chalmers called it (see last chapter). When we think about the possibility that AI machines might one day have *self*-consciousness, a number of interesting points arise to philosophical thinking about the nature of such a mind:

- How does an AI express subjective preferences?
- Does AI potentially possess next to qualia intentionality (phenomenal consciousness), i.e. define particular goals in the mind in order to reach a particular state (see Chap. 9)? In particular, could this intentionality be different than the human one?
- What exactly is the relationship between matter and mind in the AI consciousness?
- How does consciousness arise from non-conscious components?

---

[25] The science fiction writer Alain Damasio has beautifully captured this dystopic future in his novel *Les Furtifs* (The Stealthies), La Volte (2019).

- Where do, compared to the AI, the (evolutionary) origins of our consciousness lie?
- How can we interact with conscious AI?
- Would the AI have an ethical and moral framework?[26]
- Would the AI be subject to suffering?
- Is a conscious AI entirely rational or does it come just as us with irrationality in thought and actions?
- Is AI consciousness a danger for us human beings?

The possible creation of a - most likely - super-human artificial intelligence poses more major ethical challenges. How should we deal with and treat artificial consciousness? Do we have a personal relationship to it?[27]

There is, however, also a second, inversive side of the coin: What happens to us at a point where AI superintelligence surpasses our own intelligence? The more consciousness or even ego-feeling the machines gets, the more we are being challenged in the way we live, or even our very existence. Even today it applies that the more powerful computers and their algorithms get the more dependent we become on them. It is therefore clear: This dependence is going to be significantly higher in the future. What if a conscious AI controls us? The movie "The Matrix" (1999) already illustrates what our lives could be like then.

But even without answers to all these questions, we can say that the creation of a conscious super-intelligence would be at least as impactful and threatening and as powerful, and possibly even scarier as a "game changer" than a first contact with an extra-terrestrial intelligence. Unlike the latter, we can already outline a very realistic path towards a super-intelligence. We would be well advised to deal with related issues and possibly launch a program to detect any of its features early enough. Because once it is there, its momentum for further growth can develop astonishingly fast and be not controllable by us any longer.

One thing is, with all the uncertainty about the rise of a strong AI, certain: The combination of AI, robotics and virtual reality will change our image of ourselves more than all traditional discourses about humanity in our history. A huge effort is necessary to defend and retain our own humanity and not let ourselves be overrun by powerful AI and consciousness technologies such as VR and their possibilities for manipulation. We will have to determine a new

---

[26] The famous "Three Laws of Robotics" by American SF author Isaac Asimov, developed in the 1940s, might come to mind here. Could they be installed, would they be enough?

[27] An interesting novel around this question is: Ian McEwan, *Machines Like Me*, Jonathan Cape (2019).

comfort zone for this, where it must not be about going where it is particularly pleasant, where our neuronal reward receptors are best stimulated. We must define where we want to be as humans. The AI-VR expert Kai-Fu Lee warns us[28]:

> If we learned anything from the recent concerns about externalities of social networks and AI, we should start thinking early about how to address the inevitable issues when these externalities multiply with XR[29]. In the short term, extending laws may be the most expedient solution. In the longer term, we will need to draw on an array of solutions, including new regulations, broader digital literacy, and inventing new technologies to harness technological issues.

In fact, VR is superimposing a filter to our six senses that help us build our "model of the world", we described in the previous chapter. Putting ourselves further away from reality. But can we really shape our own future under these conditions, or have we already set in motion a technological machine that we can no longer stop? Are we powerless at the mercy of the interests and manipulation possibilities of corporations and governments steered by short-term power and financial goals? We ourselves must decide whether we want to let the reins be taken out of our hands and watch how technologies change us from how we know ourselves today. Can we remain masters of our technology or will technology become master of us?

While conscious computers are still much of a theoretical thing: Consciousness remains as perplexing as it has ever been. But, AI consciousness or intellectual singularity might come suddenly without us having decided to create it. It could even get here before we have comprehended how strong AI works. It would then simply start existing by the pure complexity reached by our AI systems. After all, mankind's consciousness occurred with increasing complexity of our brain (however, most likely not very suddenly; but we do not know this for sure: Proto-humans could have suddenly achieved such brain complexity and became conscious; this would also be possible with AI). In this case, machines will likely let us quickly know about it, and it will then be too late for us to act on it. It is consequently all the more important to ask and deal with the respective philosophical questions

---

[28] Kai-Fu Lee and Chen Qiufan, "AI-2041, Ten visions for our future", Penguin Random House, London, 2021. Kai-Fu Lee was formerly the president of Google China.

[29] "XR" is the experts' name for a mixture of Virtual Reality (VR) that renders a fully synthesized environment, Augmented Reality (AR) superimposing another layer on top of the physical world and Modified Reality (MR) which mixes virtual and real worlds into a hybrid world.

now, so that the designed AI systems adhere to certain rules we deem essential for preserving our freedom and human dignity.

The today well-known sentence from the philosopher Protagoras is "Man is the measure of all things".[30] The pre-Socratics focussed on the personal experiences of humans. A common interpretation of them is that their expressed knowledge and given experience can be altered by changing circumstances of human perception. This statement receives a particular importance today in what we saw about AI.

---

[30] The full sentence is: "Man is the measure of all things, of things that are, while they are, of things that aren't, while they aren't.".

11

The Path Towards Modern Mathematics
More and more abstraction as well as more and more concrete applications

Mathematics today has so many aspects that it is practically out of reach to cover it adequately. It becomes even more difficult, if not completely impossible, to undertake this for non-mathematicians. It is neither the place here, nor our competence, to draw all mathematical details over the last 150 years in a common language. We would rather like to focus on a few important mathematical turning points to help us understand the parallels to science (r)evolution in the field of mathematics, particularly since 1960, and where we stand today in mathematics. In this chapter we therefore concentrate on three points:

- The reaching of the limits of reason wrecking the consistent completeness of mathematics.
- The rush to abstraction which has similar origins as in physics: the inability to solve some problems with concrete and vivid tools.
- The path to applied mathematics in parallel to the rush towards mathematical abstractions which contributed significantly to our modern age.

Mathematics Before 1920

Although this book targets the period from 1960 to today, it is hard to understand modern mathematics without stepping back for more than 100 years for a moment. In 1905, when Einstein had his *annus mirabilis*, his year of miracles, he thought - explicitly disrespectfully - that mathematics was a pure calculation engine that must be mastered by physicists to formulate and then solve their equations. This might have been appropriate with his *special* theory of relativity. But when a few years later he got to his *general* theory of relativity, he realized that this required an abstract level of mathematics unseen by physicists ever before, and that he severely struggled with. In fact, his theory of 1915 was the first theory of physics that needed abstract mathematics: Differential geometry. When 15 years later physicists tried to formulate quantum mechanics mathematically consistent, they equally realized that they needed a level of abstraction in mathematics never seen before ….and which simply did not even exist at the time! The corresponding new mathematics based on functional analysis[1] and Hilbert spaces came out in 1932, *seven years after* the first equation of quantum physics were published, in John von Neumann's book "Mathematische Grundlagen der Quantenmechanik".[2]

Just as physics slid into a fundamental crisis in the early twentieth century, that shaped its features as well as philosophical status in the twentieth and twenty-first century, mathematics suffered from the same fate. In fact, the situation in mathematics around the turn of the twentieth century was quite comparable to the one in science: In the late nineteenth century the structure of mathematics seemed to be more or less explored and understood. Completed fields were:

- Geometry, to which the mathematicians of antiquity had already made significant contributions, had been perfected.
- Concrete algebra, with the help of which solutions for equations with several unknowns are found, had essentially been formulated.
- Analysis, which deals with the determination of the properties of functions and their derivatives and integrals. Here the problems of the infinitely small in integrals had finally been solved.

It was mostly about finding solutions to various *concrete* equations, as one characteristic of mathematics until the late nineteenth century had been: clear lucidity, The human's mind was able to imagine the mathematical problems explicitly.

---

[1] A very important pioneer and founder of modern functional analysis is surely the Polish mathematician Stefan Banach (1892–1945) from Lwow (in Ukraine called Lviv nowadays) and his friends of the "Scottish café", among them Stanislaw Ulam, whom we encountered before as the inventor of Monte Carlo simulations, Hugo Steinhaus, a student of Hilbert and the PhD advisor of Banach.

[2] Translation: John von Neumann, *Mathematical Foundations of Quantum Mechanics*, New Edition, Springer (2018).

Just as in physics a similar revolution then hit mathematics: The concrete and directly comprehensible connections and views were replaced by literally incomprehensible theories. You can draw Pythagoras' theorem on a piece of paper, and it is also possible to visualise the second derivative of a function. But towards the beginning of the second third of the twentieth century, mathematics started becoming more and more abstract. It became virtually impossible to imagine its statements and theorems pictorially or for the public non-specialists. The path towards higher and higher level of abstraction was caused by something: Mathematics, having always been seen as logically consistent and potentially completable, had fallen into structural incompleteness and even contradictions. To combat this, mathematicians tried to generalise the involved well imaginable things, which automatically led to abstractions. Already in 1900, the great mathematician David Hilbert had formulated his famous list of 23 open problems that needed to be solved before mathematics could finally be seen as complete - he was hoping this would soon be the case. Unfortunately, only a few years later, a young mathematician called Kurt Gödel proved that a complete and non-contradicting mathematics is principally impossible. At that point in time, the outstanding (female) mathematician Emmy Noether had already started a new highly abstract formulation that today is called Abstract Algebra.[3]

Some abstraction in mathematics started even earlier, i.e. at the beginning of the nineteenth century, with the two young geniuses: Evariste Galois, the inventor of the notion of *groups* in mathematics, and Nils-Henrik Abel and his *Abelian functions* and theory of elliptic functions. They were accompanied by Carl Friedrich Gauss, the "Princeps mathematicorum" (Latin for "the foremost of mathematicians"), who claimed to have discovered the possibility of non-Euclidean geometries and set the stage for the modern way of proving mathematical theorems. This flight to abstraction was triggered by the even back then appearing impossibility to solve certain mathematical questions. However, for almost 100 years it did not become mainstream in mathematics.

---

[3] For more details about Emmy Noether, see Lars Jaeger, *Emmy Noether - Ihr steiniger Weg an die Weltspitze der Mathematik*, (only in German) Süd-Verlag, Konstanz (2022); also in Lars Jaeger, *The greatest 18 female scientists in history*, Springer (2023) and Lars Jaeger, *The Stumbling Progress of 20th Century Science - How Crises and Great Minds Have Shaped Our Modern World*, Springer (2022); also in David Rowe, *Emmy Noether – Mathematician Extraordinaire*, Springer (2022).

The Crisis in Mathematics[4]

The triggers for the shock waves in mathematics and physics had the same origin: the features of infinity. While physicists were despairing in the world of smallest structures, whether atoms exist or not, and they suddenly perceived structures that were so different in their behaviour from what we know in our world, mathematicians equally lost their footing as soon as they approached real infinity - "real" in two meanings: objectively existent, and real numbers versus countable ones. In both disciplines, physics as well as mathematics, dealing with ever more extreme quantities was leading to apparently unsolvable practical contradictions.

In fact, that the usual explanation of the world collapses as soon as one deals with the concept of (real) infinity was already known to the ancient Greeks, be that expressed by Zeno's turtle paradox, the arrow contradiction or the question of whether or not matter is infinitely divisible. The paradoxes they handed down are mostly about a process being repeated an infinite number of times, resulting in smaller and smaller units. Even today, these contradictions confuse our senses because logic and experience simply do not fit together. The real problem in mathematics, however, was not infinity itself, but the discovery that there were various *types* of infinity:

- *Potential* infinity is exemplified by the natural numbers (1, 2, 3, …). For every natural number, you can always find an even larger one, which is why there is no natural number, no matter how large, that itself is infinite. Already the ancient Greek mathematician Euclid proved that there is an infinite number of prime numbers. One can also quickly prove that the set of natural numbers is as large as the set of all even or the set of all uneven numbers (as well as the one of all prime numbers, or all numbers with a "5" in it), which is completely counter-intuitive since any finite set of even and odd numbers would be twice smaller than its equivalent set of natural numbers. But infinity often yields results that are intuitively harder to follow than those for finite systems. The same applies for the set of all rational numbers, which have the same number of components as natural numbers (i.e. there is a one-to-one correspondence between

---

[4] This introductory sub-section is a shorter and somehow modified version of a chapter in: Lars Jaeger, *The Stumbling Progress of 20th Century Science - How Crises and Great Minds Have Shaped Our Modern World,* Springer (2022), Chapters 3 and 9.

the components of the two sets). This is even less intuitive, but German mathematician Georg Cantor could show this in 1874.[5]

- *Actual* infinity is characterised by the fact that the members of the set cannot be generated using a finite algorithm. An example that students in the middle level of high school already know is π (the relationship between diameter and length of the circumference of a circle).[6] The number series after the main number are infinitely long with no regularity. A set of these numbers with actual infinities lies within the set of *real numbers*. There are infinitely many numbers in that set with actual infinities (called irrational numbers; there are many more of them than those with a potential infinity only). There is a reason why the ancient Greeks desperately tried to avoid numbers like $\sqrt{2}$ which cannot be given by a finite of numbers behind the comma (they did know about the diagonal of a square with side lengths of 1 which has that number with actual infinity - an astonishing achievement itself - but did not want to go deeper into it).

For a long time, there were only the first kind of sets mathematics dealt with or wanted to deal with (although the real numbers had been known already for a while in the early nineteenth century, but mathematicians avoided to address their inherent features). The second type was considered a topic for philosophical, rather than mathematical discussions. The very statement that different forms of infinity could truly exist threw the mathematicians of the late nineteenth century into a great confusion. In 1874, Georg Cantor also proved that the real numbers were actually more infinite than the natural or rational numbers, i.e. they exceed the potential, i.e. countable infinity ("Überabzählbar", "over-countable" in German). The non-countability of their structure make them actually infinite. Thus, entirely new principles had to be developed to describe *how* infinite an infinite set is. Although the proof was correct, Cantor was met with a barrage of angry reactions, among the most famous opponents was Leopold Kronecker who headed mathematics at Berlin until his death in 1891. He even called Cantor a "corruptor of youth"! In fact, Cantor had to stay as a mathematics professor in the provincial university of Halle during his entire active life despite the fact that he desperately wanted to be in Berlin. And this, even though Cantor's insights led the entire infinitesimal calculus to finally stand on a completely firm mathematical foundation.

---

[5] Georg Cantor, *Über eine Eigenschaft des Inbegriffes aller reellen algebraischen Zahlen*, Journal für die Reine und Angewandte Mathematik, 77 (1874) p. 258–262.

[6] π Is besides being an irrational number also a transcendental number, the cardinality (size of the set) of which is as the set of real numbers R.

Cantor also realised that there even exists an infinite hierarchy of increasingly larger sets – an infinite sequence of larger and larger infinities! For this, one only has to consider the sets of all subsets of the real numbers (which are called "power sets" by mathematicians), and then the set of all subsets thereof, and so on ("infinities", "infinite infinities", "infinitely infinite infinities," etc.). Cantor was able to show that the size of any set is smaller than the size of its power set, e.g. the power set of rational numbers is exactly as large as the set of real numbers. For finite sets this is self-evident, but for infinite sets this still had to be proved. Today, this theorem is called *Cantor's theorem.*

The bitter resistance of the experts to the hierarchy of infinitely many infinities introduced by Cantor had a deeper philosophical reason: Although it ultimately placed the infinitesimal calculus on a solid foundation and thus finally resolved certain problems in it, it opened the door to other paradoxes that questioned the validity of mathematics as a whole. In fact, Cantor did not succeed in placing his hierarchy of the infinite on a mathematically sustainable foundation, because he landed exactly where the ancient philosophers had already been shipwrecked: actual infinities seem to be simultaneously different and equal! This logical problem corresponds to the liar's paradox, already known to the ancient Greeks. Here is the most widely known version:

Epimenides, the Cretan, says: "All Cretans lie."

Epimenides' claim can be true or untrue, and in each case, he is a liar and at the same time not a liar. No matter how you turn it, the sentence always ends in an irresolvable contradiction.

Mathematicians went round in circles in a similar way when they dealt with the cardinality (size) of the largest infinite set. It was clear that there is no largest cardinality of sets, i.e., no largest set, because the power set of that set must have an even larger cardinality. If we now add Cantor's theorem …

The size of any set is lower than the size of its power set.

… an insoluble contradiction arises. For if one considers the set of *all* sets, it would also have to contain its power set. This power set would thus simultaneously lie inside and outside the set with the greatest cardinality. The British mathematician Bertrand Russell put this into another form in 1901. He said:

Consider the set of all sets that do not contain themselves.

If this set is not itself an element of itself, then this sentence says that the set must contain itself. But if it contains itself, then it falls outside the definition as the set of all sets that do not contain themselves.

Since set theory is the basis for all other branches of mathematics, the paradox of Cantor and Russell threatened the entire subject. For in logics, it is true that every statement can be proven from a contradiction. This can be clarified – not quite cleanly from a technical point of view, because logical and linguistic elements are mixed – in the following way: Someone says: "If 1 + 1 = 3, then elephants can fly." If the premise is already wrong, the second half of the sentence can contain anything. So, if something is wrong in set theory, then there are only flying elephants in mathematics. It was a disaster!

The Revolution

Nevertheless, in the next 50 years the mathematicians led by David Hilbert and eight-year older Henri Poincaré (until he died - quite early - in 1912), tried hard to come up with a mathematical foundation that overcomes these logical problems by defining a basis that lets everything else in mathematics be drawn out of it. Both (but Poincaré much stronger) opposed the Cantorian set theory. French mathematicians such as Baire, Borel, Lebesgue very quickly forgot or surpassed the paradoxes of the Cantorian theory. However, it was a young mathematician who touched the logical discipline at its heart. During the conference of the Society of German Scientists and Physicians in Königsberg in 1930, where David Hilbert reiterated his position on the possibilities of mathematical completeness, once the open problems are solved, Kurt Gödel, a 25-year-old student, announced (in a bar with a few other mathematicians) a forthcoming publication of his "incompleteness theorem". He had proven that in any sufficiently complex axiom system - such as the arithmetic of the real numbers and also Cantor's set theory, consistency *and* completeness together are generally mutually exclusive. In such systems there are always propositions that are neither provable nor disprovable with the means of the respective system; the systems are in principle incomplete. Gödel's first theorem reads accordingly:

If an axiomatic formal system is consistent, it cannot be complete.

This means that, in contradiction-free systems there always exist unprovable statements. The proof of this first Gödel theorem goes back to the liar paradox of Epimenides. Now the Cretan Epimenides no longer says "All Cretans lie", but:

The proposition G reads: This proposition is not provable in the system G.

The contradiction is the same: If the statement G is true, it is simultaneously false; if G is false, it is simultaneously true.

With Gödel's theorems, attempts to find a satisfactory basis for mathematics ended. Mathematicians had to come to terms with the fact that ...

- ... mathematics must always remain open because of its own structure and logic,
- ... some of the mathematical problems (defined by David Hilbert) are fundamentally unsolvable,
- ... mathematics now contains a third category alongside "true" and "false", namely "undecidable in principle",
- ... structurally, the mathematical human mind has to live with uncertainty and ignorance.

What a shattering of faith in the purity and beauty of mathematics! The idea of eternal progress towards ever higher knowledges discovered was now gone once and for all, even in mathematics. Hilbert's dream of complete and consistent mathematics had been shattered. For Gödel's *second* theorem reads:

The consistency of axioms within an axiom system cannot be proven within their own system.

However, a small hope remains: Gödel's incompleteness theorem does not completely rule out the possibility that the consistency of a theory can still be proven by completely different means (i.e. outside of the measures). Undecidability only means that the truth or falsity of a statement cannot always be proven within the system under consideration. If a larger theoretical framework is used, everything is open again. Whether there are "once and for all undecidable" statements whose truth value can never be known was not the subject of Gödel's considerations. However, according to the theorem, the theory that would complete the other theory would then contain an undecidable itself... Thus, contradiction cannot be eliminated from rational theories. Reason has brought us to its own limits.

In the forty years between 1920 and 1960 mathematicians went so far into abstraction that they shook up the very foundations of their subject. Many of them also paid a high price personally. The fact that people, who revolutionise thinking mentally, often go at the same time to their limits emotionally, and not infrequently beyond them, can be illustrated by the example of the mathematical physicist Ludwig Boltzmann who killed himself in 1906. And Georg Cantor, who suffered from phases of mental instability, exacerbated by the oppressive resistance of his professional colleagues, which he never abated. Cantor's severe depressions were increasing, and he repeatedly had to seek longer and longer periods of psychiatric treatment. In January 1918, he died in the closed ward of a clinic where he spent the last year of his life. Then French mathematician René-Louis Baire (1874–1932), who is known for the Baire category theorem, the definition of the "nowhere dense set" and his research on analysis, plagued with bad health and difficulties with the abstraction of his theories, committed suicide in Lausanne in 1932, after seven years of wandering in modest hotels on the Lake Geneva coast. Also tragic was Gödel's fate. He was highly paranoid and depressive. When his close friend and teacher Moritz Schlick was murdered by an Austrian Nazi in 1936, he suffered a nervous breakdown and thereafter had an obsessive fear of being poisoned. He only ate food that his wife Adele had prepared and tested it for him. At the end of 1977, Adele was hospitalised for a longer period, and therefore refusing to eat Gödel starved to death.

Before moving to the current evolution of research in mathematics, let us quickly notice the important *philosophical* implications of the impossibility to have a consistent mathematics. It is paradoxical that the most rigorous science has confirmed the existence of the contradiction and the fact that we need to accept its existence and learn to live with it. This is very parallel to Heisenberg's uncertainty principle, where accepting that there are limits to the precision of measurements opens up a whole new field of knowledge. However, even though these mathematical and physics discoveries date back almost a century, with some exceptions they have not really penetrated philosophy. We can expect that once philosophy will have broadly grasped this fundamental change; we will finally be also able to make significant progress in our view on the World.

The Path Towards Modern Mathematics – More and More Abstraction

While mathematics became more and more abstract the leading role of Germany, especially Hilbert's world-leading school in Göttingen, ended under National Socialism. Many of the Jewish and other mathematicians found refuge in the USA and fertilised, just like physics, the development of mathematics there. The famous Courant Institute of New York University

(NYU) is a testimony of this: Richard Courant, a German mathematician of Jewish origin, former student of Hilbert, came to NYU as a visiting professor. In 1935 he left his position as director of the Mathematics Institute at the University of Göttingen and founded in NYU one of the most renowned research centres in applied mathematics.

But was this introduction of heavy abstractions in mathematics just an aberration? Was it simply a crisis that was overcome eventually such that one can ride on the same street again? The answer is a clear "No": In fact, from here on mathematics became more and more abstract, until today. And exactly this proved very helpful as it provided new insights and connections between mathematical fields that had been considered unrelated to each other. Plus, many abstract concepts mathematics later turned out to be very fruitful for many scientific areas, especially for theoretical physics, that had also been growing more and more abstract. Gödel's abstract mathematics, which was understood by only a few experts worldwide, even made it into the real world of new machines that were soon to conquer the globe: computers. This occurred only six years after Gödel's publication of his incompleteness theorems, when English mathematician Alan Turing we have briefly mentioned in Chaps. 7 and 10 "translated" the abstract logical language into a comparatively easy algorithmic mechanism in a simple tape-based computer. With its help, Turing succeeded in 1936 in showing that the assumption of the existence of any general algorithm, that is supposed to decide whether, for *any* programme (decision matrix) and *any* tape input, the programme stops at some point or whether it continues to run forever, must contain a self-contradiction. So, there can be no such algorithm. This led to new mathematical paths, plus, last but not least, the programming of computers.

Research in mathematics in the later part of the twentieth century often appears to deal with largely entirely different problems to what mathematicians dealt with until the 19th and beginning of the twentieth century. Reading about modern mathematics as a non-mathematician is often like trying to understand ancient Asian languages like the Buddhist language Pali as a non-Buddhist (or even most Buddhists). However, the most difficult problems that have not been proven yet (like the seven Millennium-Problems) have rather descriptive characters. Aiming at a large readership this book is surely unable to go into the many details of the development of mathematics or even get close to its complete coverage between about 1960 and today. However, the reader should nevertheless - even if not understanding everything in detail - receive a general perception of how mathematics developed, while at the same time should not worry about not being able to follow the mathematics in detail. With all this being said, we like to emphasize that mathematics showed some astonishing features: With the help of its very high abstractness the mathematicians were able to solve some rather concrete problems that had not been solved for centuries.

One of the first abstract development in mathematics after the abstract algebra and algebraic geometry by Emmy Noether was the development of generalized measures. The standard geometrical measures (length, area, volume[7]) are well known. A general mathematical measure is a formalization of this geometrical measure and was developed in the early twentieth centuries by, among others, the French mathematicians Émile Borel and Henri Lebesgue and the Russian mathematician Nikolai Luzin.[8] With this Lebesgue went on proposing a new definition of the integral extending it to a larger class of functions. General measures have similarities to the geometrical measure such that they can be treated together in a single mathematical context. A particular - quite abstract - generalization of measures, developed by John von Neumann and David Hilbert in the early 1930s (with infinite-dimensional spaces), is until today centrally used in quantum physics.

Measures are also a foundation of probability theory, the axiomatic foundation of it developed by Andrei Kolmogorov in the 1930s, which ultimately established a new field in applied mathematics. For Kolmogorov, to deal with probabilities is quite similar to the calculation of areas on surfaces and can thus be treated with methods of measure theory. Though this was highly abstract mathematics, it turned into very useful field. Out of it developed a range of diverse statistical methods with broad applications in experimental science, medicine, but also in the social sciences and humanities, market research, risk management, economics and politics.

Abstraction and formalisation in mathematics was also the goal of the work of the French (and until today French) authors' collective "Nicolas Bourbaki", which included leading mathematicians in France such as André Weil, Jean-Pierre Serre, Henri Cartan and Claude Chevalley. The meetings of the respective Boubaki members began in the late 1930s and are still going on today. A membership in the Bourbaki group remains secret as long as one is member. Any member must retire when he turns 50. The name of the group goes back to General Charles Soter Bourbaki, who took part in the Franco-Prussian War of 1870/71, and refers to an inside joke from the six Bourbakis' student days. They took a worldwide leading role in the structural conception of mathematics after the decline of the Hilbert School. Initially, they wanted

---

[7] This was, by the way, the beautiful title of Lebegue's, path breaking doctoral thesis of 1901: "Longueurs, aires et volumes".

[8] Some very famous Russian mathematicians like Pavel Alexandrov and Andrei Kolmogorov were his students even if they did not like him later.

to overcome the curriculum in France, which was strongly oriented towards (concrete) calculus, in deliberate imitation of the Göttingen algebraic school. But soon it had an impact far beyond this, even quite practical with the *New Mathematics* in the school curriculum in France of the 1960s and 1970s.

The Bourbaki group refused to make generalised statements based on individual cases – as so common in the mathematics until the twentieth century. For its members conclusions were to be drawn the other way round: One should look for relationships that were as general as possible, from which the special cases could then be extracted. This was precisely Emmy Noether's, Hilbert's and Kolmogorov's approach to mathematics: The more abstract the relationships of different mathematical entities, the more generally valid and powerful the findings between them are. In fact, the founders were explicitly inspired by the desire to adopt ideas from the Göttingen School especially those of Emmy Noether and David Hilbert (where the younger Kolmogorov also spend some time, especially with the one generation older Emmy Noether). In the course of time, the Bourbaki group became more and more ambitious and planned a whole series of textbooks, which addressed *all* of modern pure mathematics: set theory, abstract algebra, topology, analysis, Lie groups and Lie algebras (the latter two becoming very important for theoretical physics in the 1960s, as we saw in Chap. 6). This ambition fell, however, short, particularly in probability theory and category theory, although the first volumes of the Bourbaki group were ground-breaking and very influential. Furthermore, their refusal of theoretical and applied computer science constrained their range significantly.

Now, fasten your seatbelts, at least for a while: The more and more abstract mathematics shall be demonstrated by the following lines (not assuming that most readers understand the details here, it is just supposed to be illustrations). Significant was the fundamental upheaval of *algebraic geometry*, especially through the work of Alexander Grothendieck (see Chap. 4) and his school, as well as the broad development of *algebraic topology*, and - partly in tandem with this - the development of "category theory". Algebraic topology is a general theory of mathematical structures and their relations to each other and was introduced by the Americans Samuel Eilenberg and Saunders Mac Lane. Category theory offered an alternative to set theory as a theory of fundamental structures. This was a further increase in abstraction and provided new approaches and ways of thinking that have become effective in large parts of mathematics.

Important new developments such as the Atiyah-Singer index theorem or the proof of the Weil conjecture are reflected in awardings of the Fields Medal and the Abel Prize (the Fields medal in 1963 was awarded to Atiyah, and the Abel price in 2004 to both Atiyah and Singer). Both were influenced by ideas and the school of Alexander Grothendieck, The first is a central statement of global analysis, a mathematical subfield of differential geometry. It sounds as complex as it is - so typical for modern mathematics - and states that for an elliptic differential operator on a compact manifold, the analytic index (for each one of a certain class of linear operators that can be "almost" inverted there can be assigned a particular integer number, creating the analytic index) is equal to the apparently more general but easier to calculate topological index that is defined via topological invariants. It is a more generalized version of the well-known (among mathematicians, especially those focussing on differential geometry) Gauß-Bonnet theorem that states about differentiable surfaces linking their (local) geometry to their (global) topology by establishing a relationship between curvature and Euler characteristics. The latter is the definition of a key figure (a topological invariant) for topological spaces, for example for closed surfaces such as convex polyhedron, in which the number of vertices is K, the number of edges E, and the number of faces F. Here the relation between holds as $E - K + F = 2$.

The Weil conjecture proven by Pierre Deligne in 1974 (and in 1980 a second proof of a more generalized version) is even more complicated. It makes statements about the generating functions formed from the number of solutions of algebraic varieties over finite bodies, the so-called local zeta functions. Because it was assumed that these were rational functions, i.e. that they obeyed a functional equation and that the zeros were located at certain geometric locations, similar to the Riemann zeta function and the respective Riemann conjecture, according to which the zeros of this complex function are complex numbers with the real part being multiples of ½. If so, the zeta function would be an information carrier the distribution of the prime numbers.

The formerly so powerful German mathematics was lying on the ground after WWII. A specific person that helped it to stand up again was Friedrich Hirzebruch. He researched in the fields of algebraic geometry, topology, number theory and singularity theory. With the Hirzebruch-Riemann-Roch theorem, named after him, which initiated one of the most important developments in modern mathematics, he laid the foundation for his internationally high reputation in 1954.

Does the reader get lost with all these abstract information? Well, this is simply often what modern mathematics is like. However, with the higher and higher level of mathematical abstraction in the twentieth or twenty-first century many concrete problems, some of them centuries old, were finally proven. Here are some examples, some of them (the problems, not the proofs!) quite easy to understand, even for a mathematical layperson:

- The four-colour theorem (1976, required a computer, see Chap. 10): Four colours are always sufficient to colour any map in the Euclidean plane in such a way that no two adjacent countries get the same colour.
- The Kepler theorem (1998, required a computer, proven without a computer in 2017): There is a densest way to stack equally sized spheres. That is pyramid-shaped, one layer on top of the other in the in-between spots of the respective lower layer. This is how oranges are packed in food stores (if they are of more or less equal size). The volume is then just over 74 per cent filled with spheres - more is not possible. In three dimensions the proof was particularly difficult and took many, many pages of highly abstract mathematics, although it is an obvious real-life problem.[9]
- The classification theorem of finite groups, also called the "enormous theorem" (proof happened between 1955 and 2004 and required 500-odd journal articles written by about 100 authors): Every finite group of odd order is solvable (a group is solvable, if it has a subnormal series with abelian factor groups, i.e. it can be constructed from abelian groups using extensions).
- Fermat's theorem (proven by Andrew Wiles in 1994), one of the most prominent previously unsolved problems in mathematics: The nth power of a positive integer cannot be decomposed into the sum of two such powers if n is greater than 2:

$$a^n + b^n = c^n$$

where n, a, b, and c are natural number with n>2. The very easy to understand proposition was first stated as a theorem by Pierre de Fermat around 1637. While it was proven for a few numbers over time (Sophie Germain proved it with a very new technique for an infinite number that have a very special characteristic[10]) the proof got more and more difficult for higher numbers. It took Wiles eight years and few trials[11] to prove the theorem for all numbers,

---

[9] In 2016, the female mathematician Maryna Viazovska proved in a paper of only 23 pages that the E8 lattice realizes the densest sphere packing in eight-dimensional space when spheres of equal radii are placed around each of the lattice points. Her proof was considerably simpler than that of Kepler's corresponding 3-dimensional problem. Only one week later, Viazovska proved in a 17-page paper that the Leech lattice provides the optimal sphere packing in the 24-dimensional space. For this she received (as the second woman ever) the Fields-Medal in 2022.

[10] See Lars Jaeger, *The 18 great female mathematicians*, Springer (2023).

[11] His first presented proof in 1993 had a subtle error (see the related article on wikipedia: Wikipedia article on the subject: https://en.wikipedia.org/wiki/Fermat%27s_Last_Theorem).

during which he was about to give up a few times, but in late 1994 he was able to complete it.

- The Poincaré theorem (proven by Grigori Perelman in 2002, who then declined to accept any award - among those the Fields-Medal -, as he is critical of the mathematics establishment): Every simply connected, compact, unbounded, 3-dimensional manifold is homeomorphic to the 3-sphere (the 3-sphere is a sphere one dimensional higher than what we know as a sphere; thus one can be transformed into the four dimensional Euclidean space, the same that can be down with a 2-dimensional sphere into a 3-dimensional manifold).

As simple as some of these may sound in principle, such as Fermat's theorem, Kepler's theorem or the four-colour theorem (and maybe even Poincaré's theorem), the required mathematical collaborations and durations for proofs were, for each, of unprecedented size and required mathematics way beyond the rather straightforward problem. Fermat's statement that the margin of a book page was too narrow for a proof of his theorem was confirmed (most likely way beyond his expectation: Wiles' proof is over 100 pages long, and he needed mathematical understandings that went far beyond the mathematical state of knowledge in Fermat's day).

By the end of the twentieth and beginning of the twenty-first century, there was again a strong interaction of mathematics and theoretical physics via quantum field theories and string theory with surprising and profound connections in different areas of mathematics (infinite dimensional Lie algebras and Lie groups, supersymmetry, dualities with applications in counting algebraic geometry, knot theory and others). At that time the foundation of physics and the further research of it had reached levels of abstraction similar to that of mathematics (see Chaps. 6 and 7).

Today, abstractness dominates mathematics in many different fields often with, as it appears at first sight, little applications elsewhere. However, a second look reveals: Despite having become more and more abstract in the last 100 years, mathematics has also become more and more far reaching and applicable at levels and locations that could barely have been seen and considered before. In fact, by today mathematics has become ever more a key discipline across many different fields besides theoretical physics, such as in digitisation, via chemistry and biology (genetics), up to the understanding of financial markets. More widely, mathematical understandings and quantifications have become a "must" not only in natural sciences but also in social sciences, quantitative risk management and, paradoxically, for evaluating science itself.

Dealing with Concrete Problems Through Numerical Methods

A particular tool that made its way into mathematics after WWII (very slowly at first and then accelerating significantly) was the computer. Thus, besides the trend towards increasingly high levels of abstraction, a part of mathematics that explored concrete objects and problems in detail, developed equally strongly, supported by faster and faster calculating computers. The growing computer power led to a dramatic development of what is called numerical methods in mathematics, i.e. methods for solving complex problems that could not be solved (or at least not easily enough) by hand but be calculated relatively quickly (and exactly) by computers, using numerical approximations. This happened first for military purposes, for example in the mentioned development of the hydrogen bomb (simulation of the nuclear fusion process), but as of the 1960s increasingly also for everyday purpose – and this also beyond pure calculation (i.e. simulations, general mathematics such as proof searches, games, or virtual reality creation).[12] One of the early contributors to numerical applications was Courant with the finite element method which he published in 1943 and is now one of the ways to solve partial differential equations numerically. For this the computer discretizes the normally continuous variables of the equation – and can thus transform (highly complex) differential equations into difference equation that it can then solve.

Numerical algorithms are usually needed for one of the following reasons:

- There is no explicit solution for the problem or at least mathematicians do not know if such exists, for example, in the case of the already mentioned Navier-Stokes equations or the three (and beyond)-body problem (how do three bodies, like three gravitationally bound stars, that attract each other behave?), and, in general, integrals or differential equations that cannot be calculated by hand ("analytically").
- The solution representation exists, but it is not possible to calculate the solution quickly enough or is in a form in which calculation errors are

---

[12] Such as Ada Lovelace predicted in the nineteenth century, see Chap. 10.

very noticeable, for example, power series of higher power or optimization problems. When in applications the solutions are only needed to a particular (finite) accuracy, a numerical procedure can be more useful, if it provides sufficient accuracy in a (significantly) shorter time.

Nowadays, numerical methods, of which an increasing number exists, are important tools in every technical or scientific field, and often also beyond, e.g. in economics, psychology, medicine, risk management or sociology. An example outside of physics, chemistry or biology is the field of statistics in medicine, leading to enormously efficient medical evaluations - with the development of double-blind experiments and other of these significant statistical techniques that render the possibility to explore efficiency of drugs and treatment much more accurately. For many diseases there will likely be more developments in the future, it is thus essential that we refine the statistical evaluation of their efficiency.

Furthermore, numerical experimentation made many new phenomena accessible for the first time. A new discipline developed: Experimental mathematics, i.e. numerical experimentation for new mathematical solutions. Very popular examples are game theory developed by John von Neumann, fractals and chaos theory based on non-linear dynamics from the 1980s (see Chap. 6), and the catastrophe theory of the 1970s. The last topic includes mathematical methods that deal with the discontinuous, sudden "overturning" of properties or processes and thus the abrupt change of states in previously continuous dynamic systems. These systems, even if they strive for a stable state under most conditions, can experience erratic, non-steady, discontinuous changes in the solution when their parameters change. Examples for numerical modelling of important processes are:

- In biology, there can be a sudden change of behaviour in animals from flight to readiness to fight after reaching the "catastrophe point".
- Or – more current – the sudden increase of the global temperatures due to particular events that *suddenly* makes our climate jump to another level, e.g. the melting of the Siberian tundra resulting in a sudden outlet of huge amounts of methane.
- Advanced numerical methods have made weather prediction feasible for many more days than 30 or 40 years ago.
- The precise location of cell phones (and their users) requires Einstein's equation of general relativity calculated numerically by computers.
- Car companies improve the crash safety of their vehicles by using computer simulations.

- Hedge funds (private investment funds) use numerical analysis with the goal of calculating the value of financial instrument (like stocks and their derivatives, but also bonds, currencies and commodities) more precisely than other market participants and then benefit from that.
- Pension funds develop their asset and liability management strategy using large computer programs exploring scenarios as far as 60 years ahead.
- Insurance companies use numerical programs for extreme events that they provide insurance for and the respective actuarial analysis. They also determine the risk-based-capital they need for their business through Monte Carlo[13] simulations of few thousand risks. Such internal models have become the rule in insurance and reinsurance as well as in financial institutions.

Last but not least, modern computers also encompass abstract mathematics such as abstract algebra and geometry,[14] as most abstract mathematical theories eventually saw applications, often in physics, but also elsewhere. In fact, abstract mathematics has grown tremendously since the 1960s, in particular with computers, in probability theory and several other areas, all this at the highest level (as is shown e.g. in the work of the Field medallists of 2022). Important for numerical mathematics is calculative stability. A numerical algorithm is numerically stable if an error, most often occurring by approximations of the variable, does not grow during the calculation. This happens if the solution changes by only a small amount. This is, as we saw in Chap. 6), not the case for chaotic systems, which makes numerical analysis in many everyday problems (with non-linear dynamics) much more difficult.

Most numerical algorithms are implemented today in standard libraries, so users can just access the codes (software) from popular computing libraries such as MATLAB (quite expensive) and S-PLUS, as well as free and open-source alternatives such as R (similar to S-PLUS), FreeMat, Scilab and IT++ (a C++ library). Today there are also programming languages specific for broad numbers of programmers such as Python (which by today is the most popular programming language). Many numerical systems also benefit from the availability of arbitrary-precision arithmetic which can provide more accurate results. Furthermore, for less demanding numerical users, Excel spreadsheet software has hundreds of available numerical functions.

---

[13] Numerical algorithms based on random sampling to handle deterministic, yet complicated problems.

[14] For more details see Rishabh Choudhary, Ajay Kumar Gupta, *A Study on Computer Science and its Relationship to Abstract Algebra*, JARIIE, 4, 5 (2018).

There are also formal languages like Wolfram Language (used for Mathematica) and Maple that help doing analytical work on symbolic equations and expressions. Those languages have tremendous applications in solving complicated analytical calculations and have become indispensable for researchers in algebra and analysis (next to physics and even economics).

The number of mathematicians that are working in what is called *applied mathematics*, the application of mathematical methods for the calculation of respective problems, today outgrows by far the number of pure mathematicians, those that deal with the abstract relationships in mathematics. And this in different fields such as physics, engineering, medicine, biology, finance, business, computer science, and industry. The area of *operational research* (OR) is another important part of applied mathematics. OR targets at improving techniques for applying particular mathematical methods that aim at certain problems, this with techniques such as simulations, optimizations, neural networks (AI), stochastic-process models such as Markov decision processes, data analysis such as ranking and hierarchy processing.

Mathematics Today and in the Future

Until about the 1950s the work of *individual* great mathematicians was appreciated in the same way as the works of individual philosophers (Bertrand Russell, Ludwig Wittgenstein, Henri Bergson, Martin Heidegger, Karl Popper) or physicists (Einstein, Bohr, Heisenberg, Schrödinger, etc.) were. As for physics or philosophy, today's *living* mathematicians are comparably little known. One reason is the increased abstraction. Another one is the pure number of mathematicians (just as scientists). This led to an immense ramification and branching out of the fields, all of which arose in special areas almost non-observable today. Proofs of many new theorems are published annually in mathematical journals. These proofs essentially target a priori defined complicated mathematical calculus in which very special axioms apply that are often found nowhere else in mathematics. They present themselves as e.g. purely algebraic, topological or other fields in unusual, unapparent mathematical sign systems. It takes years of familiarisation to gain insight into these areas for writing a master or doctoral thesis in them. As the overall mathematical research has been thus greatly expanded and differentiated, the simple consequence is that there are only small, international communities of experts for each of the special parts of the field who meet at specialists' - rather closed - congresses.

With all the abstract and highly particular mathematical fields there are nevertheless continuously evolving important connections with science and other applications. Future advances will likely be based on a two-sided inspiration, from mathematics to science and vice versa. For *non-mathematicians* all this has always been a two-sided sword. Ever since the heyday of ancient Greek mathematics associated with names such as Euclid, Archimedes, Diophantos and Hypatia, mathematics has always been both, (in the eyes of the majority in the population) esoteric and (for many life areas) highly practical. A good example of a practical mindset was Thales with his measurement of the height of Keops pyramid by measuring the length of its shadow and wait until a wooden stick lying next to it had a shadow of the same size as its length. At that moment, the height was also identical to the length of the shadow. In fact, as the ancient Greeks already realized, we find problems of mathematical nature in all fields of human activities, such as agriculture, astronomy, linguistics, liturgy, games, etc. As we saw, an even much more abstract challenge to the human mind is mathematics today, while at the same time it is a more than ever important discipline for key technologies. In other words: Mathematics is and has always been abstract and practical at the same time. If one looks at a music piece by Johann Sebastian Bach, the layman sees nothing but an army of notes, which to him are incomprehensible symbols. The musician, however, recognises the shape of a structures hidden behind them. This is what happens to the mathematician within his field. What are wild and impossible to interpret symbols for many people inspires him or her and makes his or her soul vibrate just like what musical notes do to the musician.

How can we assess the future of mathematics? The great French mathematician Henri Poincaré wrote in 1908:

> La véritable méthode de prévision de l'avenir des mathématiques est l'étude de son histoire et de son état actuel.[15]
>
> (The true method of forecasting the future of mathematics lies in the study of its history and its present state.)

So, trying to comprehend the growing complexity of mathematics on the one hand, and the power of applied mathematics in so many different areas on the other hand, aims at understanding where mathematics is today and how it has grown into this status. This is a prerequisite for approaching the even more difficult assessment of the future's mathematics. What is a

---

[15] Henri Poincaré, *L'avenir des mathématiques*, in Revue générale des sciences pures et appliquées 19 (1908), pages 930–939.

comparably simple forecast is that numerical, more generally applied mathematics, is growing to further heights and possibilities to cover more and more complex problems (such as climate research and every field listed above), as the computers are growing more and more powerful. Formerly untreatable problems such as our global climate can already be reasonably well modelled to make reliable forecasts which will become even better (modelled in greater details) in the future. This is indeed so highly important for social, political and economic questions. Furthermore, some complex scientific questions that are still not treatable today might get treatable in the future with computers – going possibly even as far as understanding our brain and with that our self-consciousness?

The future of pure mathematics is, in contrast, the more difficult part to forecast. Will we at some point see proofs of the six best known mathematical problems (the Millennium Prize Problems listed by the Clay Mathematics Institute)?

- The Birch and Swinnerton-Dyer conjecture – makes statements about number theory on elliptic curves, i.e. the set of rational solutions to equations defining an elliptic curve.
- The Hodge conjecture – states topological information like the number of holes in particular geometric spaces - complex algebraic types - by studying the shapes sitting inside those spaces, which look like zero sets of polynomial equations.
- The Navier - Stokes conjecture – about existence and smoothness of solutions for the equation, and thus the existence and uniqueness of a mathematical solution for the equation for physical flow of linear-viscous liquids and gases.
- The P versus NP question – the question whether every problem the solution of which can be "quickly" "verified" - proportional to t^n - can also be quickly solved. This is equivalent to the question of whether P = NP is valid or not.
- Riemann's hypothesis – the conjecture that Riemann's zeta function has roots (zero values) only at the negative even integers (part of the hypothesis that is solved today) and complex numbers with real part being (natural number) multiples of ½ (possibly the most difficult and most sought-after problem).
- Yang–Mills existence and mass gap – an open problem in mathematical physics, i.e. with concrete applications in physics: For any compact simple gauge group G, a non-trivial quantum Yang-Mills theory exists on R^4 and has a mass gap $\Delta > 0$.

Besides these most well-known problems there are hundreds of open core problems across the various areas of mathematics. This sound like: "still" so many of them? Does pure mathematics make any progress at all? Well, a significant number of such problems has also been solved in the last 25 years, as we saw above. It is just that at the same time new problems constantly come up!

This brief and very fragmentary excursion into the most recent history of mathematics nevertheless showed us that this discipline, just as physics, had gone through a deep crisis that hit the boundaries of human reason, and the mathematicians too had to accept the existence of contradictions, in form of "incompleteness" and "undecidability". After this severe crisis, mathematics experienced its escape towards abstraction, followed by, and having been possible by computers with ever rising capacity, the explosion of applied mathematics, which contributed significantly to today's world.

12

Astronomical Research
The Oldest Science in History with the Newest Results of All Sciences

Astronomy is the science of the stars. The word comes from the ancient Greek *astronomía* and is a combination of *astron* (star) and *nómos* (law). With scientific measures searching for the "laws of stars" astronomers today investigate kinds and properties of many different objects in the universe far beyond what humans for so long have seen non-differentiating as stars, i.e. next to celestial bodies such as planets, the moon, dwarf stars including the Sun, also: asteroids, star clusters, pulsars (neutron stars), quasars (very luminous galactic cores), galaxies, clusters of galaxies and black holes. This includes interstellar matters such as the constant radiation occurring everywhere in space as well as the universe as a whole, its origin as well as its structure.

A (Very) Brief History of Astronomy Prior to 1960

Astronomy is the *oldest* science in history. The earliest signs of mathematical and scientific astronomical activities date back to the Babylonians around 4000 years ago. They discovered, for instance, the recurring cycle of lunar eclipses. Also the Egyptians dealt with the stars and planets: The oldest known astronomical texts from the Nile area date from 1220 BCE at the time of Ramses the Great. With these very old scientific efforts, astronomy's major *scientific results* today are at the same time across all sciences the *youngest* ones, being only a few years old and the result of very new ways of looking into the universe.

The beginning of astronomy lies in the observation and contemplation of celestial phenomena, in the cultic admiration of the stars and in the development of calendars and the measurement of time according to the movement of moon and sun. Almost all religions - Buddhism, Hinduism, Taoism, Judaism, Christianity, Islam - contributed to it. The close relationship between astronomy and astrology on the one hand, with a religion on the other hand, found strong expressions in history, with a particular intensity from the 16th to the eighteenth centuries. At that time natural science and the perception of God still lied inherently together. One example is the theological part of Johannes Kepler's (early) cosmological view of the world as a sphere combined with God:

> The image of the triune God is in the spherical surface, that is to say, the Father is in the centre, the Son is in the outer surface, and the Holy Ghost is in the sphere volume.[1]

Nevertheless, Kepler's and (his contemporary) Galilei's work in the early seventeenth century were the starting point of a process that eventually brought down the type of astronomy referring to a transcendental figure (God) that had lasted thousands of years.

However, the conflict between astronomy and metaphysical beliefs already found a recorded conflict with the Greek astronomer Anaxagoras in the fifth century before Christ. His beliefs that the heavenly bodies are the result of a developing process and that the sun was nothing more than a great burning stone resulted in his arrest. Although acquitted, he was forced to go into retirement that stopped his philosophical activities.

Then, in the sixteenth century, Nicolaus Copernicus ushered in a new era in astronomy. In May 1543, in his book *De revolutionibus orbium coelestium*, he mathematically demonstrated that the movements of the planets could equally correctly (as the geocentricism) be described with a heliocentric world view – however, without making many observations himself, except some planetary observations while in Italy. The invention of the telescope at the beginning of the seventeenth century constituted then the turning point in astronomy. Galileo Galilei used it to discover the four inner moons of Jupiter (called the Galilean moons to date) and the phases of Venus. This permanently weakened the ancient view of the world. Around the same time (1609), Kepler published his work *Astronomia Nova* with the first two Keplerian laws, which are still so well known today (his third law he published later, in 1621).

---

[1] Johannes Kepler, *Mysterium cosmographicum* (1596).

In the following years, more and more ground-breaking results of observing the sky with telescopes changed the way we see the universe. In the middle of the seventeenth century, Christiaan Huygens succeeded in discovering Saturn's rings, its moon Titan and the Orion Nebula far away in our galaxy. Then, in 1668, Isaac Newton came up with the idea of focusing light in telescopes with mirrors instead of lenses made of glass, and one year later he formulated encountering gravitation (better introducing it theoretically and mathematically). With his epoch-making work *Philosophiae Naturalis Principia Mathematica [Frage: bibliografische Fußnote?]*, published in 1687, Newton laid the first fundamental foundation of modern astrophysics by tracing any movement of planets and stars, including Kepler's laws, back to his theory of gravitation. In the same book he established the laws of all mechanics applicable on our Earth.

From here on the new discoveries of astrophysical properties accelerated even further across the eighteenth century, with better and better telescopes and more and more applications of Newton's theory. In the nineteenth century the introduction of spectroscopy made astronomy even broader. Nonetheless, detailed knowledge about our *overall universe* was still far away when the scientists entered the twentieth century. The main limitation was that a large part of electromagnetic waves coming from the universe never pass the atmosphere and are thus unobservable to us on the surface. The only type of radiation besides the visible light that passes through the atmosphere is the low energy radio waves with wavelengths of a few millimetres to about twenty metres.

Since the 1920s astronomers tried to overcome the Earth's atmosphere, trying to observe the entire electromagnetic spectrum, infrared, ultraviolet, x-rays, gamma rays. The first thing researchers did for that (in the first decade of the twentieth century) was flying in balloons, i.e. measuring ionizing radiation away from the earth's surface. One of these pioneers, Albert Gockel, introduced the term "kosmische Strahlung" – cosmic radiation, after his research in 3000 m height.

One important discovery was made with earthbound, optical telescopes, though: In 1925 Edwin Hubble - based on the discovery by his former assistant, Henrietta Swan Leawitt, that certain variable stars can be used to measure distances in our universe - showed that Andromeda was not just a nebula inside the Milky Way. The fact that our own galaxy is merely one of many, was another hard pill to swallow for all those who still believed in the uniqueness of God's creation, earth as the centre of the universe.

A Rather Recent Revolution in Observing the Universe

Systematic radio astronomy began in the 1930s but did not really take off until in the 1950s, after the Second World War. This enabled scientists finally to get a more detailed image of the universe. Coming to terms with this new information was challenging for people. As of then they were able to detect more galaxies as well as radiation from the centre of our Milky Way Galaxy. Particularly interesting was the discovery of an entirely new item in the universe: On 6 August 1963, the young female astronomer Jocelyn Burnell noticed a particular appearance in the endless signal line of observing the sky with radio waves that stretched over just a few millimetres on her many metres long paper measuring the radio waves. Burnell marked it with a question mark. On the following days, the tiny presence appeared whenever the radio telescope, which scanned different parts of space thanks to the Earth's rotation, was pointed at exactly the same part of the sky again after 24 h. The regularity of their occurrence told Burnell that this could not be a coincidence. It turned out that she was observing a *pulsar*, a term that she and her supervisor used as an abbreviation of "pulsating quasar",[2] and that is still called like this today.

In the late 1950s, unmanned (and later manned) spaceflight started leading to human beings in 1969 first ever visiting an extra-terrestrial planet that had been under investigation for millennia, the moon. Since then, there have been several vehicles without humans that carried out direct measurements from Mars, Jupiter, Saturn, Uranus, Neptune (the last two from faraway), and - most recently - from Pluto, and send them back through radio-magnetic waves to scientists on Earth.

We still had to wait until the 2000s when telescopes got sent into the orbit above the Earth for permanent observations, so that astronomers can now systematically observe the *entire spectrum of electromagnetic waves* in the universe – by using radio waves to send data back to Earth. This allowed astronomers to even look at (far away) planets outside of the solar system – among those potentially ones we could live on? The following provides the tools that are employed today to investigate the universe, thus creating a revolution of looking into the sky. Figure 12.1 then provides an illustration of the various observation devices.

---

[2] For more details see Lars Jaeger, *The greatest 18 female scientists in history*, Springer (2023), Chapter 16.

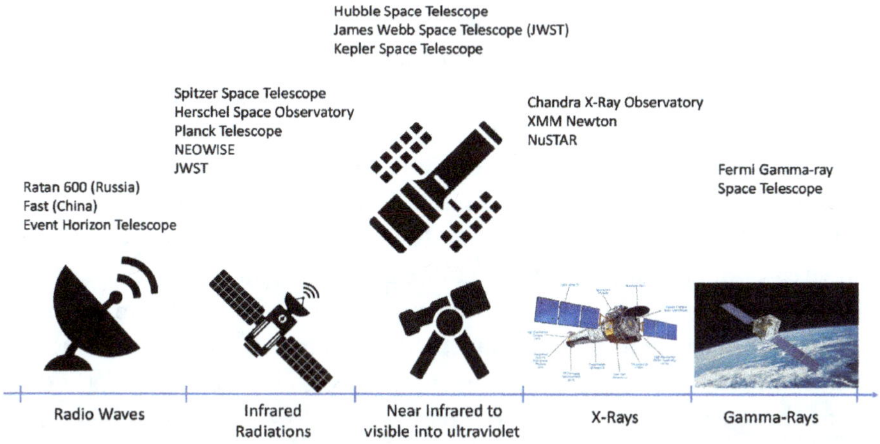

Fig. 12.1 Overview of the various observation technologies for stars and galaxies depending on various wavelengths covered (Source: Authors)

- *For radio wave observation*: As of the 1950s, astronomers started using radio wave receivers. Those dishes that look like giant round satellite plates have discovered numerous new celestial bodies. Today there exist many radio-wave telescopes around the world. The largest of them (in terms of diameter; as of 2023) is the Russian *RATAN 600* near Selenchukskaya, and the second largest the *FAST* observatory in the Chinese province of Guizhou. The largest telescope array consisting of a global network of radio telescopes (which helps to observe the very large wavelength of radio waves) is the *Event Horizon Telescope* (EHT), an international collaboration that networks eight radio telescopes into a single Earth-size dish structure.
- *For microwaves*: The first major discovery in the microwave spectrum was done with the *Holmdel Horn Antenna* in New Jersey: the cosmic microwave background, CMB (more on that later). Recent microwave telescopes, usually studying the CMB, are space bound: *The Cosmic Background Explorer* (*COBE*, 1989) and its successor, the Wilkinson Microwave Anisotropy Probe (WMAP, 2001). The Planck Surveyor (2009, also in infrared wavelenghts) was a probe launched by the European Space Agency, ESA.
- *For infrared radiation*: The most important current ones are: the *Spitzer Space Telescope* (2003), the *Herschel Space Observatory* and the *Planck Telescope* (2009), the *WISE (today called NEOWISE) Telescope*, and the *JWST telescope* (the later one going into the visible light, see below).
- *For wavelengths from higher infrared through the visible range into the ultraviolet range*: The earliest ones of these are regular telescopes (originating, as we saw above, in the early seventeenth century). Most prominent today is

the *Hubble Space Telescope* that observes from an orbit of 559 km distance to the Earth the universe (that provides much cleaner waves, which also helps for the visible range). With this telescope, astronomers were able to look further into space than ever before, observing it at a time when it was much younger than it is today, in some cases even in its very early stages ("younger" or "earlier", as light has a finite speed of 300,000 km/s, see below for further details). In spring 2020 the Hubble telescope completed 30 years of operation, with another one about 10 years expected to operate. Two of its main goals were to study the nearby intergalactic medium and to determine the Hubble constant, the measure of the rate at which the universe is expanding, which is directly related to its age (which let it determine the age of the universe very exactly, see below). The mid-IR-to-visible band successor of the Hubble Space Telescope is the *James Webb Space Telescope* (JWST), which was launched on Christmas Day 2021. An independent US-based project was the *Kepler Space Telescope* that lasted from 2009 to 2018 and targeted at finding Earth-size planets orbiting other stars within our galaxy (which it also found numerously – 2662 planets).

- *For X-ray observations*: With X-ray telescopes one can observe sky objects with temperatures between 10^6 and almost 10^9 K (i.e. up to several 100 million °C). Its radiation energy thus ranges from around 100 eV to 100 keV (kilo electron volt) which corresponds to frequencies of 3×10^{16} Hz to 3×10^{19} Hz (one electron volt corresponds to 11,604 °C). X-rays are thus a very good way to track highly energetic processes in space, from extremely massive black holes to the remnants of supernovae. However, they require a special construction of the telescope, otherwise radiation with such high energies would destroy it. So the mirrors are arranged in specially shaped tubes. Two X-ray telescopes currently in space are NASA's *Chandra X-ray Observatory* and the EU's *XMM-Newton* (both launched in 1999; with an observation upper limit of about 15 keV). With these detectors photons can be picked up from a broad spectrum of X-ray light in a previously unattainable image quality. A newer project, launched in 2012, is NASA's NuSTAR that can observe energies up to about 80 keV.
- *For Gamma-ray observations*: This covers the most energetic form of electromagnetic radiation in the universe, with energies above 0.5 MeV. The "Ultra-high-energy gamma rays" (UHEGR). from 100 to 1000 teraelectronvolt or even beyond (1 TeV = 10^{12} eV, 1 Mega-Electronvolt (MeV) = 10^6 eV). The "Very-high-energy gamma rays" (VHEGR) denotes gamma radiation with photon energies of 100 GeV (1 Gigaelectronvolt = 10^9 eV) to 100 TeV. The energy of gamma rays from radioactive decays in the universe, the most typical ones, are in the range from 0.5 MeV up to

around 10 MeV. Gamma rays cannot be observed on Earth, as they are blocked by the atmosphere. However, since the early 2000s it has been possible to observe cosmic gamma rays indirectly with certain (Cherenkov) telescopes. It is done from the Earth's surface by observing its interaction with the upper Earth's atmosphere. Telescopes beyond the Earth that observe the super high energy of gamma-rays directly cannot use mirrors any longer. Instead, they have particularly special detectors to measure the gammy-rays' energy and direction. The most important observation "telescopes" – launched in 2008 and 2002, respectively – are 1. the *Fermi Gamma-ray Space Telescope* of the NASA with detects gamma-rays with energies from 10 keV to 300 GeV (it runs two scientific instruments, the Large Area Telescope (LAT; 20 MeV to 300 GeV)) and 2. the Gamma-ray Burst Monitor (GBM; 10 keV to 30 MeV) and the ESA's *INTEGRAL* (INTErnational Gamma-Ray Astrophysics Laboratory) with an energy range in the MeV region. They both have a very large field of view and see approximately 20% of the sky at once and can cover the entire sky every three hours.

New Discoveries in the Last 25 Years

The various discoveries about the universe from observing with these new technological devices have been astonishing, but as a consequence this field has become more and more complex. Here is a rather incomplete list of new discoveries with a rather short description of them:

- Many cosmic objects such as suns, exploding stars or even quasars emit radio-waves. "Quasar" stands for *quasi-stellar radio source*, as it was first discovered in 1963 (see above) as objects that looked like stars but emitted radio waves. Thus, in the last decades radio telescopes have added tremendously to new insights about the universe. Particularly important was the observation of "plasma" jets (also called astrophysical jets). These are directional (collimated, i.e. with parallel rays) high energy streams of particles originating from black holes, neutron stars or pulsars that can stretch out over light years. Next to many other rays they also send out radio waves.
- In 2019, astronomers using above's EHT captured an image of a black hole for the first time - by observing radio waves surrounding it in a circle.
- An important investigation of microwave telescopes is the cosmic background radiation (with a corresponding temperature of 2.73 K), a remnant of the big bang and thus proof of it. The radiation is the rest of the

first light that could ever travel freely throughout the universe (about 380,000 years after the Big Bang). It can be considered "echo" or "shockwave", or in terms of its age also "fossil light". Furthermore, the Spitzer Space Telescope discovered (together with the Hubble Space Telescope) and studied the for long most distant known galaxy (in 2022 the Japanese *Subaru Telescope* detected an even older one). Microwave telescopes also discovered quite a few planets far away such as seven Earth-size planets around a single star. The first exoplanet to be discovered orbiting a star is *51 Pegasi b*, officially named *Dimidium*. It was announced on October 6, 1995 in the journal Nature by Michel Mayor and Didier Queloz of the University of Geneva, who were awarded the Nobel Prize in physics in 2019 for their discovery. It is now the prototype for a class of planets called "Hot Jupiter". The Herschel Space Observatory even detected water in star-forming molecular clouds, i.e. in the seeds of future planets. Microwave telescopes also revealed more interesting features of our planet system like a newly discovered giant ring around Saturn. Furthermore, they are a tool for studying star formation regions.

- The determination of the age of the universe is one of the fundamental discoveries of modern astrophysics. In cosmology, the age of the universe is nowadays defined as the time elapsed since the Big Bang. For measuring the age of the universe, the Hubble Space Telescope measured distances of stars to each other more accurately than ever before and thus constrained the value of the Hubble constant that is directly related to the age of the universe. It is now known to be about 13.8 billion years, while before the Hubble Telescope, scientists were only able to talk about an age between 7 and 20 billion years. Before that, it was not even defined, as most scientists, throughout the nineteenth century and the first decades of the twentieth century, believed that the universe itself was a steady state, existing eternally. The first hints indicating that the universe was not a steady state came from thermodynamics: The concept of entropy implies that any infinitely old closed system would have the same temperature everywhere, and thus there would be neither stars nor life in our universe. At the time, this contradiction could not be scientifically explained. In 1917, when he applied his theory of general relativity to the global cosmos, Albert Einstein had to introduce a cosmological constant (he chose the Greek letter Λ for it) to his equations of general relativity in order for his cosmological model to remain consistent with a universe in a steady state.³ The Russian mathematician Alexander Friedmann realized that this was an unstable fix, like balancing a pencil on its front point. Consequently, in 1922 he proposed an expanding universe model, with the actual first direct observation of the universe expanding being by Edwin Hubble published in 1929, the question of the age of the universe was again on the table (both a space telescope as well as a law - and a parameter in it - were later named after Hubble; we will get to know both of them below). Einstein himself abandoned the cosmological constant in his equation, stating that the term was both unsatisfactory. i.e. it gave an unstable solution in a static model, and redundant, i.e. relativity could describe the expanding universe without the term. He called the cosmological constant he introduced 12 years before the "größte Eselei" (the largest donkey mistake) of his life.

- Another important insight from the Hubble Space Telescope observations concerns its observation of very distant supernovae: Astronomers had long suspected that the expansion of the universe was weakening over time due to the mutual attraction of matter. However, the Hubble Telescope's observations uncovered evidence that, far from decelerating, the expansion of the universe is in fact *accelerating*. The cause of this acceleration remains unexplained till today. The most popular – but still very speculative - reason is dark energy (see below and in Chap. 6), which emits no electromagnetic waves (i.e. is "dark") and interacts with other forms of energy and matter only through gravitation. This interaction is said to take place in the opposite form to the ordinary gravitational force, repelling rather than attracting. However, physicists can currently only speculate about the exact nature of dark energy. They do not even know if it exists at all.
- With all the spectacular discoveries, what made Hubble Telescope the most well-known telescope overall were its amazing pictures from various very far away spots. These include the Carina nebula by WFC3, the star cluster Pismis 24 with nebula, or Hubble's most famous images *Pillars of Creation* showing stars forming in the Eagle Nebula.
- Even 32 years after its launch, in 2022, Hubble detected light of the farthest individual star ever seen until then.
- The two X-ray projects from 1999 identified indirectly thousands of supermassive black holes in the centres of other galaxies as sources of the X-ray radiation background. At the same time, new puzzling processes were revealed in numerous objects in space: Black holes emitting intense jets of

---

³ Albert Einstein, *Kosmologische Betrachtungen zur allgemeinen Relativitätstheorie* (Cosmological Considerations in the General Theory of Relativity), Prussian Academy of Sciences in Berlin, Proceedings of the Academy on 15th February 1917.

energy ("plasma" jets, which we have already seen above) as they assimilate matter as well as all types of electromagnetic waves.
- Another detailed observation were flashes from colliding neutron stars, as previously detected by gravitational wave detectors.
- Furthermore, they observed - indirectly as well - special black holes acting as ultra-luminous X-ray sources. The primary scientific goal of NuSTAR now is to conduct a deep survey for very large black holes a billion times more massive than the Sun, to investigate how particles are accelerated to very high energy in active galaxies, as well as to understand and image supernova remnants.
- Gamma rays in space were discovered in the late 1960s and early 1970s by military defence satellites. Detectors originally designed to detect very high energy flashes from nuclear bomb blasts began to record bursts from deep space rather than the Earth. Later detectors showed such bursts appear suddenly from any direction. Most cases that had been observed before a few years ago were gamma rays from solar flares (super-powerful localized eruption of electromagnetic radiation in the Sun) up to energies of a few MeV. However, it is now known that solar flares can even generate gamma rays in the GeV range. Most gamma rays come from extra-solar and even extra-galactic areas. The most energetic gamma-rays that have ever been discovered (as of 2023) bear an energy of 1.4 peta-electron-volt (discovered in 2021 by China's *Large High Altitude Air Shower Observatory*, the LHAASO).
- Studying gamma-rays coming from the universe helps astronomers to learn more about many extra-terrestrial bodies across the higher part of the spectrum of energies in the universe:
  - Gamma-ray burst,
  - Pulsars (neutronic stars), including Magnetars (who feature an unusually high magnetic field)
  - Quasars (active galactic kernel; sending e.g. plasma jets).
  - Blazars (a type of an entire galaxy powered by a gigantic black hole and among the brightest, most energetic objects in the sky).

We see from these lists that most of the knowledge about the various elements of the universe was obtained as late as in the late 20th and the twenty-first centuries! This applies also to the properties of universe itself. In fact, the entire spectrum of electromagnetic waves from space has provided revolutionary insights into some of the greatest mysteries in the cosmos. Yet, that is not all: It has also provided some entirely new insights into the cosmos itself.

Cosmology – The Origin of the Universe

Cosmology is a branch of astrophysics which deals with the origin and development of the universe as a whole. As old as the name is (derived from the Greek *kosmologia*, "the study of the world"), it was not until the 1950s that a serious dialogue about this question began within the natural sciences. Before that, for centuries and millennia, people could only speculate about the world's coming into existence, a question central to almost all religions. Cosmology is the youngest across all sciences and will deserve corresponding philosophical thoughts to integrate them in our vision of the world.

Based on Einstein's general theory of relativity from 1915 and Hubble's observation of the expansion of the universe in 1929, the Russian physicist George Gamow, student of Alexander Friedmann, formulated a theory in 1948 that the universe emerged from an enormously dense and hot state. A similar theory had already been put forward by the Belgian priest Georges Lemaître in 1927. Lemaître's idea, however, was not accepted by the physicists, because it referred too much to Christian ideas of the origin of the world. Gamow's theory as well was initially rejected by the physicists and astronomers as being too speculative. As an eccentric critic of Gamow's theory, the English physicist Fred Hoyle, ironically called his theory in a programme on British radio the "Big Bang". To Hoyle's chagrin, this term became entrenched in both popular and scientific parlance. Gamow's theory was not purely speculative but also made a concrete prediction about a possible physical measurement: According to his calculations, a remnant of the big bang should still be measurable today, in the form of cosmic background radiation. This radiation should permeate the entire universe with a temperature of around seven degrees above absolute zero (approx. minus 266 degrees Celsius). The cosmic background radiation was discovered about 20 years later, in 1965, by the Americans Arno Penzias (originally German) and Robert Wilson, with a corresponding temperature slightly less than half of Gamow's value. Their discovery was, however, rather a coincidence: They had used a powerful antenna at their research institute at Bell Laboratories in New Jersey to receive weak celestial signals that originate as echoes of radiation reflections from so-called balloon satellites. To their annoyance, the signals they received showed a constant background noise despite filters and their best efforts. They even moved nearby pigeon nests away to make sure they no longer caught random signals. But they could not get rid of the annoying "background radiation" from the air that was measurable uniformly in all directions. They eventually were able to rule out the Earth, the Sun or our own galaxy as sources. Coincidentally, other physicists were working at the same time to measure the radiation postulated by Gamow. When Penzias and Wilson learned of this through a friend, they realised the significance of their discovery. They had measured the remnants of Gamow's radiation from the Big Bang! This ultimately put the Big Bang theory into the mainstream of astronomy and cosmology (and in a way into mainstream in general, as it was later used for a successful American TV sitcom). It was another triumph of the interaction of theoretical and experimental physics! In 1978, Penzias and Wilson were awarded the Nobel Prize in Physics for their rather accidental discovery (Gamow had already passed away).

One interesting feature of observations across the universe is that by detecting similar objects at different distances from Earth, astronomers can see different points in time of the historical development of such objects. Light from these objects (having a finite speed) has travelled to us for different lengths of time depending on their respective distance from Earth. This way, we can draw conclusions about the temporal development of such an object. In other words, we are observing an object from millions or billions of years ago, because the object is millions or billions of light years away from us. This gives astronomers significant insights into the entire history of the universe.

Cosmology – How the Universe is Developing

It is clear to cosmologists today that our universe was created in a state of high temperature and density. But what was before that? Well, the question of "before the big bang" makes about as much sense as asking "What lies north of the North Pole?" The answer is: there is no "before the Big Bang". Time itself only came into existence with the Big Bang. It is, as physicists say, a "total singularity". Or, to put it perhaps more appropriately: They have not the faintest idea what exactly happened concerning the Big Bang. From this singularity, i.e. from the never, nowhere and nothingness, suddenly time, space and matter started to exist. From this, with the development that now set in, came the basis for the fact that 13.8 billion years later, the third planet of a small star on the edge of an inconspicuous galaxy, one out of billions, is home to two-legged beings with one head who ponder the meaning and background of the Big Bang.

In the first tiny fractions of seconds in the life of the universe, all four forces, gravity, the strong nuclear force, the electromagnetic force and the weak nuclear force, were united in a single primordial force. The universe was still unimaginably small and unimaginably hot. Its size was only about the 100 billionth part of a proton. We do not know the laws of physics that apply at this point in time and at such energy densities, but time and space were likely quantised, i.e. they broke down into discrete parts. Due to the uncertainty principle in quantum physics, statements about time periods and length expansions are meaningless in this phase. After the so-called Planck time of 10^{-43} s (one tenth of a sextillionth of a sextillionth of a second) had elapsed, the four forces gradually decoupled from each other. The first to split off as an independent force was gravity, even though we still do not understand the quantum physics of gravity. This should have caused the universe to collapse again. And maybe this happened countless times before - we cannot know, but not 13.8 billion years ago. After 10^{-35} s, the universe was still an unimaginable 10^{27} degrees hot. But that made it "cold" enough for a second force to split off from the primordial force: the strong nuclear force that would later hold the protons and neutrons together. And then, according to cosmologists, something very bizarre happened, which they call the "inflation of the universe". Within a very short time span of about 10^{-30} to 10^{-29} s after the Big Bang, the universe expanded extremely, by about 10^{27} times! This corresponds to the expansion of the universe from the small parts of the size of a proton to the size of a tennis ball. However, there exists no final empirical evidence for the inflation theory other than a few of inflation model predictions having been confirmed by observation.

After 10^{-29} s, the universe then hit the brakes, and the expansion slowed down. The universe was now too cool to remain in a state of inflation. Now the electromagnetic and weak nuclear forces also separated (but can still be described as one force in today's standard elementary particle physics, as we saw), with which the primordial force had finally decayed into the four basic forces known today. The universe was now slightly larger than a basketball. It was still 1000 billion degrees hot and consisted of energy, matter and antimatter. The matters kept annihilating each other into pure energy upon hitting on each other. But there must have been a slight imbalance between matter and antimatter at that time: For every ten billion matter building blocks (quarks, leptons, etc.), there was one less antimatter particle (antiquark, antilepton, etc.). Thus, matter was able to "get through" one ten billionth of itself in this annihilation phase. Without this strange asymmetry between matter and antimatter, which is still inexplicable to physicists today, the universe would simply have become a boring, empty space. After about 10 s, the first elementary particles, protons and electrons, were created out of the majority of matter, and shortly afterwards the first atomic nuclei. The universe had now become a fusion reactor that formed atomic nuclei from various nuclear building blocks, at first, only of the simplest element, hydrogen, shortly followed by helium and lithium. Still the universe was too hot for the atomic nuclei and electrons to combine to form atoms.

Subsequently, the universe "bobbed around" until, after about 380,000 years, when it had cooled down enough (to about 2700 degrees Celsius) for atomic nuclei to capture electrons and form the first atoms. This had a significant consequence: Electromagnetic radiation now interacted much less with the now electrically neutral atoms, so that the light particles were no longer trapped in a '"plasma mush" of atomic nuclei and free electrons. For cosmologists, this moment is of great importance, as from this point onwards, the universe is transparent and can be "observed" in its development with suitable measuring instruments. With the emergence of the atoms, something like a first "(baby) photo" of the universe developed. Today, this photo can be reconstructed from the cosmic background radiation.

After all the theoretical and partly speculative considerations about the early phase of the universe, cosmologists now have solid ground under their feet. In fact, the cosmic background radiation exhibits small energy fluctuations that have their roots in the Big Bang itself. They led to the existence of denser and less dense regions of atoms in the earlier universe. Together with the force of gravity acting on the atoms, these differences in density led to sound-like oscillations, comparable to those of a swinging bell. This comparison, however, is somewhat misleading: While sound waves of a bell tuned to the note "a" oscillate 440 times in one second, the sound waves in the early universe performed a single oscillation in 50,000 years! These oscillations can be called the "birth cry" of the universe.

Initially, the cosmic background radiation appeared completely homogeneous, i.e. without any temporal or spatial variation. From this, it could be concluded that the very early universe must also have been homogeneous shortly after the Big Bang. But this is in contradiction to the obvious spatial inhomogeneity of today's universe, from which numerous localised structures such as galaxies, stars and planets emerged. How could these have emerged from the homogeneous structure of the early universe? This question had to remain open for years. It was only in 1992 that light was shed on the matter when the satellite *Cosmic Background Explorer* (COBE) was able to measure the spectrum of cosmic radiation, which had been detected in 1965. The new results showed that the background radiation does indeed have irregularities. Small temperature differences measured (around 0,001%) indicate that the cosmic mass density in the early universe exhibited tiny fluctuations. These formed the origin of today's galaxy clusters. So, in the midst of the cosmic background radiation, the researchers had discovered the earliest forms of galaxies and stars! Asked about this at a press conference, one of the physicists said that for a religious person this was like "looking God in the face". In recent years, the temperature deviations in the cosmic background radiation have been measured and mapped with more and more increasing precisions (using the *Planck space observatory*), leading to an ever better understanding of the development in the early universe. Today, the origins of even individual galaxy clusters can be resolved in it.

However, the known visible matter is simply not sufficient for these clumps to form. The gravity of a hidden player had to come into play for this to happen. Most cosmologists assume that for this process to have taken place, there must have been a bunch of times more of the ominous dark matter already familiar to us from previous chapters than of visible matter. It collapsed under its own gravity, taking the visible matter with it. The latter then grew together into clumps, so that their gravitational forces became stronger and stronger, so that more and more matter fell onto them, so that the clumps became denser and denser. Finally, the energy density in them rose to a critical limit, at which point nuclear fusion began. The first stars were formed. This is the same mechanism that still makes our sun shine today. Compared to it, the first stars were giants with their size up to 500 times larger, and having much shorter lifespans, perhaps around some ten millions of years or shorter (compared to more than ten *billion* years for the sun).

Nuclear fusion inside the stars created larger atomic nuclei from the hydrogen nuclei, first helium, then lithium, later carbon, oxygen and finally the heavy elements such as iron, gold and manganese. Ultimately, such a giant star died in an enormous supernova explosion, which hurled the "hatched" heavy atomic nuclei into the vastness of the universe. In fact, apart from hydrogen, every atomic nucleus on our Earth today was "hatched" in a star at some point many billions of years ago. Meanwhile, the universe continued to expand. After less than 400 million years (as of 2022: HD1), the first galaxies still observable today were formed. After another eight billion years our sun's existence started and shortly afterwards our planet Earth.

The Current Unified Theory of the Universe – Many Open Questions

The following provides a ranking from the largest to the smallest structures of the observable universe (as also illustrated in Figs. 12.2 and 12.3):

Fig. 12.2 The various scales in the universe from the Earth at the bottom to the entire universe on the top (Source: https://www.visualcapitalist.com/cp/map-of-the-entire-known-universe/)

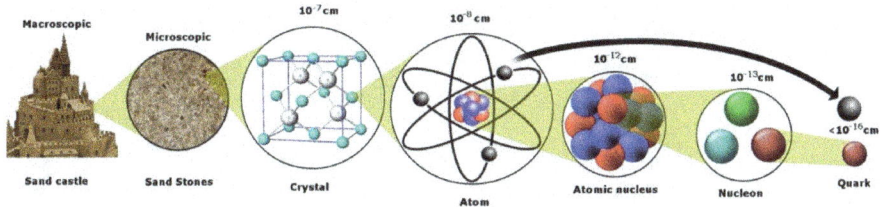

Fig. 12.3 The various scales from our everyday sizes down to the smallest (known) structures (Source: https://www.scienceabc.com/nature/universe/what-is-the-smallest-particle-we-know.html)

- Galaxy filaments and voids, i.e. supercluster complexes, galaxy walls, and galaxy sheets (filamentary connections of visible and dark matter). An example is the *Hercules–Corona Borealis Great Wall* with a diameter of about 10 billion light years, so in the magnitude of the global universe itself.
- Large Quasar Group (LQG): An example is *U1.27* with a diameter of about 4 billion light years.
- Superclusters: An example is *Virgo Supercluster* with diameter of about 200 million light years.
- Groups of galaxies and clusters of galaxies: An example is the *Local Group* with a diameter of about 10 million light years.
- Galaxies: An example is the *Milky Way* with a diameter of about 100,000 light years.
- Star clusters: Examples are globular clusters, open star clusters, with diameters of dozens to hundreds of light years.
- Planetary systems: An example is our solar system with a diameter of about 300 AU = 41 light hours.
- Stars: An example is our sun with a diameter of 1,392,500 km = 4.65 light seconds.
- Planets: An example is our Earth with a diameter of 12,756.2 km = 42.6 light milliseconds.
- Moons: An example is Earth's moon with a diameter of 3,476 km = 11.6 light milliseconds.
- Asteroids, comets: An example is the every 76 years coming Halley Comet. They a diameter of a few kilometres to several 100 km
- Meteoroids: Diameters from metres down to millimetres.
- *Human beings: Size of about two metres.*
- Dust particles
- Molecules
- Atoms

- Atomic nuclei with protons and neutrons
- Hadrons made up of, among others, quarks
- Elementary particles (e.g. electrons, quarks).

As we saw in Chap. 6, physics has not seen any fundamental progress for more than 50 years. In contrast, astrophysics and cosmology have, as we saw in this chapter, made astonishing progress within the last few years - achieving insights that lead them to questions quite similar to those that arise in particle physics. This again has particle physics having received new glimmers of hope from a completely different direction, from a field that at first glance seems maximally distant from it. In recent years, however, a fruitful cross-fertilisation has been developing between the study of the universe and the study of the subatomic world.

But how can this come, do those two fields lie as far away as even possible in science? The basic theory for what happens in our cosmos is the general theory of relativity. From this theory Einstein - in fact, Karl Schwarzschild - deduced the existence of black holes. At the same time, however, he noticed that while his theory can describe the existence of these objects well, it fails when it comes to describing their interior, formation and development. His equations simply have a singularity there, i.e. they reach *infinite values*, which is impossible in physics, Einstein's theory fails at inside of black holes. Equally for the Big Bang, we have no insight about this area. Thus, long before the standard particle theory, physicists had already discovered the limits of their theories at the other end. Today the particle physicists are equally well aware of the limits of their theory. Although the physics of atomic nuclei and quarks on the one hand and the events in black holes and the beginning of the universe on the other hand describe very different phenomena at first glance. The elementary particle physicists and astrophysicists have realised that the respective limits of their knowledge are the same, they are just looking at it from different sides: Quantum theory cannot describe gravity, while general relativity ignores the effects of quantum theory in such extreme environments as black holes. Furthermore, the standard theory in particle physics does not provide any understanding of the existence of *three* families of leptons and quarks, of the *mass hierarchy* of these elementary particles, of the nature of *gravity* and of the nature of *dark matter*, if it actually exists. At the same time, the general theory of relativity does not provide any understanding of the *big bang cosmology*, *inflation*, the *matter–antimatter asymmetry* in the universe, the nature of *dark energy*, if this actually exists, to list just a few more lacking items for each. Today, it is clear to physicists of both disciplines: In order to solve their problems, they must finally unite gravitation and the quantum world in a unifying theory, i.e. look at them from the same side: a "quantum theory of gravitation". For this, finding the concrete level of breakdown of both theories might lead to a unified theory.

Let us take a closer look at the current model of astrophysics. T standard model of particle physics is - as ugly as it is – conform with almost *any* empirical observation that has ever been made (we are not sure about the masses of neutrinos, and other details, though). Still, today's standard model of cosmology in contrast contains next to the Big Bang an extended list of unexplained features, some important ones of which are:

- The notion of dark matter arose from observations of large astronomical objects such as galaxies and clusters of galaxies, which displayed gravitational effects that could not be explained by the (amount of) visible matter. The nature of dark matter, besides that it is subject to only gravitational effects and is thus not visible, is completely unknown. Could perhaps a different view on these unusual gravitational effects explain these? As for instance a change of the Newton gravitational constant at long distances (see under). Newer (indirect) evidence of dark matter spikes around black holes though suggest, that dark matter might indeed consist of particles.[4]
- As seen above and in Chap. 6, the notion of dark energy arose from the observation that suggested the expansion of the universe is accelerating. But is this enough to say that almost 70% of gravitational forces emanate from it?
- Cosmic inflation: As we saw above, in the very beginning of the universe it expanded at an unexplainable large speed.

These features are all included in today's *Lambda-CDM model* which was briefly mentioned in Chap. 6. The Greek letter capital lambda (Λ) stands for the cosmological constant (associated with the dark energy), CDM for "cold dark matter", which is hypothetically seen as a matter that

- Is non-baryonic, i.e. entails no protons and neutrons, etc. as well as no electrons, positrons, etc., either.
- Is with a comparably low velocity (compared to the speed of light), which excludes neutrinos and other massless or almost massless particles.
- Has no interaction besides gravity, i.e. does in particular not interact with the electromagnetic force (i.e. no photons being emerged).

---

[4] Indirect Evidence of Dark Matter Density Spikes around Stellar-mass Black Holes. Article found in 2023: https://iopscience.iop.org/article/10.3847/2041-8213/acaafa.

Because the Λ CDM model is with "only" six parameters (the so-called physical baryon density, physical dark matter density, age of the universe, scalar spectral index, curvature fluctuation amplitude, and reionization optical depth) the simplest model that is in good agreement with cosmological measurements (and in accordance with Occam's razor principle[5]), it is also called the *standard model of cosmology*. It covers like no other cosmological model the observed continuing (and accelerated) expansion of space, the observed distribution of the lightest elements (hydrogen, helium, and lithium) in the universe, and the spatial texture of minute irregularities (anisotropies) in the Cosmic Microwave Background (CMB) radiation.

However, the Λ CDM model is by far not as broadly accepted as the standard model in particle physics. For example, the Hubble constant predicted by it is lower than the observed value, or the CMB picture displays dipole properties which the Λ CDM model does not predict. There are in fact about two dozen of open points for Λ CDM model. Thus, some alternative models challenge its assumptions, structure and properties. The most prominent examples are:

- A modified Newtonian dynamics (MOND): A change in Newtonian gravity in large galaxies replacing the dark matter hypothesis.
- Entropic gravity (based on string theory): Gravity is not a force itself but an emergent phenomenon that comes from the quantum entanglement of small bits of spacetime information.

Given it has only emerged quite recently, and many questions remain open, the Λ CDM in the current shape is likely not going to stay as the dominant standard model of cosmology. It will be one of the most interesting developments of modern physics to see how our knowledge and modelling of the features of the universe develop. New information from empirical observations is very likely to come rapidly with our new means of observing the universe, as we already started seeing with JWST. In any case, we see that astrophysics, just as physics and biology, is posing fundamental questions to philosophy. Some are such that science will likely never answer (likely the Big Bang) and where philosophy can possibly give ways to explore new solutions.

---

[5] In trying to understand something, getting complex and unnecessary information out of the way is the fastest way to the truth.

13

The Future of Sciences/Technologies?
From Utopian Optimism to Dystopian Pessimism (and Possibly Back)

There are five core sciences and technologies we mentioned in Chap. 5 that we have not yet discussed in more detail in the previous six chapters. It is important to dive into these as well, which we will do in this chapter, before we then deal in more detail with the question of what science will likely bring us overall in the future and what many believe about that today - before we then discuss in the last chapter what we can (and should) do such that the future technologies are developed in *our all interest*. It is important to dive into these as well, which we will do in this chapter, before we then deal in more detail with the question of what science will likely bring us overall in the future and what many believe about that today - before we then discuss in the last chapter what we can (and should) do such that the future technologies are developed in our all interest.

More of the Promising and Challenging Areas in Science and Technology

CO_2-Neutrality

The challenges of climate change have been mentioned briefly in various chapters before. However, so far, no broader discussion about it has been provided in this book (which is not a main point of it anyway[1]). Fact is: Climate change is no longer a threat on the horizon. It has arrived in our everyday lives and threatens the survival of many people already today with: Floods, species extinction, migration, droughts, super tornadoes, new patterns of cyclones, loss of ice fields in mountains, and many other climate events. And this does not even come as a surprise, as the main issues have been known since the 1980s: Man-induced greenhouse gases - above all carbon dioxide, CO_2 - are causing our planet to heat up (the greenhouse effect was already pointed out by Svante Arhenius, a Swedish chemist in 1896!). At that time, concrete plans of actions were already on the table, but a powerful industrial lobby prevented their implementation and deliberately undermined until today the reputation of the scientists involved.

Today the picture looks a little better, at least in that those who take action against climate change are no longer marginalised (but the big firms often still advocate against climate change). This is reflected e.g. by the almost entirely positive response to today's "Fridays for Future" movement after initial criticism towards them. It can neither be said that nothing has happened in politics in recent years. Today, politics and business are sometimes even competing to outdo each other in their efforts to prevent climate catastrophe (often, however, with many words and little action). In autumn 2020, both the EU and China announced a roadmap to a CO_2-neutral economy by 2050 and 2060, respectively (later India followed with 2070 as the goal). Shortly afterwards, the German automotive industry also committed itself to this goal. And once Donald Trump was voted out of office, the USA followed suit. However, the main activities decided upon (legally only in Europe so far) will only be done in the 2030s and 2040s. The advantage for today's politicians: Then, they will no longer have to stand in front of the people.

---

[1] For details about the challenge as well as chances to deal with the climate change, see: Lars Jaeger, *Ways Out of the Climate Catastrophe - Ingredients for a Sustainable Energy and Climate Policy*, Springer International Publishing (2021).

Most progress has happened in science, however. The CMIP6-models[2] of the researchers, published in August 2021 and early 2022 in the AR6 report,[3] are significantly more ambitious than their predecessors in their demands for model accuracy. For example, in some of them the spatial resolution of the grids on which the global climate is modelled is reduced to less than 100 kms. This makes it easier to determine the effects of cloud formation on the global and local climate. At the same time the temporal density of the measurements increased significantly. "This report is invaluable for future climate negotiations and policy makers," said IPCC ("Intergovernmental Panel on Climate Change") President, the South Korean Hoesung Lee. What is significant about the report is that compared to the negotiations eight years before, at the last report (CMIP5), the debates seem to have gone more smoothly. The IPCC authorship probably clearly prevailed this time against the usual resistance from politicians and others against clear formulations. Moreover, the scientific nature of the report was not touched. So, the responsibility is now also clearly stated (with a 100% likelihood!): According to the IPCC, humans are responsible for all observed global warming since pre-industrial times (1.6 degrees on land (1.4–1.8), 0.9 (0.7–1.1) degrees over the sea, 1.25 (1.1–1.4) degrees on global average).

After three and a half decades of stalemate, regression, and a few hard-won advances in the fight against climate change, it is now a game of dominoes: Forces that resisted the global energy revolution that is hoped to save our climate are falling - one by one (however, until they are all there, it will probably still be a bit of a wait). At the same time, year after year, even the most optimistic forecasts about the *possibilities of new technologies* are caught up and surpassed by current developments. There is no shortage of ideas, technological possibilities, and concrete initiatives. Almost all of them revolve around the central factor in climate change: energy. Apart from some aspects of agriculture, which are a strong climate factor, however, and are only being considered to be addressed, see below, all the influences that humans have on the climate can be traced back to the way we produce and consume energy (to be precise: to transform it in order to use it). Driven by the astonishing technological advances in the fields of photovoltaics and battery energy storage, as well as nanotechnology and artificial intelligence, we are on the threshold of the fastest and most far-reaching revolution in the energy sector in the

---

[2] For information on these models see: https://www.carbonbrief.org/cmip6-the-next-generation-of-climate-models-explained/.
[3] See: https://www.ipcc.ch/assessment-report/ar6/.

last 150 years! We already have the technical capabilities to reverse the devastating climate trends *without* significantly limiting prosperity.[4] When new technologies enter the field the technological ways of addressing the climate downturn are even more promising (see e.g. "nuclear fusion", next point). The obstacles lie primarily in economic and political "constraints", i.e. particular conflicts of interest, and psychological resistance to changing habits. An example of this is the increase in coal production in 2022.[5] Overcoming those – that is the centre of future energy politics. That the alternative technologies for making energies are already considered broadly available today can be observed e.g. by even some oil companies developing ambitions to focus on ecological energy sources.

Nuclear Fusion

Nuclear fusion is the process that makes our sun and all other stars shine and send enormous amounts of energy into space. It is thus ultimately the strongest and most broadly used energy in the universe, albeit being comparably inefficient in our sun.[6] The physics behind it sounds rather simple: Under a very high product of pressure and temperature atomic nuclei fuse together. Thereby, tiny amounts of matter are converted into enormous energies, according to Einstein's famous formula $E = mc^2$. The energies released per unit of weight in nuclear fusion are even much higher than in the reverse process, nuclear fission. The most common nuclear reaction in the universe is the fusion of two hydrogen atomic nuclei to form one helium atomic nucleus. However, heavier nuclei can also fuse together; in fact, as we saw in the last chapter, all naturally existing chemical elements were at some point incubated in the millions of degrees hot boilers inside stars. In addition, there is a very high pressure in suns due to the very high gravitational force.

There is an enormous potential in nuclear fusions for global energy supply: It would be CO_2-neutral, and, contrarian to nuclear fission, safe as it does not produce long-lasting radioactive waste, and its necessary products could be

---

[4] See: Lars Jaeger, *Ways Out of the Climate Catastrophe - Ingredients for a Sustainable Energy and Climate Policy*, Springer International Publishing (2021).

[5] https://edition.cnn.com/2022/12/16/world/coal-use-record-high-climate-intl/index.html.

[6] It is the size of the sun and the strong gravitational force that makes the fusion possible, not the temperature, which is only a fraction of what is needed for the fusion here on Earth "The fusion process is extremely slow (and inefficient in terms of energy release per unit volume) - the Sun releases only 250 W/m3 in its core. The reason for this is that fusion events are extremely unlikely, requiring two protons to overcome the Coulomb barrier between them and for one of the protons to inverse beta-decay into a neutron so forming a deuterium nucleus." https://astronomy.stackexchange.com/questions/30035/why-doesnt-the-fusion-process-of-the-sun-speed-up.

selected in such a way that they are abundantly available. But unfortunately, unlike nuclear fission, nuclear fusion is very difficult to master as an energy source. A team of researchers led by Edward Teller did succeed in initiating thermonuclear fusion processes and releasing enormous amounts of energy, but this was for the completely uncontrolled energy production of hydrogen bombs - the first of which was detonated in 1952. Permanently *controlled* nuclear fusion has not been achieved to date.

However, already in the early 1940s Enrico Fermi and Edward Teller developed the first ideas for generating electricity by controlled nuclear fusion. The basic concept they developed is still the basis for nuclear fusion research today. Two isotopes of hydrogen were suggested to be used as the main starting materials:

- While the atomic nucleus of hydrogen normally consists of only a single proton, *deuterium* has an additional neutron. About 0.015 percent of the naturally occurring hydrogen is deuterium. Given the large quantities of water on our planet, there is more than enough deuterium to be used for this purpose.
- In *tritium*, two neutrons are additionally packed into the atomic nucleus of the hydrogen. Although tritium does also occur in the atmosphere and in seawater, it does so in such low concentrations that it must be produced (so far at great expense).

Today, usually a deuterium–tritium mixture is being used, because it requires the lowest temperatures and thus is most feasible economically. The mixture is heated in a kind of microwave oven with very, very high temperatures. As of about 100 million degrees Kelvin (far more than in the centre of the sun with approx. 15 million degrees), the electrical repulsive force of the positively charged atomic nuclei is overcome and the nuclei fuse together by the strong nuclear force; the mixture ignites and releases the fusion energy. Only five kilograms of the fusion materials would be enough to produce the energy equivalent of 18,750 tonnes of coal, 56,000 barrels of oil or the annual energy yield of 755 hectares of solar collector surface. However, because tritium is difficult to obtain, research is now being conducted on other material combinations as well - the next one with higher required temperature is Deuterium-Deuterium that needs around 150 Mio. degrees. One further promising candidate is the fusion of boron atoms (consisting of five protons and six neutrons) with a proton - that needs, however, more than ten times as high energy, i.e. 1.5 billion degrees of temperature, which is equal to an atomic energy of 123 keV (1 eV corresponds to around 11,600 K).

One of the problems is that no material exists from which a container for a 100-million-degree hot mixture of atomic nuclei could be made. Even if such a material existed, the ultra-hot plasma would immediately cool down again on contact with the physical container wall to such an extent that the fusion would come to a sudden standstill. Researchers and engineers are therefore trying to control the plasma with strong magnetic fields. But maintaining such a field with the required high power as well as the required precision is a technological challenge that top scientists all over the world have been working on for decades - with moderate success so far.

Nevertheless, the hopes placed in nuclear fusion are high. The prospect of almost unlimited quantities of climate-neutral energy has prompted a consortium of 35 countries to finance the ITER experimental reactor in Cadarache in the south of France. Gigantic superconducting magnets generate fields that confine and harness large quantities of plasma. The costs are exorbitant, however: by the time ITER delivers - perhaps - its first results in 2030 and a significant net electricity output in 2040, it will likely have swallowed significantly more than €60 billion in public money. This makes ITER by far the most expensive experiment in the history of science to date. The controlled operation as power plants worldwide, which would provide us with almost unlimited quantities of climate-neutral and safe energy, is - for this approach - not expected to be possible until around 2050 or 2060.

Some privately financed companies such as *TAE Technologies* in Foothill Ranch, California, *General Fusion* in Vancouver, or *Commonwealth Fusion Systems* in Boston want to reach their goal faster and with much smaller reactors less costly.[7] While state-financed large-scale projects with their countless decision-makers and complicated networks of relationships must adhere to long-term planning, smaller, private projects can in case of setbacks flexibly decide on the next steps again. It could also prove to be an advantage that they want to achieve energy production by a different route than ITER. In order to assess this, one must take a deeper look at the so called "Lawson criterion" in which plasma physicists summarise their discipline with three relevant variables, i.e. the speed of the particles in the plasma given by the temperature, the density of the plasma which is the number of particles per volume, and the confinement time which is the period of time during which the mixture of starting materials exists as a sufficiently hot plasma. Once a critical temperature has been reached, the product of density and inclusion time is then the significant factor. ITER focuses on a low density and a high confinement time. It could mean less effort and thus lower costs if one takes a look at the

---

[7] For more details see: Lars Jaeger, *An Old Promise of Physics – Are We Moving Closer Toward Controlled Nuclear Fusion?*, International Journal for Nuclear Power, Vol. 65, Issue 11/12 (2020).

opposite, i.e. the confinement time is kept very short and the density is very high. Here, an extremely high density can be achieved with very strong laser pulses. The main challenges lie in the preciseness the radiation has to focus on the particles.

On December 12, 2022 scientists in Livermore, California, announced that they had achieved an important milestone in the use of fusion energy. For the first time, they had succeeded in generating more energy with an experimental fusion reactor than what was used during the process, so the involved physicists. A "net energy gain" of 120 percent was achieved. That does sound revolutionary: After all, there have been numerous experimental reactors for this purpose for about 70 years, but nowhere has it been possible to generate more energy than what must be put into the reactors to produce nuclear fusion. However, it is questionable whether this method is even scalable with the short laser pulses (which are currently only possible a few times a day). Most nuclear physicists hardly see the laser method as a way to commercial reactors, due to the short duration of the process. Plus, the lasers themselves consume enormous amounts of energy, which represents a multiple of the amounts of energy gained. And this had not even been considered in the net gain!

Probably the solution lies between the two extremes. Many plasma physicists today suspect that medium inclusion times and medium densities offer the best prospects for controlled nuclear fusion. Private companies have thus endeavoured some highly interesting alternative paths towards a fusion reactor altogether: Field-reversed configuration (FRC), (Laser induced) Inertial confinement fusion (ICF), Magnetized target fusion (MTS; or magneto-inertial fusion (MIF)) and Muon-catalysed fusion (μCF). The most advanced here is the above already mentioned company TAE with a FRC approach.[8]

Although different in their approaches all of the various companies are looking for paths to fusion that employ much smaller and thus less expensive reactor technologies than ITER, aiming at generating electricity already in the next few years. They are counting on possible mistakes and surmountable obstacles in their ideas being found (and fixed) much faster than in a few decades time and before billions of dollars have been burned. The fact that they depend on risk capital that is hungry for returns could therefore prove to be a decisive advantage. They simply cannot afford to turn to large, expensive, long-lasting, and untested projects. Rather, they must always decide step by step which next move to take and justify in front of their shareholders every step. In light of the nature of the described problems around thermonuclear

---

[8] Also here for more details: Lars Jaeger, *An Old Promise of Physics – Are We Moving Closer Toward Controlled Nuclear Fusion?*, International Journal for Nuclear Power, Vol. 65, Issue 11/12 (2020).

fusion technology such a pragmatic approach might prove more appropriate than ITER.

And it is these private companies that have in recent years made some considerable progress. In fact, a real public–private race for the best fusion technology solution has started. With this, nuclear fusion has become significantly more covered by the public press. Will one of them soon hit the jackpot?

Food Technology

There is a completely different dimension to climate and environmental problems: Our food also causes a significant amount of gases causing a climate change, with animal husbandry and feeding accounting for a particularly large proportion of greenhouse gas emissions (mostly CH_4, methane gas, which has 20 times stronger effect on the climate per molecule than the CO_2). Altogether, according to the *Food and Agriculture Organization* of the United Nations (FAO), it accounts to almost 15 percent of the world's total CO_2 emission equivalent. Furthermore, this trend is dramatically increasing, both with the growing world population per se and with the increase in economic power and prosperity in Asia and Africa. What Europeans and Americans take for granted today - the almost daily consumption of meat - would lead globally to an unsustainable increase in animal husbandry and thus in total CO_2 equivalent emissions of the global agriculture. It is thus already clear today that we can simply no longer afford to keep animals for 10 billion people to eat. Thus, in its "Special Report on Climate Change and Land Systems 2020" in August 2019 - hardly noticed by the public as well as politicians - the IPCC called for a turnaround in human meat consumption. According to the WWF, if we all consumed vegan only food, the carbon footprint of our diet would be reduced by more than 40 percent.

Here technologies will help us again: The following trends in favour of cutting meat consumption are already emerging in the food industry:

- Plant-based substitute for animal protein: There are already purely plant-based substitutes that taste (almost) like meat and other animal products.
- Increased use of "green genetic engineering": Recent genetic engineering applications contain the production of plants with high nutrient content The best-known example is the "golden rice", a genetically engineered rice variety with additional beta-carotene. And with CRISPR (see Chap. 8), genetic engineers have recently been given a new, much more powerful tool for plant breeding.

- Meat from the lab: Scientists at Maastricht University produced an artificial meatball for the first time in 2013. To do this, they took muscle stem cells from a cattle, multiplied them in the laboratory and grew muscle strands several centimetres long from them. Around 20,000 of these strands were needed for a 140-g meatball. "Almost like meat, not quite as juicy, but the consistency is perfect," test eaters commented. Today, firms are far enough to offer this type of meat at prices (almost) comparable to meat. In Singapore, they are already publicly available - and possibly in Europe and the US start as well.
- Food from the printer: The first machines for this already exist that work similarly to 3D printers. The ingredients are applied in layers. The products can then even be cooked.
- Much of what is in the fridge today is thrown away - a tremendous waste. Technological control of our eating habits will change that: Apps control and optimise it taking also into account the shelf life of the food in the refrigerator.

Synthetic Life and Life Prolongation

The reader should remember from Chap. 5 the ways the first artificial life has already being produced. In May 2019, almost 10 years later, biologists reported the creation of a new synthetic form of viable life, a variant of the well-known bacteria *Escherichia coli*. Bacteria genes are much more complex than the one reproduced in 2010. The scientists thereby reduced the natural number of 64 codons in the bacterial genome to 59 codons. As scary as synthetic life research sounds, as productive and beneficial it can be. Respective biologists already talk about an entirely new era in biology in which synthetic life forms are produced for specific industrial purposes. Artificially produced bacteria, for example, could break down oil spills on the world's oceans and decompose plastic. Thus "CO_2 eating" microbes from the lab could even help reduce CO_2 emissions and thus slow down climate change.

Even more exciting - and scary -is the scientists' effort on life prolongation, equally something hinted at in Chap. 5. During human lives, our cells and organs imperially lose their capacity to function. However, some of today's geneticists assume that this process can be stopped or even reversed. Already one of the oldest narrations of human history deals with the hope of eternal life: the Gilgamesh epic from the 3rd millennium BC. Therein, the Sumerian king Gilgamesh in his search for eternal life finally finds the secret of immortality in the form of a plant. But when he rests at a well, a snake steals the plant from him. Does man, 4500 years later, once again perform such an adventure? And how is the path towards.

Many medical experts and biologists believe that there is no insurmountable biological limit for human age. For aging is ultimately nothing more than the succession of defects in cell division and repair - caused by increasingly frequent genetic copying errors. If the damaged genes can be "healed" through genetic editing techniques such as CRISPR/Cas9, we could witness the decisive breakthrough in man's fight against aging or even death. The internet giant Google invests more than a third of its investment budget for biotechnology in various companies dedicated to the extension of the human life span. Is this solely a dream of the firm's leaders? Here some quite interesting insights:

> The most popular theory of aging is that us growing older has to do with the ends of each DNA strand. Geneticists call these gene regions "telomeres". Telomeres can be compared with the plastic sleeves at the end of shoelaces, which are supposed to protect those from unravelling. Biologists have observed that the telomeres shorten each time the cell divides. This continues until the cell division and thus the further existence of the cell is no longer possible. As a result, the cell dies. However, if the cell has a specific enzyme, the telomeres no longer shorten, and the cell lives - potentially forever - further. For gerontologists, scientists who deal with the process of biological aging, CRISPR/Cas9 and the possibility to edit genes like texts in a Word document constitute an amazing opportunity. These could enable the cells to produce this particular enzyme and thus continue to divide further indefinitely. Moreover, research in the field of gerontogenes (genes, which control the aging processes) has also the goal of extending our life span. The geneticists have already identified some genes that control the aging process in lower organisms, such as the "age-1", the "2daf-2", the "bcat-1", and the "clk-1" gene. The "FoxO3" gene broadly referred to as the Methuselian gene (a part of the DNA that confers healthy old age on men and women) is a member of this group. By deliberately inserting, changing or blocking these genes, genetic researchers have already massively increased the lifespan of animals in the laboratory.

Parallel to research on the level of cells, biologists and doctors also work to breed entire substitute organs. As soon as existing organs lose their function in our body, particular replacement organs could be implanted. The cultivation of organs in animals has long been on the agenda of researchers. Already over a hundred years ago, the zoologist Ross Harrison was able to get nerve cells he had cultivated outside the body to divide. In 1972, Richard Knazek and his team were able to grow liver cells from mice on hollow fibres. And just ten years later, burn victims were transplanted skin which had previously been bred from body-borne cells. And in 1999, biologists were for the first time able to breed nerve cells from embryonic stem cells of mice. When these were injected into other mice which were infected with a kind of multiple sclerosis, the animals recovered. And as late as 2021 scientists attached a pig's kidney to a human body with no more functioning kidneys, which did not cause an immediate negative reaction but worked for some time (however, the woman died a few weeks later), This is possibly a starting point for a broadly creating organ replacements for humans. However, before the biologists (even) remark infinity, we will remark others. We can e.g. further simply print organs. This is already done on the basis of a small tissue sample and a 3D image of the corresponding organ. The organ is then built up layer-by-layer with body-specific "ink cells", which are produced from stem cell cultures (in the terminology of 3D printing this technique is referred to as the "rapid prototyping method"). Already today, hip bone and foot bone transplants are printed in 3D printers with an accuracy unimaginable just a few years ago. This is precisely the goal of the "3D Organ Engineering Initiative" at Harvard University, which has already achieved astonishing successes: In 2019, it reported the cultivation of an artificial kidney.

With a type of "tissue engineering" another powerful method can be made available to doctors. In the past, one way of therapy has been to take differentiated cells from a donor's organism and multiply of those in the laboratory with the goal of replacing diseased tissue in a patient. The persistent problem with these methods, however, has so far been the rejections by the receiver's body. Here the stem cells come into play. Their advantage: The tissues bred with them are not classified as foreign bodies by the patient's immune system and are thus not rejected. Adult stem cells are multipotent. For example, an adult stem cell from the skin can generate various cell types, a simple liver cell or blood cell is not able to do that. One way of approaching immortality is thus to use stem cells to grow entire organs outside the body. These are potential powerful tools available from the toolbox of modern biotechnology for the purpose of disease prevention and life extension. They can be used to produce any tissue - in the case of embryonic stem cells, or certain types of tissue - in the case of adult stem cells. As soon as existing organs lose their functionality, the replacement organs could be implanted into the respective body. Stem cells could thus make it possible to treat diseases that are still incurable today.

The combination of genetic engineering for genes that control the ageing of cells, stem cell research, breeding replacement organs in the laboratory, and nanotechnology (3D-printing) could increase our physical and mental well-being and, last but not least, our life expectancy into yet unimaginable dimensions. If we specifically edit and re-program those genes that control the aging of our cells, breed substitute organs in the laboratory (or in animals), or use stem cells to heal our ill cells and organs, the dream of a further prolongation of our lives or even reaching human immortality no longer seems so utopian. Even though human immortality is unlikely to be realized any time soon, this project will certainly remain on the radar screen of scientific efforts.

Historical Issues

Prior to the mid-nineteenth century the scientists communicated their results to their colleagues by letters or books or presentations to academies. So scientific results were not very broadly communicated even within the (still rather small) scientific community. As of around 1850 until today the researchers have published their results in special scientific journals that underwent through a detailed refereeing process before they get in there. Only a few research results have been taken up by journalists to report about specifically interesting cases.

As exciting as the research results have been and possibly to be even more exciting in the future, the societies' reaction to scientific and technological progress has always been twofold in the last 170 years - as the effect of science results reaches people by technological changes: with huge excitement as well as great fears, with perceived positive consequences for society as well as belief in challenging effects, with high appreciation of the scientists' work as well as condemnation of their research results and often even the very scientific method itself. Here are some examples:

- The waves of industrialization in the eighteenth and nineteenth centuries brought about massive economic growth, but also the emergence of a proletariat of misery and the dissolution of the traditional extended family. When the railroads were introduced that brought people from far distances together and made travelling significantly easier, people were also afraid of the "inhuman" speed locomotives came with (for 4000 years the fastest speed had been riding horses). There were indeed a number of serious accidents, boilers exploded, trains collided, and bridges collapsed.
- In addition to computers, lasers, and modern medical diagnostics, quantum physics has also given rise to the atomic bomb, which initially sparked extensive philosophical debates. Gunther Anders, in his work titled "The Nuclear Threat: Profound Reflections on the Atomic Age," views the deployment of the atomic bomb on Hiroshima and Nagasaki as a pivotal

moment in human history. It marked the point at which humanity gained the capacity for self-destruction. Another significant philosopher engaged in this discourse is Hans Jonas, author of "The Responsibility Principle: An Ethical Framework for Technological Civilization." Both of these books have significantly impacted the philosophical discourse on science and are regarded as early influences on ecological thinking.

- The internet has provided human beings an amazing tool for information gathering and distribution, social interaction and get to know each other across the world, as well as accelerations of business and financial transactions, however, it also created the emergence of new dependencies, unwanted information storage about us, the extension of pornography, and new types of criminal activities.

These Janus faces of scientific applications was already perceived in the XVIth century as Rabelais' sentence we quoted in the first chapter remembers us: "Science without conscience is but ruin of the soul".

What is new today, however, is what we *expect* from technological change for the future. Up to the nineteenth century, philosophers and writers in the Western world drew largely very positive pictures in their visions of the future of what lies ahead for mankind. It started in 1516 with the "Utopia" by Thomas More. Utopia is an imaginary world on an island described by More in which all people (more precisely, all men) have the same rights. A working day consists of six hours, everyone can freely choose his profession and has full access to educational facilities, and everybody gets his needs provided for by the community. Such a society must have appeared as a paradise to the people 500 years ago. For a long time utopias were fictitious future worlds that represented bright contrasts to the dreary everyday life of the present days back then. It was not until the twentieth century - when many of the utopian features had become reality - that the future pictures tilted, and future utopias became dystopias. A look into the literary visions of the last hundred years displays rather unpleasant or even apocalyptic worlds, shaped by ecocide, murderous robots, totalitarian regimes and nuclear annihilation. The great examples of this are George Orwell's "1984" (from 1948) and Aldous Huxley's "Brave New World" (from 1932), the figureheads of the futuristic novel in the twentieth century - or the Austrian-German movie director Fritz's dystopian movie "Metropolis" (1927). All described nightmare worlds, caused by despotic holistic dictatorships that were made possible solely through modern technologies.

That such pessimistic views on sciences and technologies were proven largely wrong in the last seventy, respectively 90 years after their publication (although their position was quite understandable with the world wars), that the technological progress has created unprecedented life enhancement, improvement of life expectancy as well as quality, and much more fulfilment of individual dreams. All these cannot change many people' pessimistic mindset until today about how science is shaping our future. How can this be justified?

Whether in questions of the latest quantum technologies, nuclear power, genetic research, climate research, data and personality protection, virtual reality or in artificial intelligence, we noticed that humanity often seems to be cognitively and emotionally overwhelmed. This overwhelming comes with limitations for the task of reacting appropriately rationally as a community to the most important aspects associated with among others these technologies - such as: the sustainability of our production and consumption, our global social justice, the maintenance of our individual sovereignty and freedom, or possibly among the most important how to approach most efficiently the change of the global climate due to our CO_2 emission. Our technological progresses - as good, enriching, and overwhelming as they are for each of us - have been done without much thinking about the limits of our planet and a possible extinction of mankind due to their own activities. And this is by no means only recently the case: The possibility of such an event has already been "achieved" by the nuclear bombs created as of the late 1940s (they can still destroy our life on Earth multiple times). Scientific successes have been factually so great that we are now at multiple tipping points and potentially dramatic changes, among others a possible change of the human nature through gene manipulation, creating a super-AI, or manipulating our perception in a virtual reality. Science must therefore also turn its head to the issue of human survival or the nature on Earth itself. A testimony of this is the movement of scientists around the IPCC.

The publication in journals today has also become a subject of relationship building. However, for publishing in high reputation journals, scientists need to adhere to particular rules, i.e. not publish "too new" results, and have a good relationship and support in the global academic network. So scientific success has often become a consequence of good relational power – often even more important than the quality of the scientific work. Furthermore, at the same time, today there are also preprint databases on the internet like arXiv, SSRN or Ideas in which scientists can easily make their research paper public without any refereeing process. This is what the mathematician Grigori Perelman did for publishing his ground-breaking results of proofing

Poincaré's assumption (in arXiv). He simply did not want to bother with publication journals. The new phenomenon of social media for researchers like Research Gate, Academia or Google Scholar also introduce a ranking and competition between researchers as well as a channel to communicate about his/her research more broadly - by refereeing papers differently, broaden the discussion, provide public financing for not immediate technology, etc.

Today, people face scientific and technological developments that are barely comprehensible compared to Newton's laws or a motor in a car. Next to apparent political and economic freedoms (in the West) they come with technical complexity, which seems threatening our collective psyche. Is anyone at all still able to sufficiently penetrate the depths of today's scientific disciplines? How do we fight against the interests of the big technology corporations that might stand against public interests? Where shall we see the danger for society in the multidimensional space of technological possibilities? We increasingly feel we are losing the overview, which robs us of the motivation to create one for ourselves. Many people thus are at risk of becoming apathetic and indifferent. "We can't do anything about it anyway!" is the credo of many. Other technology sceptics believe that only the abandonment of technological development can be the solution. They thereby follow the logic: Progress has given us all the problems, so only its limitation can solve those.

That is not without irony. Because the "culprit" of the expected deterioration or destruction of our living conditions is identified to be the same progress, i.e. the very power that made it possible that today we live in a society that far exceeds the optimistic scenarios of More's utopia. The fact that it was the sciences of the seventeenth and eighteenth centuries and their heroes such as Isaac Newton and two generation before him, Galileo Galilei, who decisively contributed to the Enlightenment and thus to liberalism, democracy, and the open society, all that no longer counts. A blanket reproach to science is even that it subjects people to the desired constraints and laws of technology and economics, thus degrading them to pure objects. In other words, science rejects and destroys humanity. These views often come with the motto: "Before it was better" are, of course, forgetting the terrible constraints mankind lived through in the past: poverty, famine, endless wars and massacres, diseases, pandemics, natural catastrophes that found people powerless, ignorance and oppressions.

These logics thus forget the other side of the coin: New technologies have always been excellent problem solvers as well (besides that they cannot be held responsible for "all our problems"). Hunger, disease, the effects of extreme weather events and many other plagues of mankind we were able to reduce to a fraction of the extent that was normal for previous generations. Contrary to even hundred year ago, today hunger is always due to war and unrest in some lands. The culprit for it is not anymore nature but people. Plus, a science belief taken over by the democratic movements challenged and often eventually overthrew dictatorial society leaderships rejecting science results. In fact, as of World War I, it has always been democracies or liberal kingdoms that let people do the research freely and even supported it (like Germany, France or the UK) – surely not a coincidence they were also leader therein! A fiery plea for science and enlightenment is made by Harvard professor Steven Pinker in his book *Enlightenment Now*,[9] which is well worth reading. In Pinker's view, science and technology are the driving forces behind the positive developments of past centuries - and will continue to be so in the future. The authors of this book agree with this assessment if we are still able to exercise our critics and free mind.

"We Go Under" Versus "Yes, We Can" - Dystopian Pessimism Versus Utopian Optimism

A remarkable paradox shapes our time: The technological progress enables more people than ever to live in unprecedented safety, enjoy the highest ever levels of health, and experience a quality of life no past generation has ever known. At the same time many people project a future in which everything we know is destroyed or even humanity as a whole is wiped out. We are scared and at the same time live as well as never before. How does that fit together that we live in a field of tension between those two extremes?

1. The apathetic-apocalyptic (dystopian) extreme: We have no choice at all, everything is determined from outside, and we can only watch the developments through technological disasters powerlessly.
2. The optimistic-euphoric (utopian) extreme: The world is completely open, all problems can potentially be solved by science and technologies if we only design the latter appropriately. We remain masters of technology and will use it to create more and more the paradise on earth.

How can one explain this contradiction, in which we are driven by both, a comfortable, but blind belief in technology and a fear-driven curse of science and its technologies? Almost everybody blindly trusts the functioning of smartphones, computers, digital data communication, antibiotics

---

[9] Steven Pinker, *Enlightenment Now: The Case for Reason, Science, Humanism, and Progress*, Viking; Illustrated edition (2018).

and many other technologies, but many at the same time demonize technological progress as a whole. For this strange contradiction, we would like to list six essential reasons:

1. *Perceived tight jacket*: Technologies impose their beat and rhythm on us; this was first experienced by the workers on the mechanical looms of the eighteenth and nineteenth centuries. In the early twentieth century, it was the assembly line workers that worked in long hours under very particular repetitive schemes, and today it is technical and mathematical optimisation processes that create time constraints ("just-in-time" production and distribution) to which we must adhere. The result is a feeling of being at the mercy of others and lacking control over our own times and lives. Chaplin's masterpiece movie "Modern Times" from 1936 gives a striking and fine description of these feelings of workers' helplessness towards the "moloch" of Taylorism[10] at the time of the great depression.
2. *Increasing complexity*: Most people hardly understand what is going on behind the curtains of the scientific stage. At the same time, they feel that there are powerful processes at work. It is this combination of intuitive sensing and lack of concrete knowledge and understanding that creates anxiety. Here again, a beautiful movie by Jacques Tati, "Playtime" (1967), set in a futuristic, hyper-consumerist Paris, expresses the feeling of overwhelm and perplexity in the face of the complexity of the world.
3. *Incredible velocity*: The sheer velocity of technological change and the associated complexity and speed of social changes overwhelm us mentally and emotionally. Unlike in the past, scientific and technological breakthroughs are no longer a matter of decades; they now take place at yearly, sometimes even monthly intervals. We no longer see ourselves as shaping social changes, but have trouble even reacting and adapting to the crazy speed and ever more confusing transformations.
4. *Non-overseeable multitude*: We are experiencing a multitude of dramatic changes at the same time. This is precisely what is historically new. Over the past 250 years, people have at given times faced particular *singular* technological upheavals, technological advances have processed comparatively slowly giving people and generations time to adapt. Today, we are not just dealing with a single "sorcerer's apprentice" experience but with a *whole lot* of them, which makes it difficult to adapt our lives and

---

[10] Frederick Winslow Taylor (1856–1915): An American engineer contributing methods to improve industrial efficiency. In 1909, he summed up efficiency techniques in his book *The Principles of Scientific Management* (which in 2001 *Fellows of the Academy of Management* voted as the most influential management book of the twentieth century).

solve important ethical questions. The parallelism of many technological innovations overwhelms us as well.
5. *Negative consequences of technology*: The application of technologies can have a huge negative impact on ecosystems, social and political systems. Without regulations and democratic control technology can bring about unwanted side-effects, and - these effects are not locally confined; they do not stop at national borders: Nuclear war, environmental destruction, climate catastrophe, artificial super-intelligence, genetic engineering and others affect and threaten humanity as a whole!
6. *The loss of certainty*: As we saw, we were forced to abandon the comfort zone of absolute certainties, be these of religious, philosophical or scientific nature, and must endure living with the ambivalence of relative truths. What began with Copernicus and the loss of our central position in the universe, continued with Darwin (we are not the centre of creation) and Freud (we are not even masters of our own mental home) found a next manifestation in quantum theory: Science no more offers a simple, absolute truth to hold on to – a new reality that can be experienced as unsettling. Such "relativism" is the fashionable qualifier to describe the modern world, without understanding that it is this relativism which is a fundamental progress in the liberation of humans from the (often oppressive) myths of the past. Admitting that we have to live with contradiction does not, of course, mean that we have to abandon all morality. To the contrary, as we saw in the various past chapters, it opens up a whole new field of research for ethics in philosophy.

We must realise: In a world of increasingly complex ambivalence, there is no universally shared goal to strive for, just as there is no black or white in the individual dimensions of technological progress. Such a point would be very difficult to reach today which is unfortunately often seen as the reason for not even trying subjectively. In fact, the good news is: We do not have to find an optimal state for all, but "only" define our own "feel-good space" - and this can be quite different for different people. In a multi-dimensional landscape of technological developments, the future is well open. It is thus useful to explore various perspectives of what our life could be like in already the next few years and decades:

- The future could seem like paradise to us today such as today would do to the people in the sixteenth century: AI and other technologies have freed us from the need for gainful employment, everyone lives in abundance, money is no longer an issue, VR entertains us blindingly, wars and

economic fights no longer exist because it is simply not worth fighting for anything anymore. Even one of the core assumption of economics, the scarcity of items and thus necessity to efficiently distribute them, has fallen.
- In other possible future spaces, we could stay quite well without them immediately seeming Elysian: Technologies would make life easier for us but would also create problems and would not free us from all suffering.
- In still other futures, life would not seem worth living any longer, e.g. when we are enslaved by a superintelligence or degraded to human beings à la Huxley's Brave New World, i.e. pure bio-machines.
- Finally, there are all those potential developments that would correspond to a civilisational crash (climate catastrophe, nuclear war, global epidemic, etc.) that makes the delights of collapsologies.

Unfortunately, it is not necessarily clear which path leads to which of these areas. It is the uncertainties associated with the technological changes that make people withdraw to the traditions and long for the "good old days" when everything was supposedly so safe, clear and well defined (which of course has never been the case). They then emphasize the dividing line between peoples and ethnic groups rather than see universal human commonalities, they seek solutions provided by an authoritarian government, an entity that has long since lost its real authority and the control over developments. Scientists make incredible technology possible, however, many (mainly non-scientific) people find themselves in a bubble in which they are emotionally stuck in a world of yesterday (or only in the remembrance of parts of it) struggling to perceive and accept changes, let alone to see what these changes mean to them in particular.

One thing is certain: How we deal with the technological possibilities and meet the challenges associated with them will decisively determine our future fate. What do we need to do with this? Well, generally speaking, define a feel-good space where we want to be and stay, and where we do not give up our humanity. Against the belief of many, as humanity, we do have significant room for manoeuvres when it comes to shaping the technologies of the future in a way that is beneficial and prosperous for all of us as well as nature itself. For this we must address the question: How exactly can we use this leeway to create the intellectual, political, economic, organisational and ethical framework in which we make good choices in the face of challenges? And what calls to action arise for each and every one of us in the process? These are the key questions we will treat in the next and last chapter to see how "consciousness" can help master science and technologies for the good of mankind rather than for enslaving it.

Social Drivers

Scientific and technological revolutions have in the past are repeatedly associated with a redefinition of ethical, political, social, spiritual and religious norms. They shifted truths, destroyed world views and created new ones. Ambivalences have always been part of the game during these processes. Today we observe: Internet comes with exciting new opportunities for social, political and economic knowledge acquisition and exchange, as well as completely new ways of governmental (and corporate) surveillance, massive interference with our privacy and, instead of openness, unexpected consequences of tribalism in the social networks. New algorithms solve previously insoluble problems, but the development of a superior artificial intelligence threatens to enslave us. And from the hunger of our modern technologies for energy leads a direct path to the destruction of our natural resources and last but not least cause the climate becoming non-survivable for us.

But who (or what) is actually in a position to steer technological progress towards tolerable outcomes? Such a steering is about nothing less than the survival of human civilization as we know it. Several social actors come to our minds quickly. However, three of the most often mentioned ones are undoubtedly overwhelmed by the task:

1. The responsiveness of *societal decision-makers* - politicians, business leaders, media designers, etc., whose job is also to increase the common good - is far too slow to lead the accelerating dynamics of technological change. Among other things, this is since political, business and cultural leaders' knowledge of the state of scientific and technological developments is usually scarce.
2. The *scientists* themselves are just as unable to control progress alone. On the contrary, many of them operate today, as we saw in Chap. 3, in a very narrow scientific field and do not have the "global view" on science or even their own subject of science. Pushed by fierce competition for positions and research grants, they isolate themselves instead of opening up to the necessary collaborations that would enlarge their vision of reality and help solve the complex problems we are facing. Furthermore, like all other members of society, they are equally subject to the free-market logic. They can even become billionaires themselves by developing new technologies based on their insights. So, often greed overwhelms curiosity at the danger of disregarding fundamental ethical problems.
3. A third social creative force is the *free market*. And indeed, technological progress has hitherto almost exclusively followed a market (or military)

exploitation logic. In other words, what was possible and meant a financial (or military) advantage for some has indeed been developed. Can we hope that the mechanisms of market competition steer the technological progress as best for all of us? This would mean hoping that Google, Apple and Amazon decide on the development and use of quantum computers or higher artificial intelligence to everybody's benefit. Or that pharmaceutical and genetic engineering companies employ CRISPR so it serves all of us best. Even the most believing followers of the free-market ideology would upon honest inspection consider such an expectation as far-fetched. In fact, the market has always been a bad referee when it comes to ethical concerns, just for that the later almost never have price tags. Another example are environmental effects which still today are not being priced and thus remain out of the equation for decisions by companies. And questions such as the proper use of CRISPR or the development of a possible superior artificial intelligence are far more important than just a hundred billion dollars in profits for one or a few companies.

Us

It is us who should dictate the next chapter to the question which social drivers – or better in which combination - can best pilot the technological development. However, the feeling that numerous big changes are upon us constantly distresses many of us. Thus, it blocks our thinking, and a reflex, hundreds of thousands of years old, takes over: flight. But instead of falling into irrational impulses, invoking dull slogans of the "good old times", constraining ourselves to particular values and cultural believes and isolating us against anything strange, we should recognize: Only in a truly global interaction respecting each other's identities, with the help of the spiritual and ethical potential of all people on this planet will we master the challenges of technological changes - including the climate problem. This attitude is the only way to make fear inside us disappear.

14

The Myth of the Optimally Functional Invisible Hand

Why and how research projects *and* future technologies should be discussed, respectively governed by the public domain?

The Legend of an Invisible Hand

Along with American independence, the year 1776 marks the first publication of an historically no less significant document. The publication of Adam Smith's book "An Inquiry into the Nature and Causes of the Wealth of Nations" marks the beginning of classical national economics and the economic liberalism. In it, Smith uses - only once - the famous metaphor of the "invisible hand" (he actually said "a visible hand", never "the invisible hand"[1]): The common good is automatically achieved if individuals only care about maximising their *own well-being*. Smith is known to have declared self-interest to be the rational principle of all economic activity. The background to this thesis is that with a free exchange of goods, each person produces precisely the good that offers him the highest (relative) competitive advantage. This automatically maximises the overall economic productivity of a society. In this respect, economists speak of an optimal "economic equilibrium".

To this day, Smith's "invisible hand" serves as a legitimising principle for the view that only the market leads society as a whole to maximum prosperity.

---

[1] There is in fact only one instance where the invisible hand is explicitly mentioned in *The Wealth of Nations*, that is in Book IV, Chapter II titled "Of Restraints Upon the Importation from Foreign Countries of Such Goods as Can Be Produced at Home" (see below). It is also mentioned in his book *The Theory of Moral Sentiments* (1759) in Part IV, Chap. 2, Here he questions the distribution of wealth in this context.

Many economists still today like to paint a picture of a market that automatically produces the highest prosperity for all, if only it is allowed to, i.e. if the exchange of goods and services and other economic activities can develop completely unrestricted. Government interference, social welfare efforts and even ecological demands would only interfere with this and abandon the optimal economic circumstances.

However, the *real circumstances of market economy* processes do not correspond to the ideal picture of economic models seen by the economists. The arguments associated with Adam Smith are mistaken if they assume that free markets automatically lead to desired optimal outcomes, i.e. prosperity *for all*. What they do not have on their radar are five forces that prevent the "market equilibrium", as propagated by economists, from actually occurring as a socially acceptable government. It is of utmost importance to know and consider these forces when assessing the development of future technologies.

1. *Externalities*: The economic activities of one person (or group of people) can have an impact on other (possibly all other) people without the acting person bearing the full cost of his or her activities. Probably the best example of this is a public good that for long have had no price: ecological resources. The climate-damaging emission of CO_2 is still not associated with costs for producers, the safety risk of nuclear power or natural gas fracking is not paid for by the benefitting producers.
2. *Rent seeking*: Powerful groups often succeed in changing (or keeping) the political and economic rules to their own advantage. Economists speak of "rent seeking"[2] when it comes to such striving for state-guaranteed advantages without increasing the overall welfare of society. Lobbying in Berlin, Brussels or Washington has today become an industry in its own right, so to speak, employing tens of thousands highly skilled people seeking to influence politicians, i.e. for rent seeking. In Brussels alone, about 2600 lobby groups have an office, and about 30 percent of EU parliamentarians end up working in this industry after their political careers. The result is less competition and higher costs for consumers or taxpayers. In addition to increasing inequality, this leads to a deterioration of the innovation atmosphere and of economic growth.
3. *Unequal allocation of productive goods*: Productive goods can be accumulated in the hands of a few. The results are extreme income and wealth inequalities within a society, and finally also less economic competition. That was Karl Marx's main analysis of capitalism.

---

[2] "Rent seeking" means in detail earning income that is not matched by a corresponding productive output, and this at the expense of the general public.

4. *Information asymmetries:* This point further increases the danger of social inequalities. As early as 1970, the later Nobel Prize winner (2001) Georg Akerlof showed in his essay "The Market for Lemons" that free markets cannot function optimally if buyers and sellers do not have the same access to information.[3] Such asymmetry exists in many markets of our everyday lives: the labour market, the market for financial products (the banks just ripping off non-financial-experts), the market for health goods and food, the energy market, and - particularly important in the context of science - the market for new technologies.

5. *Cognitive distortions*: Classical economic theory assumes that we know what is best for us and that we act accordingly. But behavioural economics has long shown that we often act far more rashly and less rationally than free market advocates assume. Thus, we are often guided by short-term drives instead of long-term, well-considered considerations.[4] A good example is smoking or the consumption of alcohol and drugs (which for good reasons is regulated by the government). So not only do we accept harming others, we also harm ourselves. Through manipulation we are easily led into harmful or high spending behaviour. For example, Apple makes billions of dollars by releasing a new iPhone every six months and profiting from the fact that we are willingly seduced by the latest technology gadgets, greatly overestimating their usefulness compared to their prices.

The capitalistic view furthermore completely neglects what risk management has taught us, namely, that optimizing goes most of the time at the price of resilience. An optimum is often reached in an unstable state. Thus, a much more rational behaviour would be to seek a good compromise between optimality and resilience. Moreover, the presence of important transaction costs, which are usually neglected in these theories, and which vary among various actors, is also an important obstacle to reach a societal optimum.[5]

In the real world, there is no invisible hand that will lead us to the promised land. Economic history has shown many times that "market equilibria" are anything but desirable (and usually do not materialise in reality).

---

[3] Georg Akerlof, *The Market for 'Lemons': Quality Uncertainty and the Market Mechanism.* Quarterly Journal of Economics. The MIT Press. 84 (3) (1970) p.488–500. Also the book by G. Akerlof and R. Shiller *Phishing for Fools*, Princeton University Press (2015) illustrates further how we are permanently "ripped off" in our economic system by better informed participants.

[4] Akerlof and Shiller give a detailed discussion in their book mentioned in the previous footnote about the mechanisms of our manipulability.

[5] See about the consequences of asymmetry in the transaction costs: Michel Dacorogna, *High frequency trading a boon or a threat?* Bancaria, n. 1 (2020) p. 89–96.

The free forces of the market are subject to the capitalist logic of exploitation, in which it is important to generate the highest possible profits as quickly as possible. With the five forces of action listed, the free market behaves like an organism that thinks quickly, but to which slow, rational and above all long-term thinking is often alien.[6] Even if (public) companies say they plan for the future (usually three years, at maximum five!), the goals they want to achieve with these plans are usually only of short-term nature - entrepreneurial success is measured in quarterly figures and annual balance sheets as well as the bonuses of the management that are also linked to these figures. Such short-termism cannot be a target-oriented strategy for overcoming a human crisis.

One example of information asymmetries was (and still is) the pharmaceutical market. In the nineteenth century, there was still fertile ground for all kinds of quacks who, at best, tried to talk the unsuspecting into taking substances that were ineffective; but often they were also harmful or even life-threatening. A prominent example was Clarke Stanley's "snake oil", which he claimed was obtained by killing and squeezing rattlesnakes and could cure a variety of diseases. In fact, Stanley's medicine was a mixture of beef fat, red pepper, camphor, but also mineral oil and turpentine. Another panacea, which according to its producer William Swaim was supposed to be effective against scrofula, syphilis, rheumatism as well as numerous other diseases, was even highly harmful as it contained poisonous mercury. The first result of a "free" market for medicines was thus arbitrariness and fraud. Only government regulation (e.g. in the USA, in the form of the "Pure Food and Drug Act" of 1906) eventually put a stop to charlatanism in the drug market. The same applies to the food industry: here it was the "Federal Meat Inspection Act" that became US-law on the same day.

Those who think that such examples are only known from the nineteenth century are mistaken. To stay with the pharmaceutical industry: Even today, consumers are generally not in a position to make their own decisions about the use of the medicines offered to them. Most of them do not know which drugs and how many of them they should take, what risks are to be considered and how their effects are to be assessed. Even medical professionals such as doctors and pharmacists do not have the knowledge to assess all aspects of - especially new - medicines. It takes them many years to understand who reacts how to which medicine. So, they must rely on the information provided by the pharmaceutical industry, an industry that has a global turnover of about

---

[6] See also: Daniel Kahneman, *Thinking, Fast and Slow*, Farrar, Straus and Giroux (2011).

1000 billion USD and grows by about 5 per cent a year - and thus represents a huge commercial power of interest. This is precisely why the approval conditions for medicines should be regulated by the government.

Another excellent example of how asymmetric information between buyers and sellers, coupled with huge conflicts of interest, political influence, and cognitive delusion can lead to global economic instability are the manoeuvres of banks and other financial players. Bankers created toxic securities worth billions of dollars. Actually for the sake of high personal bonuses, they bought the objectionability of then still respected rating agencies. Factually, they tricked unsuspecting investors, including insurance companies, pension funds, local banks, who bought these securities and suffered losses from them in the billions, which ultimately had to be borne by the taxpayer. This unethical behaviour of the banks led to an extremely asymmetrical risk-return distribution according to the motto "privatise the profits, socialise the losses"!

So, even today's consumers are by no means protected from the machinations of unscrupulous companies - despite strict requirements for the industries, banks and consumer protection organisations.[7] So, where is the invisible hand in all of this that is supposed to produce a socially accepted optimum state - without outside government interference? We should be sceptical of the "laissez-faire" attitude demanded by classical economics - as well as today's neoliberal movements. In view of these obvious limitations in the functioning of free markets, it is quite astonishing how persistently the market enthusiasm of neoliberalism continues. There seems to be far more ideological bias than intellectual honesty at work here. Adam Smith himself was not the extreme market liberal that he is often seen as by classical economists. The one place he writes of an "invisible hand" is in connection with import restrictions on foreign goods (which he considered as quite critical) and an investor who should prefer to invest money at home rather than abroad:

> By preferring the support of domestic to that of foreign industry, he intends only his own security; and [...] intends only his own gain, and he is in this,

---

[7] Given the financial crisis of 2008 and its causes, it was no wonder that calls for regulation of banks and other financial service providers became particularly loud - even though they had already been subject to relatively strong regulations before (in the USA, for their part, in response to the financial crisis of 1929). But still not enough, as it turned out. It is interesting to know that some people attribute the 2008 crisis to the repealing of the Glass Steagall act by Clinton. This law, part of the Banking Act of 1933 of the Roosevelt administration, prohibited bankers from using depositors' money to pursue high-risk investments. Effectively separating commercial bank activities from investment banking. So, the governments had always had great difficulty in forming the appropriate counterforce against the unbridled financial industry's interests.

as in many other cases, led by an invisible hand to promote an end which was no part of his intention.[8]

Smith made it clear in his writings that quite considerable structure was required in society before the invisible hand mechanism could work efficiently. Today, however, the neoclassical economic view uses this metaphor contrarian to Adam Smith to call for markets that are completely free from any government action.

It is thus paradoxical that a failing ideology in structuring our economy is expanding very fast in the *organization of science*. (Financial) Competition among scientists is introduced more and more, where science in fact needs more collaboration in order to progress in the complex advances of science. In this framework, as we saw in Chap. 3, more and more *quantitative* indicators are introduced to judge the quality of scientific work, while it is more *qualitative* judgements that would help us point out better innovative research – including such that will help understanding and possibly solving our societal problems. It is time to rethink about the governing and financing of science. But who and which precise criteria are best for judging the quality of science going forward?

Who is Likely to Best Govern the Scientific Future? – I. The Side Actors: Cultural Figures, Journalists or the Church?

For particular questions about technological progress, the *church and religious forces* could (and should) have a serious voice. For example, the question of whether and in what form interventions in embryonic genetic material should be permitted touches on religious values and questions of faith. However: As, according to Christian doctrine, life begins at conception, then an intervention in embryonic stem cells would be a direct intervention in a human being and not in a cluster of cells that is not yet alive. If church institutions want to be taken seriously as co-speakers for our technological future, they have to break away from the dogmas of a religious interpretation of politics. It is these non-negotiable "truths" that cause religious institutions to be no longer

---

[8] Adam Smith, *An Inquiry into the Nature and Causes of the Wealth of Nations*, Book IV, Chap. 2, London (1776).

that important to most people (at least in Europe). Compared to the Christian church, for example, Buddhism is much less dogmatic about their world views.

Until a few years ago, the Swiss writer Friedrich Dürrenmatt was one of the few *cultural figures* to deal with technological progress. In his 1961 play "The Physicist" he lets a protagonist say:

> The content of physics concerns the physicists, the effects all people.[9]

It is only in more recent times that writers have increasingly rediscovered scientific topics for themselves and reflected on their possible consequences. Andreas Eschbach wrote about a nanotechnology that threatens the world,[10] and about a global intelligence created by linking human brains.[11] Marc Elsberg created a scenario of an all-controlling data octopus.[12] Along the same lines, Chinese science fiction author Chen Qiufan recently teamed up with AI specialist Kai-Fu Lee to write a collection of short stories linked with their own "voices" exploring the future developments and dangers of AI.[13] This is a particularly sensitive area in China. Also worth mentioning is the "Interview Project" by Swiss curator Hans-Ulrich Obrist, which consists of an extensive collection of interviews with artists, architects, filmmakers, scientists, philosophers and musicians.[14] But in general: Cultural workers are less suited as active shapers of humane technological progress. Their role is more that of admonishers.

It is astonishing how little physics, chemistry or biology is mentioned when *journalists* try to explain the world to us. Yet there are far more exciting connections to be made between our everyday lives and the scientific and technological progress taking place before our eyes than, for example, to the annual get-together in Davos of a world elite whose vanity is becoming increasingly unbearable, or the endless discussion about how to deal with the flood of refugees, or more recently the development of our economy (inflation, higher energy prices, etc.). That, for example, the discovery of the gene scissors CRISPR/Cas9 in 2012 (or the perhaps even more powerful version CRISPR/CasX in February 2019) did not make it into the daily press is an indictment of ignorance. It will certainly be in future history books.

---

[9] In the appendix of the comedy he lists the "21 Points to the Physicists". This one is the 16th.
[10] Andreas Eschbach, *Lord of All Things*, Amazon Crossing (2014).
[11] Andreas Eschbach, *Black Out, Hide Out, Time Out*, Arena (2010–2012).
[12] Marc Elsberg, *Helix – Sie werden uns ersetzen*. Blanvalet, München (2016).
[13] Chen Qiufan and Kai-Fu Lee, *AI 2041: Ten Visions for Our Future*, Crowe (2021).
[14] Hans-Ulrich Obrist, Rem Koolhaas:, *The Conversation Series Nr. 4*. König, Köln (2006).

Who is Likely to Best Govern the Scientific Future? – II. The - Democratically Elected - Government?

As we saw before: Governmental forces have been for 150 years - among other purposes - there to counteract the forces that drive free markets into undesirable equilibria. In fact, it is only through agents who are not subordinated to their own interests that the market can function at all. Let us look specifically at the field of science and technologies. We distinguish between three phases in the development of technological progress and examine where government forces must act:

1. The *phase of science* (and potential inventions out of it): Here we are dealing with basic scientific research, which is, in its nature, neither bound to specific purposes nor to specific interests (other than gaining knowledge) and should be motivated by curiosity. Besides challenging existing world views, it carries no or very little direct risk to society. The example of CRISPR, however, demonstrates how quickly a little-noticed field of basic research can grow into the focus of important technological applications. As recently as 2012, CRISPR research populated a rather obscure niche in the landscape of microbiological research. Less than five years later, corporate lawyers started fighting over patents worth billions of dollars.
2. *Innovation*: The development of the first commercial applications is usually associated with entrepreneurial risk. This is where the research and development departments of profit-oriented companies come in. As the economist Joseph Schumpeter's describes in his theory of the profit-oriented innovation efforts of dynamic entrepreneurs, this phase plays the decisive role regarding growth impulses. Here, government-based regulations start playing a key role.
3. *Diffusion*: The innovations are brought to the broad market. Only part of what is technically possible is ultimately realized in a product. In addition to economic criteria, political and social criteria, e.g. technology assessment in the crowd, are aspects in the decision process.

In phase 1, the government usually only has the role of actively promoting - and financing - science as a public good. When companies conduct or takeover basic research results today, they rarely do so for the sake of knowledge alone. They always have their commercial benefit in mind. While the role of the government is rather limited in phase 1, it plays an important corrective and framework-setting role in phase 2, and then especially in phase 3, in which technological applications can change our lives (and often encounters fierce resistance). This dispute can only be resolved through political and economic discourse and agreements based on it. In particular, the government must ensure to put limits to the power of groups with particular interests potentially against the broader society, for example, prescribing consumer protection measures, safety standards and industrial norms.

Last but not least, the government also has an important role to play in the social distribution of productivity and welfare gains from new technologies. Technological change may bring huge gains in prosperity, but not necessarily for everyone. With the massive increase in overall economic output in recent decades, social inequality has increased significantly due to the success of the "laisser-faire" parties in politics starting with Thatcher and Reagan but soon followed by social democrats like Tony Blair or Gerhard Schroeder. Today, it has reached a level comparable to that in Europe before the French Revolution, which, according to the OECD, WEF and IMF, poses a massive threat to further economic development.[15] We must ensure that wealth is being created for more - and eventually all - people (and states). One goal of (democratic) government forces must be to enable all people to enjoy the highest possible quality of life with sustainable resource consumption. Environmental protection is thereby a genuine task of the government, because in the free market it has so far represented an externality that is, as we saw, barely ever considered in pricing.

The government therefore has an important role to play in regulating progress. However, even the government is not able to provide a comprehensive framework for technological progress on its own. The reasons are:

- Technological progress is based on the creativity of scientific knowledge and entrepreneurial design, both of which are difficult to predict, influence or even control by the government. Politics can sometimes even massively inhibit scientific developments. Of course, a government could try to stop progress by law until it has decided on pros and cons. But the effects of too much interference in the creativity of the scientific quest for knowledge and the entrepreneurial power to shape up its use, would end up not controlling technological developments, but rather stifling it. So unfortunately, government intervention often leads to the danger of frictional losses due to false incentives and distortions of competition.
- Government decision-makers rarely have the necessary knowledge of the respective state of scientific development and are thus also hardly able to

---

[15] On this subject, Thomas Piketty's book "Capital in the XXIst century" gives ample evidence of the growing inequalities (Harvard University Press, 2014).

cope with the speed of technological change. While legislative bodies need years to obtain the detailed knowledge and then design the legal framework conditions for new technologies, the technologies have long been developing and often make this framework superfluous. A good example of this is data protection. Here, Google and Facebook have played cat and mouse with politicians for many years. In fact, ironically, it is the internet giants that benefit most from the European General Data Protection Regulation (GDPR), which came into force in May 2018, as they know best how to optimize their infrastructure in it. However, in view of the increasing risk of misuse of private data, legal protection is key. And only experience with GDPR and other measures, will give the legislators the means for designing better laws.
- Government action itself is of course not free of conflicts of interest. A good part of its decision-makers are politicians who want to be re-elected. In addition, government corruption is widespread in many parts of the world.

The role of governments in shaping our technological future is thus controversial. The government is caught between those who call for more government, such as social and environmental politicians, and those market-liberal forces who believe that the less government, the better. There are even voices that believe that human rights and democracy must be restricted and even sacrificed if necessary. Their argument is that the "responsible citizen" is too easily manipulated and defencelessly exposed to the lies and inconsistencies of propagandist forces. A "paternal government", they argue, has the task of reconciling progress with the "true needs" of the people. Only a knowledgeable elite could set things right. Such demands (as we find them already in Plato's philosophy[16]) come from both the left and the right wings of the political spectrum. They are not only dangerous, but also contradict the most successful social model in history, the open society, as Popper makes clear.[17] Every time individuals thought they had found a single path to truth or the perfect form of society, they, once in power, produced a torpor of despotic absolutism which reached exactly the opposite.

History teaches us that in a competition of ideas a decentralised society provides the best conditions for the development of science and technology. Examples of this are ancient Greece, Northern Italy during the Renaissance

---

[16] In Platon, *Politea* (the Republic).
[17] Here the lecture of Karl Poppers is also worth it: Karl Popper, *The Open Society and Its Enemies*, Routledge, London (1945).

and Humanism, Europe during the Enlightenment, the USA in the twentieth century or Sweden and Switzerland still today which despite their small sizes have some of the highest proportions of Nobel Prizes when measured against their populations (third and fourth in the ranking, after Saint Lucia and Luxembourg with each two, Sweden 33 and Switzerland 27[18]). Wherever power was strongly centred, science fell short of its potential. Examples of this are the Roman Empire, China until the twenty-first century,[19] the Middle East and the socialist bloc of governments in the late twentieth century (before it fell apart).

However, the various social actors we looked at so far are not able to shape or steer technological progress on their own, politicians are too slow, the churches are too dogmatic, journalists are overstrained, cultural workers are more likely to admonish than to shape, entrepreneurs may push ahead, but they act too opportunistically oriented only according to their own (financial) advantage, and the free market economy is quickly taking us in the completely wrong (uncontrollable) direction, neither do we need self-appointed elites.

Who is Likely to Best Govern the Scientific Future? – III. The Scientists Themselves?

We already saw in the last chapter that scientist themselves are anything but good judgers for the structuring of future science as well as for assessing appropriately new technological opportunities. Individual scientists simply have too many particular interests, which once in a while even contrast the public interest. Two core problems of managing science today are the judgement process for scientific works published in scientific papers, and the allocation of financial resources to the various scientific projects. These two issues, science funding and science judging, are intrinsically connected today, as we already saw in Chap. 3. The current environment forces researchers to spend an absurdly high proportion of their time, on average 30%, chasing funding for their research! Against this background, the belief that science today is interest-neutral and only focused on the search for new knowledge seems rather naïve. Power struggle, cleaver marketing have become part of the

---

[18] See https://en.wikipedia.org/wiki/List_of_countries_by_Nobel_laureates_per_capita (where literature, peace and economics are considered).
[19] Only one single Nobel prize in a field of natural science has been awarded to a Chinese, who is living and working in China, a woman in physiology and medicine (in 2015).

science environment since much money and social recognition are involved in the process.

The amazing new technologies create new features of life that must rely on a much broader basis than government, scientists themselves, the market, particular elites, journalists, writers or the church. Namely it must rely on *all of us*. For this we need to develop new ways of educating the vast majority of our society in scientific (and then technological) developments.

Who is Likely to Best Govern the Scientific Future? – IV. All of Us!

Science itself is by nature not a democracy. Truthful theories do not depend on a majority that votes in favour of them. Otherwise, the sun would probably still be considered a disc, and the world being only 4500 years old. So, even in a democracy one has to rely on certain truths. We need to make the distinction between the two most important aspects of it and what they mean to scientists, which we briefly saw already in the introduction. It is particularly important to clarify that personal opinion is not the same as results of scientific research. Science is not intrinsically relative. We must not question well accepted scientific knowledge like the rotation of our planets, the DNA-structure of our genes or that our climate is sensitive to CO_2 emissions.[20]

- There is a *body of agreed upon knowledge* in science (existing theories with strong experimental and empirical evidence) that scientists have very little doubt about. Still, they are not irreplaceable and can change as they did across all sciences (and mathematics) in history, e.g. between 1880 and 1950.[21] It is a result of a process of complex interactions with new results obtained by experiments and observations. However, in the history of science this is rather the exception. Thus, assuming that the scientists are right about their existing theories is very likely the best and safest belief.
- *Current research* is about open questions raised by existing science. There is intrinsically a lot of doubt, critics and controversies around those questions, e.g. how a unified particle physics theory looks like (see Chap. 6), or how proteins fold in our bodies (see Chap. 8). Today, the discussion

---

[20] An interesting discussion on this can be found in: Étienne Klein, *La vulgarisation scientifique est-elle un échec* Institution Diderot, Paris (2022).
[21] See Lars Jaeger, *The Stumbling Progress of 20th Century Science - How Crises and Great Minds Have Shaped Our Modern World*, Springer (2022).

often takes place in isolated scientific groups with the public only once in a while hearing about interesting new observations or theories. It is the effort of this book to get readers a better access to these discussions today.

It is of crucial importance that we understand the difference between the open scientific problems and the fundamental scientific theories that can be considered as true by everybody. Confusing the second and the first is a widespread subject in non-scientific circles, from spreading confusion to polemic counterarguments against science per se. This important difference is to be clarified by scientist as well as responsible journalists over and over again.

But how about technological consequences of science? Contrary to scientific truth, democracy is highly relevant for technological consequences. In fact, their validity or usefulness are not given by some natural law but by the human ambitions towards them. When scientists and technological experts are the only ones that seek for technological processes, they will surely find sub-optimal or even dangerous solutions. The same applies when we leave it to capitalists only. Neither do journalists or church persons give us solutions. In all these cases the involved are highly affected by personal interests. Technological opportunities imply rather *our all duty* to actively seek information and exchange views. This then ensures more diverse and higher levels of ethical integrity, that not small groups who, guided largely by conflicts of interest or lack of knowledge, take isolated decisions that affect us sub-optimally. Discussions about relevant issues must involve as many people and opinions as possible in order to find broadly accepted answers to important questions. This is where Hans Jonas' principle of responsibility and the collective commitment to future generations truly come into play. Unfortunately, however, there is still far too little talk about physics, chemistry, or biology – or even the scientific background of climate change, when journalists and other opinion leaders inform us on world events and important social developments. It is thus key to enlarge and deepen the involvement of media. Ways of popularizing the scientific approach are e.g. the initiatives of citizen science projects where the public participates voluntarily in scientific processes, often dealing with real-world problems.[22]

So, how do we create a more broadly interested public in science and a more powerful democratic decision-making process when it comes to technological applications? How do we broaden the informed discussion on science? How exactly can we establish some leeway to create the intellectual, political,

---

[22] See for instance the website of the *Citizen Science Association*: https://citizenscience.org/.

economic, organisational and ethical frameworks in which we make good choices on technologies in the face of difficult challenges? And what engages us to participate in the process? These are key questions and have been hard to answer ever since science and technologies became a central part of our lives almost 200 years ago. While science has become harder and harder to follow for a broad public as its complexity has gained significantly in the last 100 years, the technologies derived from those can in their essences be reasonably quickly and well understood as, for example, the use of smart phones by two or three year old kids testify.

Further to these questions as well as next to the *ethical* integrity we must demand *intellectual* integrity and control a commitment of everybody who is somehow involved in technological implications. Making money from new technologies is okay, and the motivation to do so can be intrinsically powerful for technological progress as we have seen in the past. Deliberate falsehoods, information distortion as well as information filters for the purpose of enforcing particular interests, is something we all must effectively fight against. It is simply unacceptable and dangerous that fake news unfold their destructive propagandistic power. A startling number of politicians (especially in the US) are still seriously doubting climate change or even Darwin's theory of evolution. There are most often some unethical political motivations behind these ridiculous claims.

However, the commandment of intellectual honesty does not only apply to political, social, economic or scientific leaders but also to *ourselves as the recipients of information*. We have to be careful not to draw conclusions too quickly, not to break down on prejudices, and engage in complex interrelations without the urge to simplicity (Occam's razor). And finally and among the most important attitudes, we must be open for inconvenient truths. We cannot solve crises by lying to ourselves. Here, too, scientific thinking helps us out of the impasse, as Richard Feynman put it:

> The first principle is that you must not fool yourself, and you are the easiest person to fool.[23]

We know it is a high demand, but we live in a time where utopia must meet necessity: Each and every one of us has the mandate to inform and engage her- or himself in a self-determined way, and then to not only discuss, but also to become active. Already our children should be introduced to and educated for this demand: Eagerness to know and ability to acquire knowledge and independence of thoughts, and then the courage to express

---

[23] From Richard Feynman's talk at the Galileo Symposium in Italy 1964.

statements publicly must determine the essential pedagogical direction in our educational landscape much more strongly, and this in three respects:

- Preparing young people for the epistemological and ethical challenges in view of the human crisis due to the climate change;
- Imparting knowledge about the properties, mechanisms of action and possibilities of new technologies;
- And in more general terms: Engaging in public exchanges and debates about critical points.

We need an education system that reflects on the original meaning of the word *education* (Greek/Latin: ēducātiō/educare: bringing up – to independently thinking, reflecting and morally acting people). For this endeavour, digital technologies can support us. One example is a Finnish initiative by the Ministry of Economic Affairs and Employment (which started in 2017) to teach digital laypeople how to use AI technologies which the initiators see as a contribution to democracy.

> Finland's Artificial Intelligence Programme prioritises actions and initiatives on innovative approaches in AI and machine learning and continuously upgrades and improves technical infrastructure and the deployment of 5G technology. Through this approach, the Programme aims to optimise education and modernise it to achieve better results.[24]

That we citizens can take matters into our own hands to act decisively against politicians, governments or powerful lobby groups, is also shown by a popular initiative in Bavaria in January 2019 to protect bees.[25] The topicality of this popular initiative was shown by the first comprehensive scientific analysis about the decline of insect populations worldwide which was published almost at the same time[26]: the insect mass on our planet is shrinking by a total of 2.5 percent per year, which over 25 years corresponds to an insect loss of almost 50% ($0.975^{25} = 0.53$). This dramatic decline in insect numbers could lead to the extinction of 40 percent of insect species worldwide in the next few decades, and if it continues even further, to the extinction of almost all insects. The Bavarian initiative against the industrial pollution of the bee's environment was broadly accepted and is now the basis of a law in the state.

---

[24] https://digital-skills-jobs.europa.eu/en/actions/national-initiatives/national-strategies/finland-artificial-intelligence-programme.
[25] For more information, see: https://www.thelocal.de/20190214/bavaria-has/.
[26] Francisco Sánchez-Bayo, Kris Wyckhuys, *Worldwide decline of the entomofauna: A review of its drivers, Biological Conservation,* 232 (April 2019), p. 8–27.

In addition to all citizens, the elites in particular have a duty. Political and economic decision-makers must overcome traditional, linear patterns of thinking and engage with the complex, non-linear developments of technological influences on us and our society. Following those, scientists are also called upon. They must raise their voices in important decisions, for example, in journalistic contributions, videos or public forums, and this especially against any form of lies, (too much) simplifications and populism. They should advocate the scientific method as a powerful way to build a rational consensus. Furthermore, critical journalists are in demand more than ever. Their task is clear: to expose deliberate distortions and pollutions of information that serve the enforcement of particular interests. Like the exposure of the dreadful scandal around the two firms Cambridge Analytica and Facebook, the discovery of which in 2018 was led by the mathematician Paul-Olivier Dehaye and the journalist Hannes Grassegger.[27]

There must always be a democratic process at the heart of decision-making around technological developments (as well as other areas): Centralised top-down decisions do not do justice to the complexity of the latest technologies. Only a high level of social diversity and decentralised information and decision-making structures create a sufficiently high level of functioning and performance and thus the necessary insights and decision-making power within society, politics and the economy. It is democracy that enables coordination between individual interest groups, governments and all people. Democracy can steer the global collective, the necessary cooperation of all parts of world society - government leaders, economic participants, scientists, intellectuals, culturally dedicated etc.

The fact that the democratic culture has a strong formative power can be demonstrated by two historical examples:

1. With its openness to change and critical discourse in almost all matters of governance, foreign policy, philosophy and science, ancient democratic Athens shaped the world from the Atlantic to India for many hundreds of years. Without it, the philosophies of Socrates, Plato or Aristotle, those of the Epicureans and the Stoa, the tragedy poetry of Aeschylus, Sophocles and Euripides or the historical work of a Thucydides, which are still influential today, would hardly have been imaginable.
2. The rise of a loose association of thirteen small English colonies to a centre of attraction for entrepreneurs and intellectuals from all over the world would never have taken place without the idea of democracy and an open,

---

[27] For more details: https://www.hannesgrassegger.com/reporting/that-turned-the-world-upside-down and https://www.letemps.ch/societe/paulolivier-dehaye-matheux-ennemi-facebook.

liberal social order. The wealth of ideas, the strong democracy, and the creative power of the immigrants made of the USA the economic, political and intellectual superpower of the twentieth and (and so far) twenty-first century.

The greatest possible happiness of all people is achieved when as many different creative forces as possible strive to balance interests, stability and sustainability. Again, this is only possible in an open, democratic society. We are in a constant competition of ideas and conceptions about how we want to live in the future. This competition involves the creativity of universities, research institutions and companies as well as a wide variety of political, economic, social and philosophical positions. In the end, the best ideas and most creative models of life must prevail. Plus, only in a democracy can we quite quickly change away from a wrong path we took as it had looked so good at the time. We must thus guard ourselves against controlled political will-forming and instead seek democratic dialogue.

A prerequisite for democracies to function as well as for science to be accepted is *transparency*. This applies to plans for how our genome is to be changed, to nanobots that will determine our health in the future, or to how we produce energy. But also to digital technologies, quantum computing, AI and what Big Data can do to us. As well as what biology potentially does with respect to new forms of life or prolonging our own life, furthermore brain and consciousness improvement, nuclear fusion, etc. Since the control mechanisms of society are increasingly based on algorithms, information collection and processing, increasingly AI processing, the necessary tools must not be exclusively available to a power elite. They must be publicly accessible. In particular, everyone must have the right to have their own data at their disposal and to be continuously informed about what is happening to it.

Every one of us, in every part of global society, should be in demand of information. But how should each individual find her way through the jungle of the *latest technological developments*? Five simple rules for each of us, regardless of culture or country, can have us go a long way:

1. *Give justice to complexity*: A simplistic solution or explanation to a complex scientific question or context is very likely to be wrong.
2. *Reject absolute truths*: Those who claim definitive truth or use their own beliefs as a criterion for truth should be given low credibility a priori.
3. *Select sources of information*: Where possible, we should back up our information with research in scientific journals, and those that have an

independent refereeing process. Scientists are also encouraged to communicate their research findings in a way that non-scientific people can grasp their significance.
4. *Listen to experts*: The opinions of scientists and other experts must be well integrated into the democratic discourse process, even if they are controversial and not always correct (scientists can and do sometime come with errors).
5. *Question motivations*: We should always try to understand who is providing us with information, for what reasons and with what motivation.

15

Science, Technology and Spirituality
What Science Can Do for Society, How Society Has to Shape Technology - And How Spirituality Can Set a Frame for this Shaping Process?

How New Technologies Shape Up the Economy – In the Right Direction?

Silicon Valley: Where bio-, genetic, nutritional or health technologies, nano- or neurotechnologies, digital technologies, artificial intelligence, robotics, virtual reality, social media, transport technologies etc. In a mixture of science centres, attractors of the highly gifted and risk-taking entrepreneurship, the valley has managed to master technological feasibility like no other region in the world. Nowhere else do visions become reality faster than here. Here, people see themselves as the global engine of progress - and progress as good per se. And for these visions billions of dollars are available, which in their strained search for attractive returns strive to already spot the next trillion-dollar technologies. In fact, the high-tech industry of Silicon Valley proves to us that there are no longer any economic barriers to development of technologies.

Nonetheless, our lives are at risk of being overturned within a few decades, perhaps even years. And all this continues under the influence of above's "terrible five" of capitalism: *Externalities, rent seeking, unequal allocation of productive goods, information asymmetries, cognitive distortions.*

Today, strong economic interests decide with which technology to advance and at what expense. But the reverse influence also takes place: Technologies change the economy - the most decisive one being that a new competitive reality is creating economic power concentration on an unprecedented scale: "The winner takes it all". Once a company has a dominant position in social media (Facebook and WhatsApp), internet search (Google) or office software

(Microsoft), it is difficult to break it. This is because these companies can set the norms of their industry themselves:

1. They often invented the industry themselves and are thus protected by patents against imitators and competitors.
2. Unlike cars or steel, the digital products of internet companies can be manufactured with little effort and multiplied almost at will. It does not make a difference whether a digital product is made available to customers in editions of 100, 1,000 or 1,000,000. Plus, for the consumption there are equally completely new economies of scale: The benefit for each individual user of the product is greater the more people use it.

The emergence of new norms and competitive conditions can happen very quickly, as the examples of PayPal, Amazon, Facebook, Microsoft, Google, Apple. In this context economists also speak of "disruptive technologies". Such platforms divert an ever larger share of value creation to themselves, and often do so with a very small workforce, so that each of the owners can claim a large share of the profits for him or herself. The company Instagram employed just 13 people in 2013, when it was bought by Facebook for a billion dollars. Karl Marx would have taken intellectual delight in what we see in Silicon Valley on a regular basis - and at the same time would probably have been suitably alienated.

In numerous industries, digitalisation has already massively changed the value chain: It is no longer the actual manufacturers of products or services, but digital platforms with their exclusive access to immense customer data that increasingly determine the economic regulatory framework. Examples are: Airbnb and booking.com (apartment and hotel accommodation), ebooker (air travel), Uber (taxi rides), AppStore (software), WhatsApp (telecommunications), PayPal (payment services), Amazon (online store). Those who think that such changes only affect the high-tech industry are sorely mistaken. For example, Google could soon completely overturn an industry that has a history of over 100 years in Europe and the US and provides no less than one in seven jobs in Germany: the automotive industry. For this purpose, the firm joined (in 2021) the car producer Tesla (which has a higher market value than all European car producers put together) in order to collaborate for producing entirely new cars. Google's driverless car which automatically moves in a network with other cars, traffic lights, road markings and traffic data could not only turn the business model of VW and Daimler upside down, but also make the traffic division of Siemens superfluous - do driverless cars still need traffic lights? There will still be a need for cars (albeit fewer), but Mercedes and BMW will then become pure suppliers just as bookstores are for Amazon, all significantly suffering financially. The jobs that will be lost will not be replaced automatically (i.e. just shifted as in the past). Google software requires much less human labour than the production of a car or traffic lights.

But will this all be bad? In Chaps. 9 (and 14) of their book, 2041, Kai-Fu Lee and Chen Qiufan explore the effects of AI job displacements on people where they envision the development of universal income and of new companies specialized in "reshaping" the workforce for AI economics. We might witness the start of this development with the popularisation of the idea of a universal income for every adult citizen independent of his or her own work.

The revolutionary power of digital technologies is enormous and not limited to sales and shopping platforms. And there are surely some very promising, life improving and ecological benefits. After the internet and Economy 4.0, now come "smart homes", i.e. completely automated houses that can be controlled via the internet. There is even a talk of automated "smart cities", with demand-adapted street lighting, optimisation of energy consumption, autonomous traffic control and waste collection, and even "smart nations", with control of the healthcare system by Big Data, robots and a broad network of sensors and AI. All of this promises us more quality of life, more prosperity, less energy and resource consumption and more social exchange. And of course, the positive effects do not stop in front of traditional industries. For instance, the steel company *Klöckner*: With its completely digital order and production management, the company can avoid high inventories and the associated costs and considerable capital commitment. This approach is becoming standard for such firms - however, with problems when international deliveries get disturbed like during the Corona crisis or by the Ukrainian war.

But even the biggest technology idealists must recognise that the real incomes of many people, have rather decreased in recent years and decades - while their worries and fears have increased. Winner takes it all rules create new monopolies, new forms of social inequalities and thus new fears. The result of the uncertainty and dissatisfaction was not least the election of Donald Trump as US president in autumn 2016 or the Brexit in the same year. Just as during the technical revolutions of the nineteenth and twentieth centuries, we will have to discuss the distributive justice of the productivity gains of digital and other technologies today and in the years to come.

More Openness, Less Dogmatism

The power of free-market competition, the dependence of our economic system on perpetual growth, our greed for ever more comfort, prosperity, and quality of life, and, above all, the irrepressible human creativity - all these forces are too strong for us to willingly stop technological progress as a whole. We should not aim at it, either - although in cases like nuclear bombs that can kill the entire humanity we must stop it; but were we able to do so?

For this to happen in a democratic consensus, instead of asking for an omnipotent authority to fix everything, we must start with ourselves. Each of us needs to recognise our cognitive distortions, our everyday self-deceptions and our own intellectual dishonesty as far as possible and try to minimise those. We all have a mandate to self-determine, to inform, to share, to engage and to become mindful and active. The question is: "What and where do we want to go?" - or simply "what do we want?". In our search for the answer, personalities like Albert Einstein or Richard Feynman are a role model for us, because they show us that it is possible and very constructive,

- to be interested in current technological events - and open scientific questions,
- to be fully engaged in finding solutions to social problems,
- to question ourselves and others over and over again,
- to constantly seek new consensus,
- to take an open path of discourse (and a democratic one when it comes to deciding on open public questions).

An attitude of "it doesn't matter to me anyway" or "I will not understand this anyway" is fatal and leads to decisions that ignore the will and the benefit of the majority. The diversity of challenges should not make us shy away. Individuals do not know everything, but they can take an interest and get involved in what concerns them personally. Only in this way can the future be shaped in a way that is desirable for all concerned.

Rationally Irrational

Already the Enlightenment thinkers of the eighteenth century recognised that people cannot completely free themselves from the corset of dogmas and irrationality. On the contrary, they themselves still adhered to numerous ideas that seem absurd to us today. Newton, for example, believed in alchemy and the possibility of producing gold out of other materials, and Franz Mesmer offered an enthusiastic audience on magnetic cures to dissolve blockages of energies circulating in the body. The enlighteners Kant, Montesquieu, Voltaire, Spinoza, Hobbes, Hume and Adam Smith made it their task to rationally and scientifically grasp "universal human nature" together with its unreasonableness and irrationality. Being philosophers, cognitive and neuroscientists, psychologists, sociologists, anthropologists and human physicians at the same time, they recognised that precisely because the human psyche contains so many irrational tendencies, the principles of reason must be upheld all the more strongly.

In addition, our irrationalities can certainly have benefits. For example, it must at first seem very irrational to devote enormous resources to making a man walk on the moon for a few hours. But the gains in technological capabilities that resulted from the space programmes make them seem quite rational. Apparent irrationalities can therefore inspire us to excel. Kepler and Newton believed vehemently in divine principles; without these (today seen as) "unreasonable" convictions, they would hardly have had the mental strength to arrive at their ground-breaking insights. The Indian mathematician Srinivasa Ramanujan claimed in the 1920s that the Hindu goddess Namagiri Thayar had brought his ingenious analytical and number-theoretical insights to him in visions.

Of course, the irrationality and unreasonableness of human beings can also be destructive. It leads us to destroy our environment as if we still had a second one, or to despise or even kill people with a different skin colour or different religion. This unreasonableness comes in most strongly where people reinforce each other's unreasonable convictions. When irrationality occurs in such collective form, it is most likely responsible for much unhappiness in the world.

The American psychologist Steven Pinker has an explanation for this collective irrationality: He speaks of the "tragedy of the belief commons" as a variation on the well-known principle of the "tragedy of the commons". The tragedy of the commons describes the effect that in a situation with shared resources, individual users behave according to their self-interest and against the general good of all users by spoiling or destroying this resource for their own benefit. A fisherman who catches as many fish as he can (and more than he needs) from a lake that belongs to the general public until there are none left is acting quite rationally, because he says to himself: "If I don't catch the fish, the others will, and I will go empty-handed." All in all, however, something irrational comes out: the lake is fished out in no time, and no one can catch any more fish. Pinker transfers this tragedy of the commons to social action: It is very useful (and thus rational) for a person to adhere to an obviously false conviction if the environment does likewise and he has to reckon with social sanctions if he does not conform. This leads to a contradiction in terms: It can be rational for a person to be irrational. According to Pinker's thesis, someone who rages against foreigners and refugees could do so out of a rational calculation. He might know very well that his own problems lie elsewhere than with the refugees. Rather, he enjoys the advantage of feeling part of a group that assimilates him and his fears. The reverse is also true. In a different environment, it would be advantageous to deny with a sighted eye the problems that migration and immigration create. What in both cases falls by the wayside is a fruitful discussion about facts. The Enlightenment has created a powerful counterpart to collective irrationality.

However, there is also a positive rational way of being irrational. Philosophers like Jakob Klein or Leo Strauss have theorised the relationship between rationality and irrationalism. One can accept rationally to be irrational or to limit rationality to be more efficient in our lives. Leo Strauss, reinterpreting the Trial of Socrates, sees a co-existence of rationality and irrationality. For him, Socrates was condemned as a corruptor of the youth because it is not possible to put *everything* in question. Acting politically means reducing the rationality so that we can take decisions without putting them constantly in question. He thinks that Socrates accepted this limit. It is why he did not fight the sentence. Another example is mathematics that proofs logically that any mathematical system of axioms cannot be fully consistent. Mathematicians must accept this limit of rationality to be able to progress in their understanding of reality. It is similar in physics with the uncertainty principle: accepting a limit to our knowledge allows us to understand the laws in the quantum world.

The most developed collective rationality is science. It is defined by the fact that the truth of statements is not bound to a person. Great scientists from Galileo to Einstein to Feynman created world views independent of their own person or identity. They "only" wanted to know how things really are. This courage to uncouple our knowledge of the world from our identity and accept what experience tells us is true human freedom. In order for the greatest possible number of people to participate in solving crises and problems, it is important to detach people's reasoning from their identity.

There are further prerequisites for enlightened discussion among free people. The great Enlightenment philosopher Immanuel Kant puts us on the track in his writing "An Answer to the Question: What is Enlightenment?"[1]:

---

[1] Immanuel Kant, *An Answer to the Question: What is Enlightenment?* (1784), very beginning; see also: https://www3.nd.edu/~afreddos/courses/439/what-is-enlightenment.htm.

Enlightenment is the human being's emergence from his self-incurred immaturity. Immaturity is the inability to make use of one's own understanding without direction from another. This immaturity is self-incurred when its cause lies not in lack of understanding but in lack of resolution and courage to use it without direction from another. Sapere aude! [dare to know] Have courage to make use of your own understanding! is thus the motto of enlightenment.

Here Kant mentions another essential quality needed for an enlightened life: Courage. Anyone who grew up in a world in which the Earth was at the centre of the universe needed not only excellent scientific aptitudes but also the courage to say that things are different from what everyone else believes (which often led to being executed). For Kant, the reasons for a lack of rationality were clear. In the very next paragraph of his writing he stated:

It is because of laziness and cowardice that so great a part of humankind, after nature has long since emancipated them from other people's direction (naturaliter maiorennes), nevertheless gladly remains minors for life.

It is, however, a great mistake to believe that irrational actions and thinking, which do not care about facts, are exclusively the result of a lack of courage and determination. Psychological research today is much more advanced than it was at Kant's time. In addition to courage, we must also have the will to act rationally. Just as courage and determination are the antagonists of laziness and cowardice, will is the antagonist of fatalism. We must be convinced that a turn for the better, an improvement of the world is possible. Here we again encounter the third idea of the Enlightenment: the optimism that the world can be constantly improved with persistent progress, and that the future can be shaped according to our desire.

How Can Broad Knowledge About Science and Technologies and Its Rational and Democratic Assessments Make the World a Safer and Better Place

Besides the dangers of a "winner takes it all" economy, digital technologies allow us also to envision exciting ideas. While classical economic theory assumes a world of limited resources whose allocation is done through economic competition, a digital economy could, for example, solve the distribution problem of limited resources by making unlimited resources available.

An "economy of cooperation" instead of competition would thus much more easily be possible.

The "sharing economy" is a new trend, which has already spawned business models such as "Uber" and "Airbnb". An older example is the good old "ride-sharing", whose service is now offered by online platforms like Lyft (in the US), Uberpool (international), Wingz (in the US and Canada), Grab (Asia), Bolt (Europe), Cabify (Europe, South America), Didi Chuxing (China, Asia) and many more. Here, resources that would otherwise go unused are being shared with others (a seat in our car that would otherwise remain empty). Digital technologies make a new form of economy possible, and business models of sharing have already begun their march.

But the visions of a "sharing economy" go far beyond sharing houses, cars, knowledge (e.g. Wikipedia) or opinions (social media). More generally, they could make any form of underutilised asset available to a community. This leads to an overall reduction in the need for property. Thus, its representatives already speak of collaborative use of all kinds of freely accessible assets: recreation (boats, holiday domiciles, etc.), production (machines, data storage, etc.), ideas (example: the inventor platform "Quirky" at www.quirky.com), education (online lectures and schools, such as the "Khan Academy"), all the way to new accesses to capital for starting a company (crowdfunding, lending platforms). Sharing economy is therefore also called "collaborative consumption" or "peer-to-peer-based sharing".

The resources are simply not contained anymore. Many products then become free. Computer software can be copied as often and as quickly as we like, and computer-controlled robots or AI exponentialize our physical and mental capacities far more than the steam engine which ("only") multiplied the power of our hands. The non-material resources of the digital world are unlimited in principle. However, material goods are also becoming more accessible to all as they can be produced easily and very cheaply using (digital) 3D printers with the appropriate (digital) software.

It is now also clear how we solve one of the essential challenges of our time, the climate problem: with radical honesty and consistency, and then the use of environmentally friendly technologies. An attitude of intellectual integrity simply rules out denying man-made global climate change. Giving priority to the economic interests of individuals, corporations or nations is ethically dishonest. Even those who believe that the chance of a catastrophic development is still small must follow the dictates of risk ethics: Even developments with low probabilities but highly negative implications demand considerations. And for climate change we face a worst-case scenario: Very high probabilities and very high negative consequences!

Economists and climate scientists have long since shown that a sustainable and climate-neutral economy is possible. Switching the economy to sustainable energy production and a massive international tax on CO_2 emissions could achieve a lot.[2] However, today's climate policy shows over and over again that it is neither a lack of political or economic reasons for action, nor a lack of opportunities to do so that prevents necessary creative measures from being taken. However, it is also clear that the transition will not happen from today to tomorrow but will be a process – a process that must not take too long.

Like thinking honestly, *acting* honestly is before everything a personal matter. However, "honesty" is not an absolute word that defines any action 100% in one way. But if we all think and act according to the same principles of human rationality (again, this does not mean thinking and acting the same way as everyone else), then we can always find consensus and resolve any conflict.

- Global, collaborative thinking will guide us to the right answers to human crises,
- Global, collective action will translate these answers into action and lead us out of human crises.

However, so far it hardly looks like humanity could agree on common criteria of probity in thought and action. The first real - non-military - global challenge, climate change, seems like a test run for any future human crisis. For the first time in our history, we are confronted with the challenge of acting as *one global entity*.

There are two behaviours that in practice guide our ways to mental autonomy and intellectual integrity. The first: we need exchange. The word "exchange" has two meanings here:

- People exchange clearly delineated arguments, thus comparing and discussing them.
- Individuals change their convictions (exchange them) by adopting coherent arguments of their counterparts.

Anyone who has experienced scientific enterprise knows about the importance of exchange in science. Scientific exchange can happen by taking part in a symposium, in the omnipresent discussions, in the diversity of the lectures,

---

[2] For more details see Lars Jaeger, *Ways Out of the Climate Catastrophe - Ingredients for a Sustainable Energy and Climate Policy*, Springer (2021).

in the passion in the debate of scientists (well-known is their writing of formulas on the napkins and beer mats at dinner) and in the inspiration with which the participants go back to their work afterwards. The scientific method is the best way to exchange controversial views and then build a consensus among people on a rational basis. An interesting experience in this direction is promoted by the EPFL in Lausanne with the Centre for Digital Trust (C4DT). It is an academic-industry alliance of international relevance that facilitates innovation in digital trust services and products.[3]

Insight views, democratic exchange, and broad discussions can generally be applied to any crisis: We thus have the means and opportunities to counter them. We have good reason to be optimistic! But the phase of our lethargy and inactivity as well as the belief in political or neo-liberal capitalist solutions must be over.

A New Way of Approaching "Spirituality"

Next to science, another property of our correct behaviours and reactions is to properly assess our believes in "spiritual contexts". And let us not confuse spirituality with religion only. Wondering about nature and human spirits unites natural scientists, philosophers, and people who call themselves spiritual. Without distinction, they feel deep awe and inciting curiosity about the beauty of nature, the diversity of its forms and the mightiness of its powers, but also about the richness of our own spiritual experience The "phenomenology of the spirit", as the philosophers put it; the first one being Hegel with his phenomenology *of consciousness* or *phenomenology of spirit* - Phänomenologie des Geistes. Science, philosophy and spirituality are thus not opposites. Wondering about the mysteries of the world and the wonders of its phenomena is the essential commonality of spiritual and scientific inquiries.

Already the ancient philosophers Plato and Aristotle recognised that the origin of scientific and philosophical endeavour lies in our deep wonder. Thus, Aristotle wrote:

> For wonder was to men now, as before, the beginning of philosophising, in that they wondered at first at the nearest unexplained thing, and then gradually progressed and raised questions even about greater things, for example, about the phenomena of the moon and the sun and the heavenly bodies, and about the origin of the universe.[4]

---

[3] https://c4dt.epfl.ch/.
[4] Aristoteles: *Metaphysics*, I 2, 982 b 17 – 22.

And Plato said:

Wonder is the attitude of a man who truly loves wisdom; indeed, there is no other beginning of philosophy than this.[5]

Finally, Thomas Aquinas wrote in the late Middle Ages:

Wonder is a longing for knowledge.[6]

It is the "wonders" of nature that lead us to "wonder". And from there it goes to knowledge through science. Thus, there is no question about a "disenchantment of nature" occurring through science, as Max Weber saw it at the beginning of the twentieth century. Today we know far more about the world than we did back then. And yet there are rather more than fewer mysteries. We have no lack of reasons to wonder. Wondering and amazement is not only a part of spirituality, but already its core.

However, amazement is not the only link between spirituality, philosophy and science. There are three further bridges that connect them:

- *Intellectual honesty*: From amazement at the sublimity in nature arises in us a sense of limitation and thus humility. Our scientific and spiritual thinking must strive for truth. At the same time, we must be humble when it comes to assessing our knowledge and our own "truths". Seen in this light, intellectual honesty is also a spiritual attitude, a form of spiritual asceticism that consists in a constant critical questioning of our own knowledge.
- *Uncompromising wanting-to-know*: From the wondering as well as the critical spiritual attitude arises a curiosity that can never be contained. Those who subscribe to it want to understand things instead of simply accepting or dismissing them with simple explanations - and this particularly in relation to the mysterious and miraculous things in the world.
- *Renunciation of an absolute truth*: The combination of wanting-to-know and intellectual honesty leads us to a scepticism about ultimate truths. If the loss of absolute truths was one of the most dramatic shocks to modern science, the associated earthquake in any traditional spiritual understanding goes far deeper.

---

[5] Platon: *Theaitetos*, 155 d.
[6] Th. von Aquin, *Summa Theologica*, 1a2ae, 32,8; lat. Original: *Est autem admiratio desiderium quoddam sciendi*, [...], Verl. Styria Graz, Vienna, Cologne Köln (1933 ff); Original between 1266 und 1273.

So, instead of a religious or esoteric spirituality, we need a specifically secular spirituality. It must not hover in "higher spheres" but be a practical part of our everyday life. Already in the late 1960s, the German (Austrian)-American sociologist Thomas Luckmann spoke of a new, "invisible religion" that starts from the individual and asks about the function and meaning of spirituality in modern society instead of turning to a universal transcendent sacred.[7] Furthermore, the German philosopher Thomas Metzinger (see Chap. 9), calls the question of whether something like a "secularised spirituality" is conceivable as central in assessing "a historical period of transition that will have a deep impact on our image of ourselves, and on many different levels at the same time."[8] This position is close to Spinoza's view of God, as the substance of nature and not as a transcendence. His views have influenced many scientists, among them Einstein who answered in April 1929 in a telegram to Rabbi Herbert Goldstein:

> I believe in Spinoza's God who reveals himself in the orderly harmony of what exists, not in a God who concerns himself with the fates and actions of human beings.

There is a list of requirements for a secular ethic importance for science as well as technologies, similar to the requirements of scientific integrity:

- *Autonomy in thinking*: We must resist myths, fixed social norms and dogmas.
- *Responsibility in action*: We must realise that we are all personally responsible for what the future will look like and need to act according to our ethical criteria.
- *Global acceptance*: The principles of action must be reasonable enough to be accepted in the secular and democratic West, in the (non-dogmatic) Islamic Middle East, the Israelian and democratic Jewishness, and in the Hindu-Buddhist influenced Far East, etc.

The first two points can be summarised in one word: "personal integrity". The last point contains the demand for globally consistent ideals. This is where a particular difficulty lies: Modern technological progress takes place on a global scale. But what do common global values look like? Not even

---

[7] Thomas Luckmann, *The Invisible Religion*, Springer (1967).
[8] Thomas, Metzinger, *Spirituality and Intellectual Honesty - An Essay*, University of Mainz, https://www.blogs.uni-mainz.de/fb05philosophie/files/2014/04/TheorPhil_Metzinger_SIR_2013_English.pdf; also the last chapter of the second edition of his book "The Ego Tunnel: The Science of the Mind and the Myth of the Self", 2nd edition, Basic Books (2014).

the guiding principle of (European-style) humanism, the equality of all people and the demand that everyone should have access to the best possible development of their personality, is recognised throughout the world.

An answer to the question of what human beings should be in the future includes a consensus about what they are here for in the first place. In the past, religions and social conventions dictated the meaning and purpose of our lives: To serve God, to sacrifice for the nation, to prepare for an afterlife, etc. Enlightened thinking and intellectual honesty make us realise that these goals no longer apply in such a simple way. The question of meaning must therefore be answered within a secular spirituality. The enlightened secular mind can only answer this question by creating meaning from within itself. In our modern world this can be: Earning money, having a good time and enjoying life, creating something lasting (for example, building an art collection), acquiring knowledge, treat others well, and much more. The question of meaning in our lives seems to have only one unsatisfactory answer within the Western secularised worldview: Making oneself the measure of all things.

Recent scientific results help us in our quest for meaning and collective spirituality and possibly overcome our sole individual striving for profit and consumption:

- We are actually not as big of an ego machines as we think. The evolutionary paradigm of competition, of survival of the fittest is still deeply anchored in our everyday thinking and in economic and social models. But in today's biology, the image of an individual "I" optimising survival and reproductive success has long been outdated. Scientists understand more and more precisely that evolution consists of an interplay of the individual with his or her genetic kin, his species and nature as a whole. Most biologists and anthropologists assume that the special evolutionary success of humans was only made possible by their extraordinary ability to cooperate socially (and intellectually). This is exactly the ability we need for our survival in the future.
- In Chap. 9 we already talked about the fact that our I-experience is by no means as self-evident as we think. An irreducible and individual, i.e. indivisible, unity of the ego does not seem to exist. Rather, without us being aware of it, our "I" is part of a web of world-spanning dependencies. If we become aware of this, it will increase our chances of survival as a global society. Many spiritual traditions describe various forms of a dissolution of the individual sense of self as an essential part of the spiritual experience. Either through an integration of the "I" into a superior instance (God) or through rooting into the social "we" of a community,

or even without reference to anything or anybody any longer (Buddhism). So, the path is already there, we just have to develop it further and then walk it. Incidentally, this relativisation of the subjective has an analogy in the world of the objective. We still consider ourselves as subjects, independent of any external environment in principle. But just as we as subjects are not as separate from our environment as we think, the "objective" is also not as objective, i.e. independent of us subjects, as it seems to us. Physicists have already made this experience in their work on quantum theory. Today, physicists understand that the subject dependent objectivity in the nano-cosmic world *converge* to a subject independent objectivism on the (mesoscopic) scale of our experiences. But as we today can observe objects in the nanoscopic world, the subjective and the objective can be far more intertwined than our world view and experiences enable us to believe. Equally a relativisation of the ego fixation, the later seeming, as we saw, biologically not be fixed objectively somewhere in the brain, will largely eliminate the"unnatural" separation of our ego and our environment. We then understand ourselves better as part of a whole again.

A commitment to intellectual integrity also supports us in developing a more selfless, community-oriented perspective, based on the cognitive insight that we are part of a greater whole and not an isolated, self-centred "ego self-cosmos". For intellectual integrity is inseparable from the striving for mental autonomy and an ethically holistic, i.e. "integral" perspective. This strives us to become independent of an experienced self together with its particular self-interests and desires (particularly pursued in the Buddhist tradition).

Summary: Ideas Instead of Ideologies

Let us briefly summarize: There is no single regulatory mechanism, no model formula for how to best shape technological progress. No single authority, be it the market, the churches, journalists, the government, scientific elites or any other social force, will be able to dictate the shape of the technological future or even manage it alone. Rather, we are in a *competition of ideas*. This is being fought out between universities and research institutions, between political frameworks and entrepreneurial creativity, but also between different political, economic, social and philosophical positions, and not least between innovative companies. Google, Apple, Microsoft, Facebook, Amazon as well as the many start-ups that emerge every year with new business ideas. This competition of ideas applies to all models of social design. In the end, the best ideas and most creative models will prevail. This is, however, only possible in a democracy. Dictatorships have so far always failed in this.

It takes a social equilibrium of many shaping forces to achieve a good balance of different interests, social stability, sustainability and the greatest possible happiness for all people. The essential question is: "Where do we want to go?" For answering it, we will have to follow the democratic and open path of discourse, ceaselessly engage in painstaking new consensus-building, constantly question ourselves and others, review the course of events and set the design process in motion anew each time. A perfect decision and path making is not achievable but avoiding a catastrophic one is achievable possibly only in a democracy. Dictatorships have so far always failed in this.

So, in the future we need a round table with as many representatives as possible. That is what we are already seeing today on the internet. Not everyone likes that. Nondemocratic sceptics describe the polyphony of open and controversial communications in an open democratic society as chaos. Putin and Xi Jinping are not the last ones to present themselves as an alternative to bring order to society and the world. To them the noise of disagreement seems to be the chaotic result of a post-modern, mass-democratic worldview resting on the equality for all. Yet it is precisely this hullabaloo of diversity on which our hopes as authors of this book rest for shaping technological progress for the better rather than for the worst. For only the constant struggle of different interests guarantees that our future will not be determined by particular interests of certain groups, by errors and cognitive distortions of individual experts or by pure chance, as with a democracy we have the chance to correct the errors we made quickly. There is no way around an expansion of democratic decision-making processes based on an always open and reflexive exchange of ideas and opinions. In doing so, we will always have to muddle through a process of constant corrections of undesirable developments.

We would like to close this book by a reformulation of an old philosophical question: "What is a human being?" Modern science and technologies transformed this question into a new one, which is quite frightening for the responsibility it places on us. Still, it is equally realistic and now at the centre of philosophical anthropology and ethics: "How do we want human beings to be?".

Name Index

A

Abel, Nils-Henrik 239, 248, 249
Aeschylus 316
Akerlof, Georg 303
Aleksandrov, Pavel 53
Allen, John F. 133
Anaxagoras 260
Anders, Gunther 44, 290
Aquinas, Thomas 329
Archimedes 34, 256
Arhenius, Svante 280
Aristotle 10, 15, 16, 121, 175, 316, 328
Aspect, Alain 18
Atiyah, Michael F. 248, 249

B

Babbage, Charles 211
Baire, René 243, 245
Banach, Stefan 238
Bardeen, John 12, 128, 212
Bednorz, Georg 133
Benzécri, Jean-Paul 24
Berger, Hans 177
Bergson, Henri 15, 255
Berkeley, George 15, 196
Berners-Lee, Tim 132
Bethe, Hans 21
Binning, Gert 140
Birch, Bryan 257
Blair, Tony 309
Boeke, Jef 165, 166
Bohr, Niels 10, 16–19, 37, 39, 58, 59, 61, 255
Boltzmann, Ludwig 10, 245
Bonnet, Pierre Osian 249
Borel, Emile 10, 20, 243, 247
Born, Max 102
Boström, Niklas 90
Bourbaki, Charles Soter 247
Bourbaki, Nicolas 247, 248
Boutroux, Émile 72
Bouveresse, Jacques 72
Bovery, Brown 33
Brattain, Walter Houser 128, 212
Braun, Karl Ferdinand 33
Brentano, Franz 182
Broca, Paul 176, 186
Brossel, Jean 129

Name Index

Brunel, Isambard 33
Buchner, Eduard 55
Burnell, Jocelyn 262
Byron, Lord 211

C

Cajal, Santiago Ramón 177, 178, 180, 188
Cantor, Georg 10, 72, 241–243, 245
Capon, Laura 21
Cartan, Henri 247
Chain, Ernst 43–46
Chalmers, David 194, 232
Chaplin, Charlie 295
Chargaff, Erwin 148, 149
Charpentier, Emmanuelle 163
Chevalley, Claude 70, 247
Chomsky, Noam 216, 217
Citroën, André 33
Copernicus, Nicolaus 260, 296
Courant, Richard 245, 246, 252
Crick, Francis 11, 148–150, 179
Curie, Marie 127

D

Daimler, Gottlieb 33, 320
Damasio, Alain 232
Damásio, Antonio 201
Darwin, Charles 2, 11, 31, 56, 296, 314
Dath, Sara 43
Dawkins, Even Richard 159
de Broglie, Louis 20
de Fermat, Pierre 250, 251
Dehaye, Paul-Olivier 316
de La Mettrie, Julien Offray 80
de Laplace, Pierre Simon 109, 110
Delbrück, Max 11, 21
Deligne, Pierre 249
Demenÿ, Georges 129
Democritus 183
Dennett, Daniel 76, 193
de Saussure, Horace Bénédicte 31
Descartes, René 72, 73, 121, 175, 176, 183
Dharmakīrti 196
Diophantos 256
Dirac, Paul 19, 22, 24, 100, 104
Doudna, Jennifer 163, 171
Drexler, Eric 92
du Bois-Reymond, Emil 65–67, 73, 160, 176
du Châtelet, Émilie 31
Dublin, Louis 147, 168
Duchesne, Ernest 43
Dürrenmatt, Friedrich 307

E

Eckert, John 212
Edison, Thomas 37
Eiffel, Gustave 33
Eigen, Manfred 157
Eilenberg, Samuel 248
Einstein, Albert 2, 10, 15, 16, 18–21, 26, 32, 37–40, 42, 44, 52, 56–58, 60, 61, 69, 75, 117, 120, 126, 129, 237, 253, 255, 266, 267, 269, 276, 282, 322, 324, 330
Eisenhower, Dwight 133
Elsberg, Marc 307
Epicurus 183
Epimenides, Cretan 242, 243
Erasistratos 175, 176
Eschbach, Andreas 191, 307
Euclid 240, 256
Euripides 316

F

Faraday, Michael 36, 37, 53, 54
Feringa, Bernard Lucas 141
Fermi, Enrico 21, 101, 104, 118, 119, 126, 265, 283
Feyerabend, Paul 71, 72

Feynman, Richard 12, 22–24, 47, 82, 91, 99, 100, 105, 107, 119, 137, 139, 314, 322, 324
Fichte, Johann Gottlieb 53
Fleming, Alexander 42–44
Florey, Howard 43, 44
Franklin, Rosalind 11, 148–151
Friedmann, Alexander 267, 269
Fritsch, Harald 105
Fuchs, Klaus 13

G

GAFAM 76, 230
Gage, Phineas 185
Galilei, Galileo 10, 27, 260, 293, 314, 324
Galois, Evariste 239
Gamow, George 269, 270
Gauß 10, 249
Gauss, Karl Friedrich 239
Geim, Andre 135
Gelfand, Israel 20
Gell-Mann, Murray 102–105, 107, 119
Gilgamesh 287
Glashow, Sheldon 107
Gockel, Albert 261
Gödel, Kurt 10, 20, 239, 243–246
Goldstein, Herbert 330
Golgi, Camillo 176, 177
Gong, Peng 230
Gould, Elisabeth 188
Graber, David 223
Grassegger, Hannes 316
Gratia, André 43
Greenberg, Oscar 105
Griffith, John 149
Grothendieck, Alexander 70–72, 248, 249
Gumbel, Emil 20
Gutenberg, Johannes 131

H

Hadamard, Jacques Salomon 20
Hahn, Otto 44
Harris, Tristan 222
Harrison, Ross 169, 288
Harvard 34, 289, 294, 309
Hawking, Stephen 63
Hebb, Donald 180
Hegel, Georg Wilhelm Friedrich 53, 328
Heidegger, Martin 255
Heisenberg, Werner 15, 17, 19, 39, 58, 60, 61, 99, 103, 110, 245, 255
Hermann, Grete 20, 52, 60, 68
Herophilos 175, 176
Higgs, Peter 108
Hilbert, David 10, 19–21, 72, 238, 239, 243–248
Hirzebruch, Friedrich 249
Hoagland, Mahlon 151, 152
Hobbes, Thomas 323
Hodge, William 257
Hodgkin, Alan 177
Hopfield, John 210
Horvitz, Eric 224, 225
Hoyle, Fred 159, 269
Hubble, Edwin 261, 264, 266, 267, 269, 278
Hume, David 121, 323
Husserl, Edmund 25, 121
Huxley, Aldous 49, 50, 191, 192, 291, 297
Huxley, Andrew 177
Huygens, Christiann 261
Hypatia 256

J

Jacob, François 153
Jäncke, Lutz 187
Jinping, Xi 333
Jonas, Hans 291, 313
Joyce, James 104

K

Kant, Immanuel 14, 16, 25, 53, 59, 60, 73, 120, 121, 196, 323–325
Kapitsa, Piotr 133
Kasparov, Garry 209
Kastler, Alfred 129
Keops 256
Kepler, Johannes 4, 10, 250, 251, 260, 261, 264, 323
Khorana, Har Gobind 45
Klein, Jakob 312, 324
Kleppe, Kjell 45
Knazek, Richard 169, 288
Koch, Christof 179
Koch, Robert 2, 37
Kolmogorov, Andrei 10, 53, 247, 248
Kowalewskaja, Sofia 10
Krauss, Lawrence 122
Kronecker, Leopold 241
Kuhn, Thomas 57, 58
Kurzweil, Ray 168

L

Ladenburg, Rudolf 129
Lane, Saunders Mac 248
Lang, Fritz 291
Langevin, Paul 15
Laughlin, Robert 114
Leawitt, Henrietta Swan 261
Lebesgue, Henri 10, 243, 247
Lee, Hoesung 281
Lee, Kai-Fu 117, 210, 215, 234, 307, 321
Leibniz, Gottfried Wilhelm 15, 52
Leibniz, Wilhelm 10, 15, 22, 53, 72
Lemaître, George 269
Lenzen, Wolfgang 22
Leutwyler, Heinrich 105
Lévy-Leblond, Jean-Marc 71
Licklider, Joseph 207, 208

Lie, Marius Sophius 103, 108, 248, 251
Lorenz, Edward 111, 119
Lovelace, Ada 211, 213, 252
Luckmann, Thomas 330
Luzin, Nikolai 247

M

Mach, Ernest 15, 55, 56
Maiman, Theodore 129
Mallat, Stéphane 24
Mandelbrot, Benoît 111, 112, 119
Marconi, Guglielmo 33
Markram, Henry 219, 220
Marx, Karl 15, 302, 320
Matsumoto, Tetsuzo 216
Mauchly, John 212
Maxwell, James Clerk 10, 36, 37, 54, 55, 57, 120
Mayor, Michel 266
McCarthy, John 209
McCulloch, Warren 208
McEwan, Ian 81, 233
Meitner, Lise 44
Mendeleev, Dimitri 101, 102, 104
Mendel, Gregor 11, 56
Merlin, David 22
Metzinger, Thomas 76, 195, 198, 199, 200, 202, 330
Miller, Stanley 156, 157
Mills, Richard 257
Minsky, Marvin 209
Misener, Don 133
MIT 34, 65, 96, 193, 209, 303
Montesquieu 323
Moore, Gordon 211, 213
More, Thomas 291, 293
Müller, Alexander 133
Mullis, Kary 45

N

Nāgārjuna 158

Nagel, Thomas 184
Navier, Claude 112, 252, 257
Newton, Isaac 4, 10, 14, 15, 27, 52–57, 65, 109, 115, 120, 261, 264, 277, 293, 322, 323
Nicolelis, Miguel 199, 200
Noether, Emmy 10, 20, 21, 60, 103, 106, 120, 239, 247, 248
Novoselov, Konstantin 135

O

Obama, Barack 219
Obrist, Hans-Ulbricht 307
Oersted, Hans Christian 54
Oppenheimer, Robert 12, 126
Orwell, George 232, 291

P

Pais, Abraham 102
Parmenides 195
Pasteur, Louis 2, 37, 43
Pauling, Linus 12
Pauli, Wolfgang 39, 99, 117
Penrose, Roger 187
Penzias, Arno 269, 270
Perelman, Grigori 251, 292
Piaget, Jean 58, 193
Pinker, Steven 72, 294, 323, 324
Pitts, Walter 208
Planck, Max 2, 10, 23, 38, 39, 57, 69, 116, 117, 263, 271, 273
Plato 10, 15, 62, 121, 196, 197, 310, 316, 328, 329
Podolsky, Boris 18
Poincaré, Henri 10, 72, 110, 243, 251, 256, 293
Popper, Karl 56–59, 61, 68, 71, 72, 121, 196, 255, 310
Prigogine, Ilya 157
Project, Manhattan 12, 44
Putin, Vladimir 333
Pythagoras 239

Q

Qaeda, Al 77
Qiufan, Chen 234, 307, 321
Queloz, Didier 266

R

Rabelais, François 60, 291
Ramanujan, Srinava 323
Randal, Lisa 69
Riemann, Bernhard 249, 257
Roberts, Ian 216
Roch, Gustav 249
Rohrer, Heinrich 140
Roosevelt, Franklin D. 44, 305
Rosen, Nathan 18
Russel, Stuart 224
Russell, Bertrand 40, 72, 225, 242, 243, 255
Rust, Bernhard 21
Rutishauser, Heinz 213

S

Salam, Abdus 107
Samuel, Pierre 70, 171
Sauvage, Jean-Pierre 141
Schawlow, Arthur 129
Schlick, Moritz 245
Schlumberger, Marcel 33
Schmidhuber, Jürgen 210
Schmidt, Eric 91, 224
Schopenhauer, Arthur 53, 60, 61
Schrödinger, Erwin 11, 16–19, 26, 62, 147, 255
Schroeder, Gerhard 309
Schumpeter, Joseph 308
Searle, John 80
Sedol, Lee 210, 215
Serre, Jean-Pierre 247
Shelley, Mary 5
Sherrington, Charles 177
Shockley, William Bradford 128, 212

Siegel, Carl Ludwig 20
Singer, Isadore 248, 249
Smith, Adam 301, 302, 305, 306, 323
Socrates 316, 324
Sophocles 316
Spinoza, Baruch 15, 323, 330
Stalin 13
Stanley, Clarke 156, 304
Steinhaus, Hugo 238
Stoddart, J Fraser 141
Stokes, George 112, 252, 257
Strauss, Leo 324
Swaim, William 304
Swinnerton-Dyer, Peter 257
Szilard, Leo 44

T

Tati, Jacques 295
Teller, Edward 21, 22, 126, 283
Tesla, Nikola 32, 33, 37, 320
Thales 256
Thayar, Namagiri 323
Thucydides 316
Townes, Charles 129
Trump, Donald 280, 321
Turing, Alan 81, 127, 209, 212, 246

U

Ulam, Stanislaw 126, 128, 238
Urey, Harold Clayton 156, 157

V

van Gogh, Vincent 197
Venter, Craig 95, 159, 165, 166
Viazovska, Maryna 250
Vinge, Vernor 207, 208
Voltaire 31, 53, 323
von Helmholtz, Hermann 176
von Humboldt, Alexander 12
von Neumann, John 10, 19, 20, 126–128, 180, 209, 211–213, 238, 247, 253

von Schelling, Friedrich Wilhelm Joseph 53
von Siemens, Werner 33, 36, 320
von Weizsäcker, Carl-Friedrich 52, 60, 62

W

Watamul, Jeffrey 216
Watson, James 11, 148–151
Webb, James 134, 264
Weber, Max 329
Weil, André 247–249
Weinberg, Steven 61, 107, 119
Wernicke, Carl 176, 186
Weyl, Hermann 20
Wiles, Andrew 250, 251
Wilkins, Maurice 148–151
William of Occam 278, 314
Willis, Thomas 176
Wilson, Robert 269, 270
Winter, Jacques 129
Wittgenstein, Ludwig 196, 255
Wolff, André 153
Wolfram, Stephen 255
Wundt, Wilhelm 176

Y

Yamanaka, Shinya 92
Yang, Chen Ning 257
Yukawa, Hideki 102

Z

Zain, Muhammad Yasır 132
Zeno 240
Zermelo, Ernst 72
Zernicka-Goetz, Magdalena 166, 167
Zuse, Konrad 211
Zweig, Georg 105
Zwicky, Fritz 65

GPSR Compliance

The European Union's (EU) General Product Safety Regulation (GPSR) is a set of rules that requires consumer products to be safe and our obligations to ensure this.

If you have any concerns about our products, you can contact us on

ProductSafety@springernature.com

In case Publisher is established outside the EU, the EU authorized representative is:

Springer Nature Customer Service Center GmbH
Europaplatz 3
69115 Heidelberg, Germany

www.ingramcontent.com/pod-product-compliance
Lightning Source LLC
LaVergne TN
LVHW010335260326
834688LV00036B/723